SIDE BIAS:
A NEUROPSYCHOLOGICAL PERSPECTIVE

SIDE BIAS: A NEUROPSYCHOLOGICAL PERSPECTIVE

Edited by

Manas K. Mandal
Indian Institute of Technology, Kharagpur, India

M. Barbara Bulman-Fleming
University of Waterloo, Ontario, Canada

and

G. Tiwari
Banaras Hindu University, Varanasi, India

SPRINGER-SCIENCE+BUSINESS MEDIA, B.V.

A C.I.P. Catalogue record for this book is available from the Library of Congress.

ISBN 978-90-481-5599-6 ISBN 978-0-306-46884-1 (eBook)
DOI 10.1007/978-0-306-46884-1

Printed on acid-free paper

Contents

Handedness: Measurement and Observations

Side Bias: Foot, Cradle, Face and Attention

Contributors

Hari S. Asthana, Ph.D.
Department of Psychology
V.K.S. University, Ara
India

Alan A. Beaton, Ph.D.
Department of Psychology
University of Wales, Swansea
U.K.

Braj Bhushan
Section of Psychology
Krishnamurti Foundation India, Varanasi
India

M. Philip Bryden, Ph.D.†
Department of Psychology
University of Waterloo, Waterloo,
Canada

Pamela J. Bryden, Ph.D.
Department of Kinesiology and Physical Education
Wilfrid Laurier University, Waterloo
Canada

Carl Gabbard, Ph.D.
Department of Health & Kinesiology
Texas A & M University, College Station
USA

Gina M. Grimshaw, Ph.D.
Department of Psychology
California State University San Marcos, San Marcos
USA

Susan Hart, Ph.D.
Department of Physical Education, Recreation & Dance
New Mexico State University, New Mexico
USA

Kenneth Hugdahl, Ph.D.
Department of Biological & Medical Psychology
University of Bergen, Bergen
Norway

Yukihide Ida, Ph.D.
Psychology Unit
Osaka-Gakuin University, Osaka
Japan

Syoichi Iwasaki, Ph.D.
Psychology Unit
Fukushima Medical University, Fukushima
Japan

Jocelyn M. Keillor, Ph.D.
Defence and Civil Institute of Environmental Medicine, Toronto
Canada

Marilyn D. Lucas, Ph.D.
Department of Psychiatry
University of Witwatersrand, Johannesburg
South Africa

Manas K. Mandal, Ph.D.
Department of Humanities & Social Sciences
Indian Institute of Technology, Kharagpur
India

Michael E.R. Nicholls, Ph.D.
Department of Psychology
University of Melbourne, Melbourne
Australia

Michael Peters, Ph.D.
Department of Psychology
University of Guelph, Guelph
Canada

Philip Ray, Ph.D.
Department of Psychology
University of Wales, Swansea
U.K.

Lesley J. Rogers, Ph.D.
School of Biological Sciences
University of New England, Armidale
Australia

Steven C. Schachter, Ph.D.
Beth Israel Deaconess Medical Centre
Harvard Medical School, Boston
USA

Üner Tan, Ph.D.
Department of Physiology
Blacksea Technical University, Trabzon
Turkey

Geetika Tiwari, Ph.D.
Department of Psychology
Banaras Hindu University, Varanasi
India

Oliver H. Turnbull, Ph.D.
School of Psychology
University of Wales, Bangor
UK

Jan W. Van Strien, Ph.D.
Department of Clinical Neuropsychology
Vrije Universiteit Amsterdam, Amsterdam
The Netherlands

This book is dedicated to M.P. Bryden†

Preface

The beginnings of the idea about a book on 'side bias' began in the year 1994 during the senior editor's research association with late Professor M.P. Bryden and colleagues at the University of Waterloo, Canada. Over many discussions with Professor Bryden, it was clear that the concept of 'side bias' encompasses all aspects of motor behaviour within the context of human (and non-human animal) laterality. The tendency to favour one side or limb over the other is important not only from the perspective of understanding the functional asymmetries of the cerebral hemispheres, but also to an understanding of a myriad of aspects of human behaviour, as the contributions to this volume will attest.

By side bias, most people would think of bias in terms of hand preference or performance. The phenomenon of side bias, however, is more general and influences motor behaviour of all kinds, ranging from simple hand movement to complex behaviours like facial expression and attention. Therefore, the concept has been operationalized in terms of bias reflected in the motor expression of paired (such as hands, feet, eyes, or ears) or nonpaired organs (such as the face) as a function of preference, performance or attentional/intentional factors.

The book has become a reality by virtue of getting many of the ideas that were discussed with Dr. Bryden together in the form of chapters on diverse areas of side bias written by distinguished scholars in this field. The need for students and researchers to have a conceptual foundation on this issue has also been given due consideration. The emphasis of this book is on peripheral or motoric indices (such as hand, foot, face, etc.) rather than on central or sensory indices of side bias (such as vision, audition, or touch). To examine these indices, the authors of this volume have sometimes taken experimental, and sometimes developmental approaches. Regrettably, very few researchers have dealt with forms of side bias other than handedness. We believe that the biopsychosocial aspects of side bias are yet to be explored fully in the current literature. The present book is intended to cast light on some of the unresolved issues of side bias. Although attempts have been made in this book to cover all forms of side bias, some, such as eyedness and earedness, are absent. Paucity of research has prevented us from getting contributions in these areas. It is hoped that a future book on this topic addresses these important areas.

The chapters in this book are divided on the basis of the primary areas of side bias. Handedness being the most studied area, the first two sections are devoted exclusively to this subject. The third section deals with some other forms of side bias.

The first section, 'Development Of Side Bias And Handedness', contains five chapters. The first chapter deals with the evolution of side bias. Professor Rogers traces population lateralization (sensory as well as motor) in lower (e.g., fish, amphibia, reptiles) and higher (e.g., birds, mammals) vertebrates. By population lateralization, the author refers to bias reflected in the whole species rather than to bias observed in a group of individuals of the same species. She believes that lateralization in humans is not unique either in its nature or extent, and she provides an excellent critical discussion of the continuity of the phenomenon of functional lateralization from animal to human being.

Van Strien, in the second chapter, deals with the genetic, intrauterine and cultural origins of human handedness. The concept of handedness is elucidated, and the developmental and cultural factors that determine handedness are discussed. Available theories explaining handedness are then critically examined with an aim to search for similarities rather than differences amongst these theories. The author concludes with a notion that 'no single (theoretical) model yet put forward explains all aspects of the origin of human handedness'. He nevertheless emphasizes the salience of genetic and intrauterine factors in the determination of handedness.

The third chapter, by Tan, examines the (palmar) grasp reflex of the left and right hands in human neonates in order to explain the phenomenon of handedness in relation to sex and familial sinistrality. An empirical study conducted by the author, with right minus left reflex strength as the dependent measure, is discussed before these issues are taken up critically.

The findings suggest that (a) females tend to be slightly more asymmetric than males (due to a stronger reflex from the left hands in males than in females), and (b) a positive history of familial sinistrality (FS+) induces a left shift in right-minus-left reflex strength. The author favours a genetic theory of human lateralization to explain the grasp-reflex asymmetry.

Iwasaki, in chapter 4, records the age and generation trends in handedness with an emphasis on an eastern perspective. Although it is expected that the prevalence of left handedness will be far less in the eastern world than in the western world, a reexamination of the issue by analysis of the relation between writing hand and the reported prevalence of correction of left hand use suggests a relatively stable index of left handedness (approximately 10%) across the world. The author believes that intervention by adults, if started at an early age (before five), can exert a significant influence on handedness at maturity. The cross-cultural consistency in handedness therefore is attributed to a genetically predisposed trait with the added feature that human beings are highly adaptive, especially at the early age, as a result of which environmental influences (such as social pressure) can alter the manifest characteristics of handedness.

Beaton and his co-authors Hugdahl and Ray argue that most manual activities involve the coordinated use of both hands rather than one. They have examined the role of the corpus callosum in bimanual performance and cite evidence of interhemispheric integration in those with an intact brain and in acallosal participants . The effects of age on handedness, unimanual asymmetry, bimanual co-ordination and hemispheric function, as revealed by dichotic-listening techniques, is reviewed in some detail. These authors also present data from their own study that examined whether there is an unequal hemispheric decline as a function of age. It was found that with increasing age there is a decline in the (a) performance of both the left and right hands, (b) tactile information transfer across corpus callosum, and (c) dichotic-ear asymmetry, as a result of a reduction in the normal right-ear advantage. These issues are discussed in chapter five.

The second section includes three chapters that deal with measurement issues and with contributions of imaging methodologies to understanding handedness. The first chapter in this section, by Schachter (which is chapter six of this book), examines the problems of measurement and quantification based on questionnaires. The author believes that handedness studies should be conducted with appropriate controls for age and other factors (such as sex, education, family history of handedness, etc.) coupled with a sufficient number of experimental and control participants to achieve the statistical power to detect a significant effect. The author further remarks that, in his opinion, the Edinburgh Handedness Inventory is the most sophisticated handedness questionnaire and that "laterality score" rather than "laterality quotient" is the more powerful index to determine the degree and direction of handedness.

Ida, Mandal, and M.P. Bryden, in chapter 7, deal with the issue of statistical artifacts that emerge out of handedness data based on questionnaires. For example, conventional analysis of hand items by factor-analytic procedures reveal a two-factor structure (skilled/unskilled). It is posited that these factors consistently emerge across cultures because of violation of the assumption of multivariate normality in factor analysis. The authors conduct a meta-analysis of data obtained in two different studies that have been reported previously, and suggest that each item in a handedness questionnaire should be normalized by transformation before multidimensionality is examined. In addition to questioning the efficacy of conventional factor analysis of handedness data, the authors discuss the cultural confounds of skilled/unskilled hand-preference factors.

In view of the increasing attention being paid to recent technological advances, Professor Michael Peters, in chapter 8, provides a thoughtful and critical examination of the contributions of brain-imaging techniques, both new and relatively older, to the understanding of handedness. Rather than attempt a massive, uncritical listing of such studies, Professor Peters singles out several exemplary studies each of which has addressed an investigation of the neuroanatomical or functional correlates of handedness, and discusses each in considerable detail. Several studies relating to the degree to which both short- and long-term experience can influence neuroanatomical asymmetries are also discussed. The concordance in findings obtained from more direct anatomical methods and brain-imaging techniques of handedness is presented, which suggests that functional and anatomical asymmetries exist between the left and right primary cortices of right handers. For left handers, the findings, as is usual, show inconsistency.

The third section of this book deals with side biases that are observed in human behaviour other than handedness. Four forms of side bias are reported : footedness, cradling bias, facedness, and pseudoneglect. The issue of footedness has been taken up in two chapters. P.J. Bryden reviews the role of skill and task complexity in lateral performance and preference of hands and feet in chapter 9. The research on expression and measurement of these lateral preferences and manual performance asymmetries is reviewed in the beginning. Next, the relation between hands and feet in terms of preference and performance is discussed, followed by a review of studies that examined the influence of skill and task complexity on the expression of lateral preference and performance. The author believes that individuals have stronger lateral preferences for more complex tasks.

Chapter ten is devoted exclusively to the issue of footedness, which is a little-studied area in comparison to handedness. The notion of foot dominance has been examined by Gabbard and Hart with operational definitions and theoretical perspectives that explain the phenomenon. The authors present their own perspective and an argument is put forward based on empirical data, which suggest that functional asymmetry of the foot can be best explained from a contextual perspective. The functions of mobility (motor action) and stability (postural control) of the lower limbs are discussed in both the unilateral and bilateral contexts.

The chapter that follows (chapter 11) takes up the interesting issue of cradling bias. By such bias, Turnbull and Lucas refer to human females who prefer to cradle infants to the left side of their body midline. The authors trace the history of cradling bias and raise an important controversy concerning the cardiac and cerebral involvement in the emergence of such a bias. The cardiac account gets its boost from the fact that heart-beat sound is essential for the development of human neonates in terms of weight gain and growth. The cerebral account of the cradling bias has been extended because of the available evidence of right-hemispheric involvement in expression, understanding, and experience of emotion. It is argued that mothers interact more closely with their babies when the infants are aligned to their left (a contralateral function of the right hemisphere) rather than to their right (a contralateral function of the left hemisphere) hemispace. The leftward bias was also found to induce greater body contact. The authors critically discuss these issues and present experimental evidence.

Unlike other forms of side bias, biases in facial expressions are less clearly noticeable. Although studies have indicated that observers are subconsciously aware of the variation in the expression of the two sides of the face, the bias can be made pronounced by preparing facial composites of a photograph displaying an emotional or non-emotional expression. In this method, the left- and right-side composites are prepared by using the lateral half of one side of the face and its mirror-reversal. Other methods that measure the hemifacial bias include measurement of the differences in the electrophysiological activation or muscle movements between the two sides of the face. These methods are discussed in the introductory section of chapter twelve by Asthana, Bhushan, and Mandal and are followed by a review of literature on facedness. The review of literature is organized to explain the phenomenon during emotional and non-emotional expressions. The left side of the face has been found to be more pronounced in all kinds of expression, suggesting the involvement of the right hemisphere by the way of contralateral muscle action. The chapter concludes with a general comment on the relations between facedness and other forms of side bias, such as handedness.

As Nicholls points out (chapter 13), asymmetry in facial expression is also detectable in portraits. He observes that despite apparent symmetry in the human face, artists often choose to portray themselves asymmetrically. Giovanni Bellini's portrait of Leonardo Loredan, Doge of Venice, is presented to illustrate his viewpoint on this issue. Four forms of asymmetry have been identified in portraiture : the expression of the face, the turning of the head, the direction of illumination, and the horizontal position of the eyes. After discussing these forms of asymmetry, the author investigates the role of handedness in depicting the leftward bias observed in portraits or profile drawing. Asymmetry in attentional bias is also discussed, as such biases could play an important role in the perception of portraits. Nicholls thereafter searches for the theoretical accounts that explain the bias. These accounts substantiate a neuropsychological viewpoint suggesting that the emotive qualities are better expressed via the left face or better perceived due to a leftward scanning bias of the observer. Artists most often take clues from such a bias and the desire to portray features contained on the left side of the face plays an important role in the bias, Nicholls concludes.

Finally Grimshaw and Keillor raise the issue of pseudoneglect, a leftward bias exhibited by some normal individuals, as a form of side bias. They first review numerous examples of hemispatial neglect in focal brain-damaged patients with an observation that the deficit occurs more frequently in right-brain-damaged patients. On the basis of these studies of clinical populations, the presumed anatomical loci of the attentional and intentional factors contributing to the neglect syndrome are identified. The authors report an experiment to show the efficacy of the line-bisection paradigm in the understanding of the syndrome of pseudoneglect. A leftward bias is documented in the attentional as well as the intentional system. They conclude that attentional factors appear to contribute more strongly to pseudoneglect than do intentional factors, but that line length is a critical factor.

Altogether, the authors whose work is contained in this volume have established that side bias is present in many forms of human behaviour as well as in the behaviour of non-human animals. The subject matter therefore provides a wide canvas to researchers interested in pursuing further study in this general area. Nevertheless, with the notable exception of handedness, most other examples of side bias have remained relatively unexplored.

Clearly, much remains to be done to develop reliable measuring tools and appropriate research methodologies. It is our hope that this volume will engage young researchers and serve to encourage them, as well as more experienced researchers, to further investigations in these domains.

Thanks to Steven Smith, Bill Tays, Brandon Wagar, and Jan Willwerth for help with some of the final editing of the figures. As the editors of this volume, we would like to express our deep thanks to our families and colleagues for their encouragement, patience and support, and most importantly, to the contributors for their cooperation and commitment to the project.

Manas K. Mandal
M. Barbara Bulman-Fleming
Geetika Tiwari

I

DEVELOPMENT OF SIDE BIAS AND HANDEDNESS

Chapter 1

Evolution of Side Biases: Motor versus Sensory Lateralization

Lesley J. Rogers
University of New England, Australia

A brain is said to be lateralized if the left and right sides (for example, the left and right hemispheres) differ from each other in either structure or function. This lateralization can often be seen in the whole animal as side biases in motor behaviour or differential perception of stimuli located on the left and right sides. In other cases lateralization may not be evident unless a region of the left or right side of the brain is damaged. In such cases, the effect of the lesion differs depending whether it is on the left or right side.

Examples of lateralization are widespread among the vertebrates, even among lower vertebrates as I will discuss in some detail, and many of those forms of lateralization are similar to lateralization in the human brain. The idea that lateralization might increase in its extent and pattern in higher species to reach its pinnacle in humans, as suggested by Corballis (1991), was an attempt to take into account the fact that nonhuman animals are lateralized without entirely rejecting the earlier notion that lateralization is unique to humans and the biological basis for human language and tool use. I will show that lateralization in humans is not unique either in nature or extent. Even the pattern of lateralization in humans shares a number of features with other vertebrates. It is true that the presence or strength of

M.K. Mandal, M.B. Bulman-Fleming and G. Tiwari (eds.), Side Bias: A Neuropsychological Perspective, 3-40.

different kinds of motor and sensory lateralization varies from one species to another but, as I will argue, this may depend on the particular environmental demands on species rather than being a reflection of evolving higher levels of cognition.

I will consider the evidence that refutes the notion of discontinuity from animals to humans and I will do so for two important reasons: 1) to ask the question whether lateralization has evolved many times over, each time in different species, or whether at least its basic pattern has been conserved ever since it first evolved, and 2) to attempt to decide whether sensory or motor lateralization has a higher priority in terms of evolutionary selection. Then I will consider the potential advantages and disadvantages of being lateralized and how they might related to an animal's survival. Principally, I consider why an animal might retain a brain that makes it more responsive to a potential predator on one side than the other. The disadvantage of this side bias is so obvious that one assumes lateralization provides advantages far outweighing this impediment to survival.

1. POPULATION & INDIVIDUAL LATERALIZATION

The first examples of lateralization in animals included avian species (Nottebohm, 1971, 1977; Rogers & Anson, 1979) and rodents (Denenberg, Hofmann, Garbanati, Sherman, Rosen, & Yutzey, 1980; Denenberg, 1981). These were examples of group (or population) lateralization, meaning that the lateralization was in the same direction for the majority of participants tested as representatives of a species. Denenberg found, for example, that emotional responses were controlled by the right hemisphere in most rats. He also found that rats that have been handled in early life have a bias to move off in a leftwards direction when placed in an open field, a side bias indicating dominant control by the right hemisphere (Sherman, Garbanati, Rosen, Yutzey, & Denenberg, 1980).

This consistent specialization of the right hemisphere in most, if not all, individuals is referred to as a 'population bias', similar to right handedness in humans. There is another form of lateralization that is present in individuals but not the population. Paw preference in rodents is an example of the latter since individual rats and mice have either a left or right paw preference for retrieving food from a tube but there is no consistent group bias for use of either the left or the right paw by all or most individuals in a population (Collins, 1985). This means that rodents do not exhibit handedness, despite the hemispheric specialization for emotional and other responses. For a species to show handedness, the majority of individuals must prefer to use the same hand and also they must show consistent use of that same hand in most manual tasks (McGrew & Marchant, 1993). The same criterion can be applied to preferred use of the hind limbs, preferred eye used to view stimuli and preferred ear used in auditory orienting and dichotic-listening tasks.

In this chapter I will deal with lateralization primarily at the population level but I will have occasion to mention the relations between 'individual' and 'population' lateralization. Population bias refers to a whole species and, of course, such a bias can only be assumed by extrapolation from results obtained by testing a group of individuals of the same species. It is recognized that group biases may not always be extrapolated to population biases, especially since past experience may alter the expression of lateralization. Nevertheless, for the most part, group and population bias will be used interchangeably in this chapter since little is known of group differences in lateralization and my aim is to attempt to draw parallels and connections between species.

2. THE FIRST APPEARANCE OF LATERALIZATION

The discovery of population lateralization in mammals and birds raised the possibility that these two types of lateralization had arisen separately in the avian and mammalian branches of evolution. Thus, apparent similarities would have come about by convergent evolution. The idea that lateralization in birds and mammals could have stemmed from a common ancestor seemed initially to be unlikely. Given that lateralization had once been considered unique to humans, it was already a stretch of the imagination to encompass the idea of lateralization in birds and rodents, let alone vertebrates lower on the evolutionary scale than these species. Consequently it was not until quite recently that researchers began to investigate lateralization in reptiles (lizards), amphibia (toads and frogs) and fish (a number of species). I will discuss examples of lateralization in these species below.

Discovery of population lateralization in lower vertebrates has shown that there may be a commonality in the lateralization in birds and mammals. In fact, recently it has been shown that lateralization appeared as early as teleost fish and appears to have been retained in divergent lines of evolution. Hence lateralization appears to be a homologous characteristic of the brain rather than a feature that has evolved in parallel but independently in different species.

The chief evidence in support of lateralization being retained, rather than reappearing afresh in each species, is similarity of the types of lateralization seen in the different species, as I will discuss now. First I will discuss lateralization of motor behaviours and then side biases that depend on lateralized processing of perceptual information. By this comparative approach I hope to show that lateralization is an important aspect of a broad range of behaviours in animals and that understanding the evolutionary origins of lateralization may shed some light on the large array of lateralized behaviours known for humans (Bradshaw & Rogers, 1993).

3. MOTOR LATERALIZATION IN LOWER VERTEBRATES

Each side of the brain controls the musculature on the opposite side of the body and receives sensory input for the opposite side of the body. Thus use of a limb, or limbs, on the left side of the body engages the right hemisphere and vice versa. The same may be said of the body musculature used in turning in animals without limbs or with vestigial limbs. In species with well developed limbs it is possible to look for lateralized limb use in touching or holding objects or performing other acts with the limbs. In species without well-developed limbs, direction of turning of the body can be assessed and, of course, turning can also be measured in species with limbs.

Turning biases are known to be characteristic of mammals, even baleen whales (Matthews, 1978) and dolphins (Ridgeway, 1986; Sobel, Supin & Mislobodoski, 1994). They have been studied in most detail in rodents (Glick, 1985). Laboratory-bred rats and mice turn in circles spontaneously at night or in the daytime after they have been injected with amphetamine (Glick & Shapiro, 1984, 1985). Half of them circle clockwise and the other half circle anticlockwise. It was not until quite recently, however, that turning biases were discovered in teleost fish and in the species tested there was a population bias for all, or most, members of the group to turn in the same direction.

3.1 Fish

Cantalupo, Bisazza, and Vallortigara (1995) examined the direction of turning in a species of poeciliid fish, *Girardinus falcatus*, when escape behaviour was evoked by showing them a model predator. The predator model was a face resembling a larger fish. It was lowered rapidly to one side of the fish tank at a time when the fish was looking directly at it. The direction of turning to escape was scored. In the first session, when the stimulus was entirely novel, there was a significant group bias for the fish to turn rightward. This meant that, although the decision to turn could have been made by viewing with either the left or right eye, once turning had been initiated, the fish could continue to view the stimulus with the left eye as it turned away, a consideration that will become relevant later when perceptual lateralization is discussed. The turning bias of the group decreased on successive presentations of the stimulus until, on the fifth presentation, there was a tendency for the fish to escape by turning leftwards, viewing the stimulus with the right eye as they did so.

It will be noted that different directions of turning were elicited by presenting novel (rightwards) or familiar (leftwards) stimuli to the fish. This differs from the circling biases measured in rats and mice. The latter were repeated rotations that occurred in the home cage without the introduction of a stimulus. Apart from the distinct possibility of species difference, the presence of a population bias for turning in fish and not rodents may depend on the situation in which turning is measured. Visual perception was essential in the test used for the fish but probably not in the test used for rodents. Although circling in rodents may involve perceptual input, possibly olfaction, it is essentially a stereotyped behaviour that may depend on processing in lower regions of the brain. In fact, side biases may emerge in rodents tested on different tasks or in different conditions: for example, Alonso, Castellano, and Rodriguez (1991) found a population bias for right-side turning in rats tested in a T-maze.

Although escape behaviour in fish is not likely to depend on sophisticated evaluations, it is a rapid response that may depend on dominance of processing on one side of the brain. Therefore, lateralized perceptual processing may impose a population bias on the motor response. Escape responses are mediated by the Mauthner cells, a pair of giant reticulospinal neurons with axons that decussate and so innervate muscles on the opposite sides of the body. The Mauthner cell on one side triggers a C-shaped contraction of the fish's body, which initiates turning (Canfield & Rose, 1993). It would appear to be sensory connections to the Mauthner cells that determine the population biases seen in the escape behaviour of the poeciliid fish. But, once the C-start turn has been initiated by the Mauthner cells, the turning response is ballistic, meaning that it requires no further sensory input. Nevertheless, visual input may be used for continued monitoring of the predator during turning even though the direction of turning has been decided already.

The first response of the fish to the model predator is likely to be a rapid avoidance response and to involve output from the Mauthner cell on the left side, because the fish turns rightwards. The left Mauthner cell would apparently be triggered by visual input from the left tectum, which receives its input from the right eye. Hence the right eye and left tectum might have a dominant role in initiating the escape response although, during the turn and perhaps thereafter, the left eye and right tectum may continue to monitor the stimulus. With repeated presentations of the predator, stimulus habituation may occur and a rapid response mediated by the left Mauthner cell may be no longer involved. This may be the explanation for the change in turning bias.

Some studies on the spontaneous rotation of fish have been carried out in the laboratory of G. Vallortigara and A. Bisazza. These studies are more equivalent to the research on spontaneous or amphetamine-induced circling in rats. Rotational swimming in mosquito fish, *Gambusia hoolbroki*, was tested by placing them in circular tanks. The researchers found that females turned in a clockwise direction in the morning and in an anticlockwise direction at night (Bisazza & Vallortigara, 1996). Males did not show any population (group) bias. The circling bias occurred only when the fish tanks were well-illuminated by light and disappeared when the fish were tested under very low intensities of light. This suggested to Vallortigara and Bisazza that the female fish were using a sun-compass as a means of orienting (Goodyear & Ferguson, 1969). In the natural environment, this species of fish uses the position of the sun to locate the coast and shallow waters for relaxed basking and feeding. They also use the sun-compass to locate the deeper waters into which they can swim to escape predators. To do this they must relate the sun's position to an internal clock that allows for the changing position of the sun throughout the day and at all times they would need to relate this to their own orientation with respect to compass directions. The calculation of the sun's position relative to the fish's own orientation might be made easier if the fish travel along fixed routes in the same direction. In the case of fish caught by Vallortigara and Bisazza this appeared to be so. But, irrespective of how complicated the calculation made by the fish, the direction of turning would have to be opposite in the morning and afternoon, as seen in the females tested by Vallortigara and Bisazza. It will be noted that this switch in the direction of circling according to time of day took place with the source of illumination placed directly overhead, and so always signalling midday. The switch in direction of circling therefore took place according to an internal, diurnal clock and not in response to a change the direction of illumination. This fact might be important in the natural environment as light refraction at the water-air interface might confound the fish's ability to detect the exact position of the sun.

The fact that females and not males showed a population bias may be explained, Vallortigara and Bisazza suggest, by higher levels of predation on female poeciliid fish than on males (Britton & Moser, 1982). Not disconnected from this, it may also depend on the fish's state of arousal because the population bias in females tended to disappear with repeated testing. Also, males tested with a predator placed in a separate compartment in the centre of the tank displayed a population bias to circle clockwise (Bisazza & Vallortigara, 1997). The predator was another familiar species of fish living in the same habitat as the mosquito fish and frequently preying on them. It would therefore be expected to raise their levels of arousal. The males tested in this way circled clockwise in both the morning and the afternoon and so showed no evidence of using a diurnal clock. However, it is possible that the visual presence of the predator overrides any influence of the diurnal clock and forces the males to swim clockwise so that they can always view the predator with their right eye, as did *Girardinus falcatus* once the model predator had become familiar. If this is the explanation, the circling bias results from perceptual asymmetry rather than being caused by a purely motor bias dependent on the fish's state of arousal (for more details see review by Bisazza, Rogers, & Vallortigara, 1998).

Another known motor bias in fish involves fin preference. The channel catfish (*Ictalurus punctatus*) produces sound by rubbing one pectoral fin against the other. Fine et al. (1996) filmed 20 catfish as they produced their sounds and, by slow-motion playback of the recordings, they found that 10 of the fish had a fin preference, 9 of these preferring to rub the right fin against the stationary left fin. Therefore, although only half of the group displayed a fin preference, of those with a preference the right fin was used to perform the activity. Data from more individuals are needed before one can be confident of this right-fin preference but the result is interesting in the context of limb preferences in amphibians, discussed next. In fact, the bone and muscle structures used by the fish to move the pectoral fins are precursors to the amphibian limb and a right bias for control presents itself as a possible evolutionary precursor to right forelimb preferences in amphibians. I mention this only tentatively at present because the right-fin bias would need to be present in a number of species of teleost fish before one would have any confidence in this proposal.

3.2 Amphibians

The Pacific tree frog, *Hyla regilla*, has a bias to jump leftwards to avoid a predator. Dill (1977) placed a frog on a pedestal above which a ball was suspended and he scored the frog's direction of jumping when the ball was released from a position in front of the frog and allowed to swing towards it. The result obtained was a slight but significant population bias for jumping leftwards. The same frogs were found to have a longer right than left leg and so this could have been the reason for the motor bias, although Dill found no significant correlation between the direction of jumping and the degree of leg asymmetry.

Asymmetrical turning with a population bias is present in the male newt, *Tritus vulgaris* (Green, 1997). Mating behaviour in newts involves the female following the male with her snout touching his tail. After he deposits a sac of spermatophores on the substrate, he turns through 90° to form a barrier to the advancing female and his tail is folded along the flank of his body facing her. The female moves forward and then stops. The female's cloaca is positioned over the spermatophore sac and the latter is transferred into her cloaca. Males have a population bias to turn leftwards during this transfer of the spermatophore sac. This turning bias may have a motor cause, although that is not known. Once the male has turned, he can observe the female with his left eye, which could be causally related to the preferred direction of turning or merely an outcome of a primarily motor turning bias.

Frogs and toads have well differentiated limbs and hands that they use in a number of ways, during feeding to clean prey and push it into their mouths and to wipe unwanted material from the head, mouth and eyes. These anurans, therefore, can be compared with higher vertebrates in terms of handedness (Bisazza, Cantalupo, Robins, Rogers, & Vallortigara, 1996; Bisazza, Cantalupo, Robins, Rogers, & Vallortigara, 1997). The results show that these modern representatives of some of the first tetrapod vertebrates have a population bias for right handedness. The first experiment tested the European toad *Bufo bufo* by placing a small elastic balloon as a hood over the head. The procedure was repeated several times for each toad. The toads had a preference to remove the balloon with the right forepaw. The strength of the right handedness was close to 60% and therefore weaker than that in humans. In fact, in all tests of handedness, to be described now, the population bias is statistically significant and in the region of 60 to 70%.

Bufo bufo was also tested by placing a small strip of paper on the toad's snout and scoring the limb used to wipe it off. Again the procedure was repeated several times for each toad in order to gauge each individual's true preference. A population bias for preferred use of the right paw was exhibited again. But a similar test of a smaller species *Bufo viridis* tended to show a bias of left-paw preference shifting towards a right-paw preference as testing was repeated. This species appeared to be very distressed during testing in the laboratory and stress levels may have contributed to, at least, the initial left-paw preference. Another species, *Bufo marinus*, did not respond by wiping the paper strip from its snout and so could not be tested in this way. *Bufo marinus* was, instead, tested by allowing the toad to clasp the experimenter's hand (alternated between left and right) and then turning the toad upside down under water. The paw used by the toad to pivot itself into the upright position was scored. A right-paw preference was found. The toads released their grasp with the left paw first and used the right paw to apply a force on the experimenter's hand and so act as a pivot. The hind limbs were splayed and did not touch the experimenter's hand.

The forelimb around which toads and frogs pivot can also be scored by laying the subject on its back on a flat surface, not underwater. This method cannot, however, be used for species with suction feet unless they are very large and less able to cling to the experimenter's fingers and hand. I am presently testing different species of free-living Australian frogs using this technique. A sample of 24 *Litoria latopalmata* has shown a population bias to pivot around the right forelimb, only 2 participants displaying a preference to pivot around the left forelimb (unpublished data; Fig. 1). The median bias to the right is 70 percent. Although the forelimbs play a part in this righting response, the hind limbs are also involved and perhaps also other muscles. Pawedness in this test therefore should not be equated to pawedness in snout wiping or even underwater pivoting since the latter scores represent exclusive use of the forelimbs, or at least almost so.

Naitoh and Wassersug (1996) have suggested right pawedness may result from use of the right paw to wipe toxic materials from the everted stomach. After ingesting toxic material, some anuran species vomit by everting the entire stomach. As the stomach is located to one side of the body and has shorter mesentery on the right than the left side, when everted, it prolapses on the right side of the animal. This anatomical bias might therefore elicit right-pawedness, they suggest. Although this is an interesting observation, it is unlikely to be the reason for the biases in righting responses when not under water because the hindlimbs are involved (Robins, Lippolis, Bisazza, Vallortigara, & Rogers, 1998) and it is certainly not the explanation for lateral biases in visual responding discussed below.

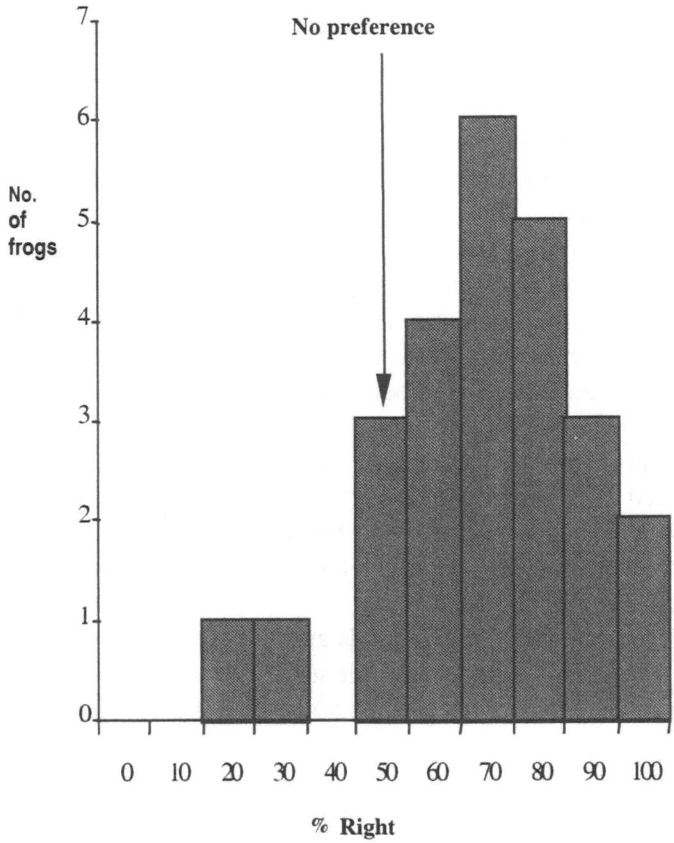

Figure 1. Frequency distribution of percent right hand preference in the Australian frog, *Litoria latopalmata*

There has been one report of motor lateralization for control of vocalization in frogs. Bauer (1993) lesioned the left or right vocal tracts in the hindbrain of frogs, *Rana pipiens,* and then attempted to elicit their alarm calls. The frequency and quality of the calls were reduced by lesions of the left side but not the right side.

3.3 Reptiles

To my knowledge there have been no reports of motor bias in reptiles, apart from a report of some data suggesting that constricting snakes have laterally biased coiling when feeding on live prey (Heinrich & Klaassen, 1985). This absence of information on motor biases in reptiles is somewhat surprising, especially in light of the clear lateralization of eye use in lizards, as discussed below.

4. SENSORY & PROCESSING LATERALIZATION IN LOWER VERTEBRATES

4.1 Fish

The possible role of visual perception in determining the direction of turning when fish see a predator has been mentioned above. Tests of visual lateralization in fish have been carried out by Miklósi, Andrew, and Savage (1998), who scored the eye used by zebrafish (*Brachydanio rerio*) as they examined familiar and unfamiliar stimuli, and before turning was initiated. The zebra fish were first trained to swim from one end to the other of a long, narrow tank. Then various visual stimuli were presented at one end of the tank. The fish were videotaped from overhead and, when a fish was within close proximity to the visual stimulus, its body angle was determined using frame-by-frame analysis. Typically, the viewing angle is between 0° (direct facing) and 20° to the left or right side. That is, either the left or right frontal field is used. Hence, eye preference could be determined.

Unfamiliar stimuli (objects or a visual scene) were first viewed with a preference for the right frontal visual field. Once stimuli had become familiar, and that was as soon as the second presentation of the same stimulus, the left eye was used preferentially. A familiar fish was also viewed by the left field and the left eye was also used to view an empty, semicircular area with grey walls. The researchers interpreted their results as showing that the fish views with the left eye when the stimulus is either neutral or familiar and therefore requires no decisions to be made about responding, and the fish uses the right eye when decisions about responding have to be made. In other words, the left side of the brain (right eye) is used when visual information has to be processed and a "considered" response must be made (e.g., whether to approach or withdraw from a novel stimulus). As will be discussed later, a similar pattern of preferences occurs in the chick; viz., the left eye is used to view familiar stimuli and the right to view unfamiliar stimuli (Dharmaretnam & Andrew, 1994).

In order to give a "considered" response, the fish would have to delay the escape response briefly. The immediate activation of the Mauthner cell, which would lead to escape, must be very briefly inhibited. This might be possible only when the right eye is in use. As the left Mauthner cell also appears to trigger the C-shaped escape response, according to the experiments with *Girardinus falcatus* discussed earlier, the left Mauthner cell (and right eye) may have a dominant role in motor responses. This possibility could be tested by electrophysiological recording of the activity of the Mauthner cells.

Detour tests in which fish have to swim around barriers may also be reveal viewing preferences. Bisazza, Pignatti, and Vallortigara (1997a) found that male mosquito fish (*Gambusia hoolbroki*) would detour leftwards to swim around a barrier with vertical bars to reach a group of conspecific females. They also swam leftwards when a simulated-predator was placed on the other side of the barrier. In both of these cases, therefore, the fish is able to view the stimulus with its right eye as it makes the detour.

Although the direction taken in making the detour could be considered to result from a motor bias only (i.e., clockwise circling), this is unlikely because the same males showed no directional bias when they had to make the detour to reach a group of conspecific males or when no stimulus was present. It would seem that the right eye is used preferentially only when viewing biologically important stimuli (the females and the predator) about which the fish has to make a response decision. However, in my opinion, it is not clear why they would not have to make such a considered response when approaching conspecific males. These results would be consistent with the above explanation for zebrafish only if the male mosquito fish have to give a "considered" response to females but not males. Although this is a conceivable explanation, no conclusion can be reached because different stimuli were used in the different testing conditions. If the hypothesis of Miklósi et al. (1998) outlined above does apply, one can predict that male mosquito fish will detour to the right when they are tested with a familiar member of their own species.

Another study of detouring behaviour by Bisazza, Pignatti, and Vallortigara (1997b) found species differences among males. They tested the males' direction of detour to reach conspecific females and found that three species (*Gambusia hoolbroki*, *Gambusia nicaraguensis* and *Poecilia reticulata*) detoured leftwards and two species (*Brachyrhaphis roseni* and *Girardinus falcatus*) detoured rightwards. A single species was selected from each of these groups (*Gambusia hoolbroki* and *Girardinus falcatus*) and new detour tests were performed using a simulated predator, which elicited leftwards detouring (right eye) in both cases, and an opaque screen, which elicited rightwards (left eye) detouring in both cases. Therefore, despite species differences in response to the females, biologically relevant stimuli are viewed by the right eye and neutral, irrelevant stimuli are viewed by the left eye. The researchers concluded that sexual motivation may have affected the direction of detour when females were used as the stimulus. The two rightward-detouring species were more disturbed by being placed in the novel environment and this suppressed their sexual motivation and, presumably, the biological relevance of females to them in that context. This is an important consideration for measuring lateralization in wild compared to captive species.

Recent evidence shows that a mosquito fish will approach a predator to inspect it more closely when it has a conspecific on its left side than when a conspecific is on its right side. Bisazza, de Santi and Vallortigara (1999) demonstrated this by placing the mosquito fish in a rectangular tank with a predator fish in a separate tank placed at one end of the 'swim-way'. A mirror was placed on one of the longer sides of the tank with the mosquito fish, on either its left or right side. Thus the image would swim along with the mosquito fish on its left or right side. With the image on its left side, the mosquito fish approached more closely to the predator than it did with the image on its right side. This result would appear to reflect the preference to monitor familiar stimuli using the left eye.

All of these experiments allow us to conclude that fish have both sensory and motor biases and that, in at least some situations, the sensory bias determines the motor bias. Forms of lateralization caused by preferential use of an eye to view different stimuli, therefore, appeared early in evolution and might be widespread among vertebrates.

4.2 Amphibians

I have mentioned above that the male newt *Tritus vulgaris* turns leftwards during mating and that this might afford him use of the left eye to view his female partner. Such a viewing preference would be consistent with use of the left eye and right side of the brain to view a familiar stimulus, provided that mating in newts involves prior familiarization. Whether visual preferences or motor lateralities are the cause of the asymmetry in turning by the courting male newt has yet to be determined experimentally.

Visual preferences, however, are now known to be present in toads (Robins et al., 1998; Vallortigara, Rogers, Bisazza, Lippolis, & Robins, 1998). The left and right visual hemifields have complementary specialization for predatory and agonistic responses. *Bufo bufo* and *Bufo marinus* frequently direct agonistic tongue strikes at each other when they are housed in groups and supplied food that elicits competition, whereas *Bufo viridis* toads tend to avoid each other. The agonistic behaviour of *Bufo bufo* and *Bufo marinus* was recorded on videotape made by placing the camera directly overhead. By playback at slow motion it was possible to determine whether the strikes were in the attacker's left or right hemifield. A population bias for striking at targets in the left visual hemifield was found (mean of 60% left and standard error of 2% for *Bufo marinus*, and 65% for *Bufo bufo*, although fewer scores were recorded for the latter species). This bias to attack targets in the left hemifield was complemented by a population bias to strike at moving prey in the right hemifield.

Prey-striking responses could be measured in all three species by placing the toad inside a transparent cylinder through which it could see a live worm or cricket suspended on a thread from an arm that rotated the prey in either a clockwise or anticlockwise direction. By videotaping from overhead and recording from playback it was possible to determine the number of strikes at the prey in the left and right hemifields. When the prey was rotated clockwise, and thus entered first the left and then the right field, almost all of the strikes occurred in the right hemifield. When the prey was rotated anticlockwise, entering first the right and then the left field, *Bufo bufo* and *Bufo viridus* directed strikes at the prey even before it crossed the midline. They gave a more symmetrical distribution of strikes in the left and right hemifields. These results show that the initial detection of the prey in the left field does not elicit prey-catching responses and that the prey must move into the right half of the binocular field before the toad will strike at it. By contrast, initial detection of the prey in the right field allows the toad to orient towards it and strike at it anywhere in the binocular field. One might say that the toad shows a form of 'stimulus-specific visual neglect' in the left field. *Bufo marinus* gave a somewhat different result: it was much less responsive to prey rotated anticlockwise than to prey rotated clockwise.

Recent experiments in my laboratory have found that *Bufo marinus* strikes preferentially at novel stimuli in the left lateral visual field (Robins & Rogers, paper in preparation). At a time when the toad was attending to a worm-like image moving on a computer screen, two small novel stimuli, resembling flies, attached to each end of a Y-frame were introduced from behind the toad into the lateral visual fields. The toad had to choose between striking to the left or right side and there was a preference for the left on the first presentation of the stimulus but not on subsequent presentations. Changing the stimulus slightly (e.g., by adding a white stripe) reinstated striking leftwards.

Overall, it might be concluded that toads, like fish, direct considered responses to prey in the right hemifield and rapid attack strikes at conspecifics in the left hemifield.

4.3 Reptiles

Lateral preference to use the left eye in aggressive interactions is as characteristic of the lizard, *Anolis carolinesis*, as it is of toads. Deckel (1995, 1996, 1998) videotaped agonistic encounters made by the lizards and found that encounters with high levels of aggression were more likely to involve use of the left eye. A high level of aggression was indicated by headbobbing with extension of the coloured, throat dewlap while the lizard was moving towards another lizard and threatening to bite. Less aggressive encounters, 'assertion displays' as opposed to 'challenge displays', showed no consistent bias in eye use and there was a trend towards use of the right eye during motionless observation of an aggressive conspecific. As these lizards have laterally placed eyes and binocular vision is obstructed by their snouts, monocular viewing is, essentially, used at all times but this requirement, in itself, would not mean that there should be a population bias in eye use. Rather, because most of the visual input from an eye goes to its contralateral hemisphere, it seems that specialization of the right side of the brain (right hemisphere) to control aggressive behaviour is the reason for the population bias.

Aggression in *Anolis carolinesis* involves a change in colouration. Nonaggressive lizards are dark brown in colour and aggressive lizards are lighter and greener in colour. Green-coloured lizards were twice as likely to use the left eye (Deckel, 1995). As melatonin has been reported to bind asymmetrically in the diencephalon (Wiechmann & Wirsig-Wiechmann, 1992), Deckel and Jevitts (1997) suggested that the left eye and right hemisphere may be involved in inhibiting melatonin release and enhancing serotonin release. If lateralized control of melatonin and serotonin release is associated with the lateralized aggressive displays of *Anolis sp.*, the same may be true for toads and higher species. However, the direction of causation has not been established: higher levels of aggression and left eye – right hemisphere use may, for example, change melatonin and serotonin levels rather than be caused by them.

5. MOTOR LATERALIZATION IN HIGHER VERTEBRATES

Various forms of motor lateralization are now known to occur in birds and mammals. Here I will make comparisons to lateralization in lower vertebrates; detailed discussion of lateralization in birds and mammals has been made previously (Bradshaw & Rogers, 1993, 1996; Rogers & Bradshaw, 1996).

5.1 Birds

Foot preferences at both the individual and population level have been reported in a number of difference avian species. In particular, many species of parrots are known to have a strong population bias to hold food in one foot while feeding (reviewed by Harris, 1989). From the data collected so far, left-footedness predominates in African (Friedman & Davis, 1938) and Australian (Rogers, 1980, 1981) parrots. A total of fourteen left-footed species have been reported so far, whereas only three species have been found to have a population bias for right-footedness (Cannon, 1983; Rogers, 1980).

In the case of parrots, the footedness involves manipulation of food objects and this requirement for manipulation may be a factor determining preferential use of a limb (Walker, 1980). A population bias for right-footedness has also been found in the goldfinch (*Carduelis carduelis*) when they were tested on a task requiring manipulation of doors and catches using the beak and foot to obtain a food reward (Ducker at al., 1986). This result supports Walker's hypothesis, but at least one species that does not use its feet to hold and manipulate food, the chicken (*Gallus gallus*), has been found to have footedness. Chicks use their feet to scratch the ground while feeding and they show a significant tendency to initiate a bout of scratching with the right foot (Rogers & Workman, 1993; Tommasi & Vallortigara, 1999). This right-footedness at the population level is also apparent for using the foot to remove a small piece of adhesive tape from the beak (84 percent right-foot preference in the group tested).

The chick may not use its feet to perform holding and fine manipulation of food objects but it does use the feet during feeding and, therefore, some form of manipulation occurs by using the feet. This may explain the foot preference. Pigeons and budgerigars do not use their feet to scratch at or hold food during feeding and they show no foot preference, to remove a piece of adhesive tape from the beak, either at the population or individual level (Rogers & Workman, 1993; Güntürkün, Kesch, & Delius, 1988).

The strength of the population bias of footedness in parrots is at least equivalent to that of right handedness in humans, if not stronger. The right-foot preference may also suggest that the left hemisphere is dominant in these feeding situations. However, it is not clear whether the left of right foot is performing the most active role. Certainly, the foot holding food is used to manipulate it to some extent but the parrot's beak is used to carry out the finer manipulations and the other foot is used in a skilled manner to balance the bird. The hemispheres would need to control different aspects of the motor output required but one may be no less important than the other. Also, the beak is a central structure controlled by both hemispheres and it may often be used for the finer motor manipulations. I have observed this particularly in Australian parrots feeding on banksia cones from which they extract seeds.

By testing chicks monocularly, Tommasi and Vallortigara (1999) have obtained insight into which hemisphere may be most active in the control of the limbs during feeding. Binocular chicks and chicks using only the left eye (right eye occluded) had a right-foot preference to initiate bouts of ground scratching, whereas chicks using only the right eye used the left foot to ground scratch. This result suggests that the hemisphere activated by visual input controls the limb used to maintain balance, and not the one to perform scratching (at least at the initiation of the scratching bout). The authors concluded that footedness may have arisen from the limb used to maintain postural and positional control rather than from the limb used in motor activities. This conclusion may, of course, apply to other species.

The avian brain also displays lateralization of motor control of the syrinx during singing. In a number of species of songbirds, centres in the left hemisphere, including the higher vocal centre (HVC) and the nucleus robustus archistriatalis (RA), are involved in singing, whereas their equivalents in the right hemisphere are not (Nottebohm, Stokes, & Leonard, 1976; Nottebohm, 1980). RA has primarily a motor function although it does receive auditory inputs (Konishi, 1994; Vicario & Yohay, 1993). This nucleus sends inputs to some of the motor neurons in the nucleus of the XIIth cranial nerve and these innervate the musculature of the syrinx (summarized in Rogers & Bradshaw, 1996). Singing of canaries, for example, is disrupted by lesioning HVC or RA in the left hemisphere but not by lesioning equivalent regions in the right hemisphere (Nottebohm et al., 1976; Nottebohm, 1977). Following lesioning of the left RA, the canary sings with a reduced frequency range and with fewer syllables. HVC has both sensory and motor functions (McCasland & Konishi, 1981) and lesions of the left HVC leave the canary with the ability to produce no more than one of its song syllables, the rest of the song being a monotonous succession of simple notes. Lesions of the right HVC have little to no effect on singing (Nottebohm, 1977).

The specialization of centres in the left hemisphere for controlling song is characteristic of six out of nine species investigated so far, two species having no lateralization (Suthers, 1990) and one species, the zebra finch, having the reverse lateralization (Williams, 1990). The exceptions to the 'rule' for left-hemisphere specialization for control of song may indicate that there has been no gradual evolutionary elabouration on left-hemisphere specialization for vocalizing already present in amphibia. Alternatively, there may be something unusual about the zebra finch's song, possibly in terms of the context in which singing occurs or the bird's state of arousal. Since the song nuclei are present in both hemispheres in all species, and auditory inputs are processed on both sides (Cynx, Williams, & Nottebohm, 1992), it is possible that lateralization for the controlling vocalizations is specific for the type and context of the vocalization measured. This point will be relevant in the discussion of motor control of vocalizations in marmosets, to follow.

In summary, birds have strong lateralization of motor responses. As discussed below, these are matched by lateralization of sensory processing.

5.2 Mammals

Hand preferences and turning biases in rodents have been mentioned above and they have been reviewed in detail elsewhere (Bradshaw & Rogers, 1993; Glick & Shapiro, 1985). Here I will discuss mainly hand preferences and not other motor biases in primates, although I will do so only briefly because there is a large amount of literature reviewing this topic (for example, Hook-Costigan & Rogers, 1996; MacNeilage, Studdert-Kennedy, & Lindblom, 1987; Ward & Hopkins, 1993) and my aim is to make comparisons to the lateralization in lower vertebrates and birds.

As an overall pattern, it can be said that prosimians have a population bias for left handedness to reach for and hold food (Ward, Milliken, & Stafford, 1993) and there may be a shift towards right handedness in apes, although there is disagreement on the latter. There have been reports of right hand preferences and other motor biases in chimpanzees raised in the laboratory (Hopkins & Bard, 1993) but no evidence of this bias was found by Marchant and McGrew (1996) when they examined hand preferences in wild chimpanzees performing a number of tasks. On the other hand, Rogers and Kaplan (1996) have found strong left handedness in orang-utans for manipulating parts of the face, as when cleaning the eyes, ears or teeth.

New World primates present a variable prevalence of handedness, although right handedness may be a relatively common feature (reviewed by Hook-Costigan & Rogers, 1996). The tamarin, *Saguinus oedipus*, has been reasonably well studied and this species is right handed when reaching for and holding food (Diamond & McGrew, 1994; King, 1995). The common marmoset, *Callithrix jacchus*, however, shows no obvious population bias although individuals all have quite strong hand preferences and there may be slightly more left handers than right handers (Hook-Costigan & Rogers, 1996; of the 21 marmosets in the colony at the University of New England, 13 are left-hand preferring). At least one New World species, *Ateles geoffroyi*, may be left handed at a population level (Laska, 1996). More data on hand preferences in New World primates need to be collected before conclusions should be made, but it does appear that the presence and direction of handedness varies with species and, perhaps, habitat as well as the task for which the hands are being used. In this sense, hand preferences in primates may be more variable than the paw and foot preferences of amphibians and birds. Primates may, however, show more consistent motor biases in muscular activity not involving the limbs, as in the facial musculature, for example.

Primates, like humans, have lateralization of control of facial expressions. Rhesus macaques (*Macaca mulatta*) express fear more strongly on the left side of the face (Hauser, 1993) and so do marmosets (*Callithrix jacchus*) (Hook-Costigan and Rogers, 1998a). The left half of the mouth opens sooner and wider than the right half. This motor bias reflects the right hemisphere's role in emotions since each hemisphere controls the musculature on the opposite side of the face. The same left-side biased lateralization is seen when marmosets open the mouth to make the 'tsik' vocalization, used in mobbing predators. The opposite lateralization appears to occur for social contact calls, 'twitters'; when this call is produced, the right side of the mouth opens to a larger extent than the left. Therefore, it seems that vocalizations used in situations eliciting high arousal and fear may be produced by centres in the right hemisphere, whereas more relaxed, contact calls are produced by centres in the left hemisphere.

6. SENSORY & PROCESSING LATERALIZATION IN HIGHER VERTEBRATES

6.1 Birds

Lateralization of visual processing in birds, mainly the chick and the pigeon, has been discussed in detail previously (Rogers, 1995, 1996). There are three main features that are relevant here; viz. lateralization of aggressive and feeding responses and of responses to novel and biologically relevant stimuli.

As in toads, the left-eye and right-hemisphere system of the chick is specialized for aggressive responses. Young chicks that have had their levels of aggression enhanced by treatment with testosterone respond to a moving hand by attacking provided that they are tested either binocularly or monocularly with the right eye occluded (Rogers, Zapia, & Bullock, 1985). The same chicks do not attack when they are tested with the left eye occluded. This form of lateralization is also present in adult chickens (Rogers, 1991).

Processing of topographical information is also a function of the left eye and right hemisphere of the chick. Rashid and Andrew (1989) tested chicks monocularly on a task requiring them to find food buried under sawdust in a large arena. The chicks used spatial cues to locate the hidden food when using the left eye but not when using the right eye. In line with this result, Vallortigara, Zandforlin, and Cailotto (1988) showed that chicks could use spatial cues to locate a small box containing food when it was on their left side but not the right.

The left eye and right hemisphere of the chick are also specialized to detect novel stimuli. A young chick interrupts feeding to pay attention to a small novel stimulus introduced into its peripheral visual field from behind and it does so earlier when the novel object advances on the chick's left side than when it does on the chick's right side (Rogers & Anson, 1979). This bias is also revealed by scoring the eye that the chick uses to view stimuli. The chick displays preferential use of the left eye to view familiar and neutral stimuli and the right eye to view attractive and biologically important stimuli (Dharmaretnam & Andrew, 1994).

The right eye and left hemisphere of the chick are specialized to control pecks during feeding and to direct pecks away from inedible pebbles to grain (Mench & Andrew, 1986; Rogers, 1997; Zappia & Rogers, 1987). This ability of the left hemisphere may stem from being able to inhibit responding when necessary and so give a considered response. Consistent with this explanation, McKenzie, Andrew, and Jones (1998) have shown that adult chickens using the right eye and left hemisphere only, are able to inhibit approaching a novel, social partner and pecking at a familiar partner.

I have provided only a brief outline of the now-comprehensive list of studies showing lateralization in the chick in order to highlight the main features. The similarity of the types of lateralization in young and adult chicks indicates that the basic adult pattern is present in early life. In addition, it resembles the pattern of lateralization in fish, toads and lizards. Lateralization of visual responding has also been reported for pigeons (Güntürkün, 1985) and marsh tits (Clayton & Krebs, 1993). Pigeons, for example, have dominance of the right eye and left hemisphere for controlling pecking and discriminating complex patterns, similar to the chick. In general, it appears that lateralization is a marked feature of the avian brain and one that is present in auditory (Miklósi, Andrew, & Dharmaretnam, 1996) and olfactory (Rogers, Andrew, & Burne, 1998) processing as well as visual processing. In the auditory modality it appears that the right ear and left hemisphere attend to important cues, whereas the left ear and right hemisphere attend to relatively unselected auditory inputs (Miklósi et al., 1996). Hence the specializations of the hemispheres for processing auditory inputs have much in common with their functions for visual processing.

Insufficient data on lateralization of olfaction are available to say whether processing in this modality matches the pattern known for visual and auditory lateralization but the initial data point in this direction. A day-old chick, presented with a blue bead from which clove oil odour is released, first pecks the bead and then shakes its head, as a disgust response. The same response is given if the chick is tested using its right nostril, the left nostril being blocked with wax (Rogers, Andrew, & Burne, 1998). Head shaking does not occur in response to presentation of the blue bead and clove oil odour if the right nostril is blocked with wax, but the chick still pecks at the bead. As input from the nasal epithelium in each nostril goes to its ipsilateral hemisphere, this result indicates that the right hemisphere responds to the novel odour and generates the immediate, stereotyped response of head shaking. The left nostril and left hemisphere attends to the visual cues only (pecking the bead) and over-rides any response to the odour. Apparently, the left hemisphere makes a considered decision not to respond to the odour and to respond to the visual cues alone. This explanation assumes that a chick using the left nostril detects the odour and decides to ignore it. Another experiment shows that this is the case. Presenting the clove oil odour together with a red bead elicits pecking and head shaking irrespective of which nostril is occluded. In this case, the left hemisphere decides not to ignore the odour, probably because red beads are less attractive to chicks (Andrew, Clifton, and Gibbs, 1981). Hence, the left hemisphere is able to inhibit the immediate response to the olfactory input and make a considered decision whether to respond to odour or not, whereas the right hemisphere gives an immediate response to the odour as well as the visual cues.

6.2 Mammals

There are a considerable number of studies that have examined effects of unihemispheric lesions on performance in primates, and others that have revealed lateralization following sectioning of the corpus callosum (Hamilton & Vermeire, 1991; reviewed also by Bradshaw & Rogers, 1993). Presentation of visual stimuli in the extreme peripheral field has also been used to reveal lateralization (Hopkins & Morris, 1989) and also monaural stimulation with species-specific vocalizations. The first evidence of auditory lateralization in a nonhuman primate was found by presenting 'coo' calls monoaurally to Japanese macaques (*Macaca fuscata*): a right-ear advantage was found for discrimination of coo types (Petersen, Beecher, Zoloth, Moody, & Stebbins, 1978). Therefore, the left hemisphere of these monkeys is specialized for processing species-specific calls. It is worth noting that similar specialization of the left hemisphere has been shown, not for processing, but for producing vocalizations in the frog and several species of songbird (discussed previously). Of particular relevance, the right ear - left hemisphere of the rat is specialized to process the calls produced by rat pups; Ehret (1987) found that a maternal rat would retrieve her distressed pups when she was tested with her left ear occluded (i.e. using the right ear) but not when her right ear was occluded.

The right-ear advantage in processing species-specific calls has also been demonstrated by playing recorded calls from a speaker placed behind the subject and scoring the ear turned to listen to the call. Using this technique, Hauser, Agnetta, and Perez (1998) found that rhesus monkeys have preferred use of the right ear to listen to a number of their natural calls. Changing interpulse interval of the 'grunt' and 'shrill bark' calls to produce a call outside the normal range either eliminated the ear bias or shifted it to the left ear. The latter shift may indicate attention to novelty by the right hemisphere.

Research using the same playback technique in my own laboratory has shown that the marmoset, *Callithrix jacchus*, turns the left ear to attend to the 'tsik' mobbing call, reflecting use of the right hemisphere (Rogers, Shuster & Hill, in preparation). It therefore appears that the hemisphere used to process conspecific vocalizations may depend on the meaning of the call, fear-inducing calls being porcessed by the right hemisphere rather than the left. Further research with different species and different vocalizations is needed to test this hypothesis.

The results of various experiments looking at the effects of lesions on visual lateralization in primates have demonstrated group biases suggesting specialization of the hemispheres at the population level. As would be expected, each study involves testing rather few participants but overall there are definite indications of population biases. There have, however, been few attempts to demonstrate that these hemispheric lateralizations are manifested in side biases for responding to visual stimuli. The reason for this, when considering visual specializations in primates at least, is the large binocular overlap of the visual fields and the incomplete decussation of the optic nerve fibres, which means that lateralizations can be revealed only in the extreme peripheral visual field. Nevertheless, primates do display eye preferences for viewing stimuli through a small hole. Bushbabies have been found to prefer the left eye to view interesting but nonarousing stimuli (Rogers, Ward, & Stafford, 1994) and marmosets prefer the right eye (Hook-Costigan & Rogers, 1998b). These biases are lost, and may be even reversed, when the participants are highly aroused. The eye preference in the relaxed state is as strong a group bias as is present in humans. For example, twenty out of the twenty-one marmosets in the colony at the University of New England have a strong right-eye preference (Hook-Costigan & Rogers, 1998b). Eye preference is unlikely to be determined by motor lateralization as there is no correlation between eye and hand preference. Instead, eye preference would seem to be the choice of a specialized hemisphere for processing the visual input since primates do retain some aspects of the evolutionary past in their visual pathways: each eye relays input to the contralateral hemisphere more rapidly and in more detail than to the ipsilateral hemisphere (summarized in Rogers et al., 1994).

The side bias in visual responding most relevant to this chapter is that shown for aggression in the baboon, *Therithecus gelada*. Casperd and Dunbar (1996) discovered that baboons are more likely to display agonistic responses to conspecifics on their left side than on their right side. This bias, presumably, stems from first detection of the conspecific in the extreme peripheral visual field and reflects control by the right hemisphere, as in toads, chicks and lizards. Thus we may conclude that right hemisphere involvement in controlling aggressive responses has been highly conserved during the evolution of species.

The same conservation across species may have occurred for processing topographical information in the right hemisphere. Right-hemisphere specialization for spatial processing is known in humans (summarized in Hellige, 1993). It is also present in the chick, as already discussed, and in the rat. Cowell, Waters, and Denenberg (1997) tested rats monocularly in the Morris water maze and found that they could locate the escape platform when using the left eye (and right hemisphere) but not when using the right eye (and left hemisphere). To my knowledge, there have been no similar experiments carried out to test for lateralization of spatial performance in intact monkeys.

7. SURVIVING WITH A LATERALIZED BRAIN

Lateralization of both motor responses and sensory processing has a long evolutionary history and, as discussed, there has been remarkable conservation of the basic pattern since it first appeared in lower vertebrates and in the divergent avian and mammalian lines (figure 2). There is no way of deciding the primacy of sensory or motor lateralization and, in fact, this might be irrelevant to consider because both might be different manifestations of the same essential lateralizations.

Lateralization at a population level may be adaptive if it involves sensory and motor processes used in some forms of social behaviour (Rogers, 1989). For example, individuals might use it to predict, and therefore control, aggression within a social group. Given the population lateralization of lizards, toads, chicks and baboons to attack using the left eye and right hemisphere, an individual might increase its chances of not being attacked by approaching others on their right sides. This ability to predict aggressive responses in a group may stabilize the social hierarchy and, in fact, it has been shown that young chicks with lateralization of aggressive responses form more stable hierarchies than those not so lateralized (Rogers & Workman, 1989).

Population lateralization, however, might have some severe disadvantages (see also Rogers, 2000). A predator could, for example, exploit the population bias to predict the direction in which its prey might escape, or even attack. A predator might also exploit the fact that it is less likely to be detected if it approaches its prey on one side compared to the other. No matter how slight the bias might be, it could be a disadvantage to the population as a whole. Individual lateralities that are not present at a population level could not be exploited in this way. Thus, population lateralization presents itself as a special case of natural selection, irrespective of what factors cause it to develop. Even if environmental stimulation is the cause of population lateralization, as is known to be so for some of the visual lateralities in the chick (Rogers, 1990), selective pressures will operate at a population level because lateralization is unlikely to be an inconsequential characteristic.

If population lateralization is advantageous for some aspects of social behaviour, it should be more common in social species than in solitary, or at least less social, ones. Some evidence supporting this hypothesis has come from studies of lateralization in shoaling and nonshoaling fish by Bisazza, Cantalupo, Capocchiano, and Vallortigara (2000). They have now tested 16 different species of fish in the detour test described above and found that some species were lateralized at the population level, whereas others showed individual lateralization but no population bias. In addition, they have devised a new test for measuring shoaling behaviour in each species and related that to the index of lateralization in the population. Six species showed shoaling and all of these were lateralized at the population level. The ten remaining species did not shoal and more than half of these were not lateralized at the population level. This finding indicates that population lateralization may be an essential characteristic of shoaling species. For each individual fish in a shoal, the optimal direction in which to turn when avoiding a predator is in the same direction as all other members of the shoal. A population lateralization in escape turning would maintain cohesion in the shoal and be a great advantage for individual survival.

In his study showing population lateralization in a species of newt, Green (1997) argued that turning in the direction characteristic for the population (leftwards) did not enhance success of transfer of the spermatophores to the female. Hence the result did not support my hypothesis that population lateralization serves a social function. Reproductive success is, of course, an outcome of social interaction but this was not the kind of social function to which I was referring. I was applying the suggestion to social ways of controlling aggression by being able to plan the direction of approach to a conspecific, as mentioned above.

There is another, different aspect of lateralization that might affect survival and that concerns the immune system. Geschwind and Galaburda (1987) hypothesized that left handedness in humans might be associated with depressed immune responsiveness. There are some data in support of this in humans although they are controversial and far from entirely convincing. In rodents, however, lesioning studies have shown that the left and right hemispheres are differentially involved in immune responses and, of particular interest here, that immune competence is related to circling bias and hand preference (Neveu, 1988). A recent study of laterality and immune response has found that mice with a left (anticlockwise) turning preference have a weaker innate immune response than mice with a right (clockwise) turning preference (Kim, Carlson, Seegal, & Lawrence, 1999). The complete picture is complicated by the various aspects of immune function. For example, left-pawed mice have been shown to have higher mitogen-induced T lymphocyte proliferation than right-pawed ones (Neveu, Barnéoud, Vitiello, Betancur, & LeMoal, 1988). Consistent with this result, a recent study has found that right handed chimpanzees have lower lymphocyte counts than chimpanzees designated as left handed or ambiguously handed (Hopkins & Parr, 1998). Also, rats that circle to the left (anticlockwise) have been found to have higher lymphocyte stimulation indices than rats that circle to the right (clockwise) (Neveu 1988). One can conclude that, despite the complications awaiting further research to be clarified, motor biases are associated with immune competence. This has obvious implications for survival. The direction of motor lateralization might even be a factor in sexual selection (i.e. females choosing partners with better immune systems). If left handedness and left turning biases are associated with weaker immune responses, there could be selective pressure for right handedness and right-side turning biases. It is worth considering whether this might explain the predominance of right limb preferences in amphibian and mammalian species. It does not, of course, explain the exceptions in which left handedness is present at a population level, unless these species experience, or have experienced, different demands on their immune systems. The hypothalamic-pituitary-adrenal axis is very likely to have a role here, also (Kim et al., 1999).

Left- and right hand preference might also be related to more general behaviours such as exploration. A recent study in my laboratory (Cameron & Rogers, 1999) tested exploration in left- and right hand preferring marmosets by placing them singly into an unfamiliar environment. The right-handed participants explored the novel environment more actively than the left handed ones. Hopkins and Bennett (1994) found a similar result in chimpanzees. Therefore, hand preference may reflect a bias towards hemispheric dominance and a consequent bias in general behaviour or temperament. Left hand preference may reflect right-hemisphere dominance and higher levels of avoidance, consistent with the role of the right hemisphere in fear and other emotional responses. Right hand preference and dominance of the left hemisphere may lead to positive responses and approach. It is possible, therefore, that selection acts on a general characteristic such as exploration and that hand preference manifests itself as a reflection of this. Population biases in hand preference might therefore occur when it is advantageous for the majority of individuals to be more exploratory (right-handed) or more cautious (left-handed). Thus, for example, right-handed populations might have benefited from colonizing new environments.

8. DISCONTINUITY OR A CONTINUUM?

I would like to return to the notion of discontinuity in the evolution of lateralization from animals to humans. In the preface of his book "The Lopsided Ape" published in 1991, Corballis wrote, "Although I was fearful that evidence on animal asymmetries would overtake me, it still seems to be the case that right handedness and cerebral asymmetry are unique to humans

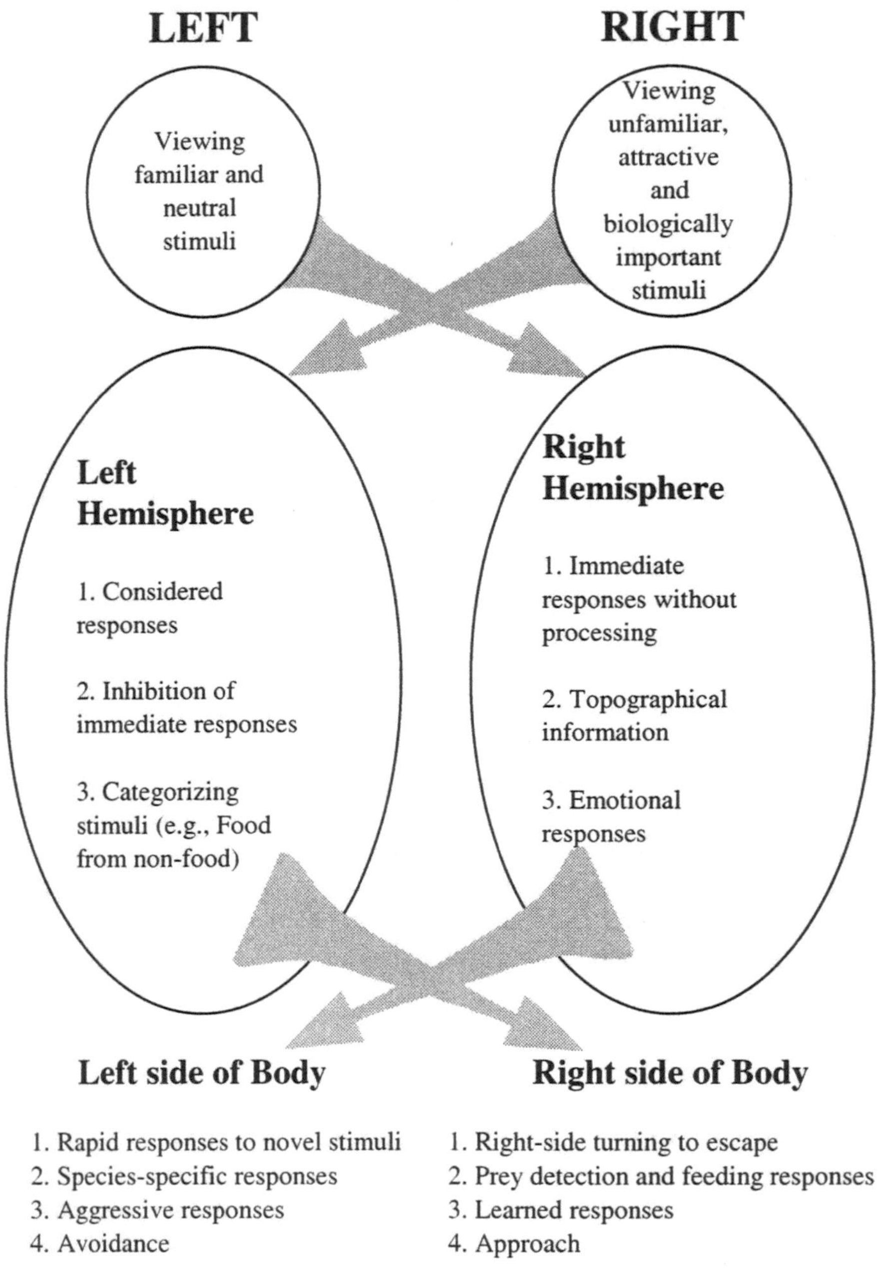

Figure 2. A summary diagram of lateralization in vertebrates

– unique not so much in their presence as in their extent, pattern, and population bias. Moreover it is not laterality per se that is critical so much as the nature of functions that are lateralized, which themselves seem to capture much of the essence of what it is to be human. Handedness is related to our extraordinary ability to manufacture and manipulate, and cerebral asymmetry is most pronounced with respect to that putatively unique faculty, language." (Corballis, 1991, p. vi).

At this time, one can say that there is sufficient evidence to refute most of the points raised in this quotation. Lateralization is an essential characteristic of language in humans but that lateralization is no more pronounced than many other forms of lateralization in other species and it is shared by communication systems in other species. In other words, lateralization and handedness are no longer attributes on which we can base a claim for human uniqueness. Population biases are as common in animals as in humans, and many forms of lateralization in animals are of the same strength, or extent, as found in humans. This chapter has not documented the strength of most of the forms of lateralization discussed but it has mentioned the 84 percent right-foot preference in the chick. The various forms of lateralized visual processing in the chick are of at least this degree of asymmetry. Footedness in Australian cockatoos is as strong as the right hand population bias in humans.

9. CONCLUSIONS

Lateralization of both sensory processing and motor function is characteristic of a broad range of species. There are many examples of it being present as a population bias. Figure 2 summarises the general pattern of hemispheric lateralization now known to be present in a number of vertebrate species and for visual, auditory and olfactory processing.

The earliest vertebrates tested, teleost fish, have lateralized turning biases and eye preferences for viewing familiar and unfamiliar stimuli. Amongst fish, we now know that the population biases for turning, at least, are present only in species that shoal. It is tempting to suggest, therefore, that an association between a population bias for turning and shoaling was the first evolutionary step for lateralization to be present at the population level. In other words, it evolved in response to the pressures of group aggregation or social behaviour.

Some forms of lateralization are notable for their persistence across species and thus evolutionary time. Here I have shown that this is the case for the specialisation of the right hemisphere for agonistic responses, and perhaps other immediate social interactions, and the left hemisphere for feeding and other responses based on making a considered decision. Once certain forms of lateralization had evolved, they were retained as a highly conserved feature of the vertebrate brain. This does not imply that they are solely genetically programmed characteristics: experience is known to establish at least some forms of lateralization (e.g., agonistic and feeding responses in the chick; reviewed by Rogers, 1996). In fact, the development of lateralization in the chick is influenced by experience and hormonal condition (Rogers, 1996) and the same appears to be true for short-term fluctuations in lateralization (Rogers, 1998). Whether a relation exists between experience in one sensory modality and lateralization in the same or other sensory modalities is one direction for future research in the field.

Rather than being an esoteric aspect of brain function, lateralization is a fundamental characteristic of the vertebrate brain essential to a broad range of neural and behavioural processes. The fact that lateralization is not unique to humans in its presence, extent or population bias makes it no less interesting. In fact, it makes lateralization an excellent basis for examining principles of brain evolution, as well as providing animal models for studying the factors that lead to and modify the development of brain lateralization.

10. REFERENCES

Alonso, J., Castellano, M. A., & Rodriguez, M. (1991). Behavioural lateralization in rats: Prenatal stress and effects on sex differences. *Brain Research, 539*, 45-50.

Bauer, R.H. (1993). Lateralization of neural control for vocalization by the frog (*Rana pipiens*). *Psychobiology, 21*, 243-248.

Andrew, R.J., Clifton, P.J., & Gibbs, M.E. (1981). Enhancement of effectiveness of learning by testosterone in domestic chicks. *Journal of Comparative Physiology and Psychology, 95*, 406-417.

Bisazza, A., Cantalupo, C, Capocchiano, M., & Vallortigara, G. (2000). Population lateralization and social behaviour: A study with sixteen species of fish. *Laterality*, in press.

Bisazza, A., Cantalupo, C., Robins, A., Rogers, L. J., & Vallortigara, G. (1996). Right-pawedness in toads. *Nature, 379*, 408.

Bisazza, A., Cantalupo, C., Robins, A., Rogers, L.J., & Vallortigara, G. (1997) Pawedness and motor asymmetries in toads. *Laterality, 1*, 161-175.

Bisazza, A., de Santi, A., & Vallortigara, G. (1999). Laterality and cooperation: Mosquitofish move closer to a predator when the companion is on their left side. *Animal Behaviour, 57,* 1145-1149.

Bisazza, A., Pignatti, R., & Vallortigara, G. (1997a). Detour tests reveal task- and stimulus-specific behavioural lateralization in mosquitofish (*Gambusia holbrooki*). *Behavioural Brain Research, 89*, 237-242.

Bisazza, A., Pignatti, R., & Vallortigara, G. (1997b). Laterality in detour behaviour: Interspecific variation in poeciliid fish. *Animal Behaviour, 54,* 1273-1281.

Bisazza, A., Rogers, L.J., & Vallortigara, G. (1998). The origins of cerebral asymmetry: A review of evidence of behavioural and brain lateralization in fishes, reptiles and amphibians. *Neuroscience and Biobehavioral Reviews, 22,* 411-426.

Bisazza, A., & Vallortigara, G. (1996). Rotational bias in mosquitofish (*Gambusia hoolbrooki):* The role of lateralization and sun-compass navigation. *Laterality, 1*, 161-175.

Bisazza, A., & Vallortigara, G. (1997). Rotational swimming preferences in mosquito fish: Evidence for brain lateralization? *Physiology and Behavior, 62*, 1405-1407.

Bradshaw, J. L., & Rogers, L. J. (1993). *The evolution of lateral asymmetries, language, tool use, and intellect.* San Diego : Academic Press.

Bradshaw, J. L., & Rogers, L. J. (1996). Tool use and evolutionary development of manual asymmetry. In D. Elliott and E.A. Roy (eds.), *Manual asymmetries in motor performance* (pp. 33-54). Boca Raton : CRC Press.

Britton, R. H., & Moser, M. E. (1982). Size specific predation by herons and its effecton the sex-ratio of natural populations of the mosquitofish Gambusia affinis. *Oecologia, 53*, 146-151.

Cameron, R., & Rogers, L. J. (1999). Hand preference of the common marmoset, problem solving and responses in a novel setting. *Journal of Comparative Psychology, 113*, 149-157.

Canfield, J. G., & Rose, G. J. (1993). Activation of the Mauthner neurons during prey capture. *Journal of Comparative Physiology, 172*, 611-618.

Cannon, C.E. (1983). Descriptions of foraging behaviour of eastern and pale-headed rosellas. *Bird Behaviour, 4*, 63-70.

Cantalupo, C., Bisazza, A., & Vallortigara, G. (1995). Lateralization of predator-evasion response in a teleost fish (*Girardinus falcatus*). *Neuropsychologia,* 33, 1637-1646.

Casperd, J. M., & Dunbar, R. I. M. (1996). Asymmetries in the visual processing of emotional cues during agonistic interactions by gelada baboons. *Behavioural Processes, 37*, 57-65.

Clayton, N. S., & Krebs, J. (1993). Lateralization and unilateral transfer in marsh tits. *Journal of Comparative Physiology, 171*, 799-806.

Collins, R. L. (1985). On the inheritance of direction and degree of asymmetry. In Glick, S.D. (ed.), *Cerebral lateralization in nonhuman species* (pp. 41-71). New York : Academic Press.

Corballis, M. (1991). *The lopsided ape: Evolution of the generative mind*. Oxford : Oxford University Press.

Cowell, P. E., Waters, N. S., & Denenberg, V. H. (1997). The effects of early environment on the development of functional laterality in Morris maze performance. In Fagot, J., Rogers, L. J., Ward, J. P., Bulman-Fleming, B., & Hopkins, W. (eds.), *Hemispheric specialisation in animals and humans* (pp. 221-232). Hove : Psychology Press.

Cynx, J., Williams, H., & Nottebohm, F. (1992). Hemispheric differences in avian song discrimination. *Proceedings of the National Academy of Sciences,* U.S.A., 89, 1372-1375.

Deckel, A.W. (1995). Laterality of aggressive responses in Anolis. *Journal of Experimental Zoology, 272*, 194-200.

Deckel, A. W. (1996). Behavioural changes in *Anolis carolinensis* following injection with fluoxetine. *Behavioural Brain Research, 78*, 175-182.

Deckel, A.W. (1998). Hemispheric control of territorial aggression in *Anolis carolinensis*: Effects of mild stress. *Brain Behaviour and Evolution, 51*, 33-39.

Deckel, A.W., & Jevitts, E. (1997). Left- vs. right-hemisphere regulation of aggressive behaviours in *Anolis carolinensis*: Effects of eye-patching and fluoxetine administration. *The Journal of Experimental Zoology, 278*, 9-21.

Denenberg, V. H. (1981). Hemispheric laterality in animals and the effects of early experience. *The Behavioural and Brain Sciences, 4*, 1-49.

Denenberg, V. H., Hofmann, M., Garbanati, J. A., Sherman, G. F., Rosen, G. D., & Yutzey, D. (1980). Handling in infancy, taste aversion, and brain laterality in rats. *Brain Research, 200*, 123-133.

Dharmaretnam, M., & Andrew, R. J. (1994). Age-and stimulus-specific effects on use of right and left eyes by the domestic chick. *Animal Behaviour, 48*, 1395-1406.

Diamond, A. C., & McGrew, W. C. (1994). True handedness in the cotton-top tamarin (*Saguinus oedipus*). *Primates, 35*, 69-77.

Dill, L.M. (1977). "Handedness" in the Pacific tree frog (*Hyla regilla).* *Canadian Journal of Zoology, 55*, 1926-1929.

Ducker, G., Luscher, C., & Schultz, P. (1986). Problemlöseverhalten von stieglitzen (*Carduelis carduelis)* bei "manipulativen" aufgaben. *Zoologisches Beitr, 23*, 377-412.

Ehret, G. (1987). Left hemisphere advantage in the mouse brain for recognizing ultrasonic communication calls. *Nature, 325*, 249-251.

Fine, M. L., McElroy, D., Rafi, J., King, C. B., Loesser, K. E., & Newton, S. (1996). Lateralization of pectoral stridulation sound production in the channel catfish. *Physiology and Behavior , 60*, 753-757.

Friedman, H., & Davis, M. (1938). "Left handedness" in parrots. *Auk., 80*, 478-480.

Geschwind, N., & Galaburda, A. M. (1987). *Cerebral lateralization: Biological mechanisms, associations, and pathology*. Cambridge, MA : The MIT Press.

Glick, S. D. (1985). Heritable differences in turning behavior of rats. *Life Sciences, 36*, 499-503.

Glick, S. D., & Shapiro, R. M. (1984). Functional and neurochemical asymmetries. In N. Geschwind & A.M. Galaburda (eds.), *Cerebral dominance: The biological foundations* (pp. 147-166). Cambridge : Harvard University Press.

Glick, S. D., & Shapiro, R. M. (1985). Functional and neurochemical mechanisms of cerebral lateralization in rats. In S.D. Glick (ed.), *Cerebral lateralization in nonhuman species* (pp. 158-184). New York : Academic Press.

Goodyear, C. P., & Ferguson, D. E. (1969). Sun-compass orientation in the mosquitofish *Gambusia affinis. Animal Behaviour, 17*, 636-640.

Green, A.J. (1997). Asymmetrical turning during spermatophore transfer in the male smooth newt. *Animal Behaviour, 54*, 343-348.

Güntürkün, O. (1985). Lateralization of visually controlled behaviour in pigeons. *Physiology and Behavior , 43*, 575-577.

Güntürkün, O., Kesch, S., & Delius, J. D. (1988). Absence of footedness in domestic pigeons. *Animal Behaviour, 36*, 602-604.

Hamilton, C. R., & Vermeire, B. A. (1991). Functional lateralization in monkeys. In F. L. Kitterle (ed.), *Cerebral laterality: Theory and research: The Toledo symposium* (pp. 19-34). Hillsdale : Erlbaum.

Harris, L. J. (1989). Footedness in parrots: Three centuries of research, theory, and mere surmise. *Canadian Journal of Psychology, 43*, 369-396.

Hauser, M. D. (1993). Right hemisphere dominance for the production of facial expression in monkeys. *Science, 261*, 475-477.

Hauser, M. D., Agnetta, B., & Perez, C. (1998). Orienting asymmetries in rhesus monkeys: The effect of time-domain changes on acoustic perception. *Animal Behaviour, 56*, 41-47.

Heinrich, M.L. & Klaasen, H.E. (1985) Side dominance in constricting snakes. *Journal of Herpetology, 19*, 531-533.

Hellige, J. B. (1993). *Hemispheric asymmetry: What's right and what's left.* Cambridge: Harvard University Press.

Hook-Costigan, M. A., & Rogers, L. J. (1996). Hand preferences in New World primates. *International Journal of Comparative Psychology, 9,* 173-207.

Hook-Costigan, M. A., & Rogers, L. J. (1998a). Lateralized use of the mouth in production of vocalizations by marmosets. *Neuropsychologia, 36,* 1265-1273.

Hook-Costigan, M. A., & Rogers, L. J. (1998b). Eye preferences in common marmosets (*Callithrix jacchus*): Influence of age, stimulus and hand preference. *Laterality, 3*, 109-130.

Hopkins, W. D., & Bard, K. A. (1993). Hemispheric specialization in infant chimpanzees (*Pan troglodytes*): Evidence for a relation with gender and arousal. *Developmental Psychobiology, 26*, 219-235.

Hopkins, W. D., & Bennett, A. (1994). Handedness and approach-avoidance behaviour in chimpanzees. *Journal of Experimental Psychology, 20*, 413-418.

Hopkins, W. D., & Parr, L. A. (1998). Lateralized behaviour and lymphocyte counts in chimpanzees (*Pan troglodytes*): A cross-sectional and longitudinal assessment. *Developmental Neuropsychology, 14*, 519-533.

Hopkins, W. D., & Morris, R. D. (1989). Laterality of visuospatial processing in two language trained chimpanzees (*Pan troglodytes*). *Behavioral Neuroscience, 103,* 227-239.

Kim, D., Carlson, J. N., Seegal, R. F., & Lawrence, D. A. (1999). Differential immune responses in mice with left- and right-turning preference. *Journal of Neuroimmunology, 93*, 164-171.

King, J. E. (1995). Laterality in hand preferences and reaching accuracy of cotton-top tamarins (*Saguinus oedipus*). *Journal of Comparative Psychology, 109*, 34-41.

Konishi, M. (1994). An outline of recent advances in bird song neurobiology. *Brain, Behaviour and Evolution, 44*, 279.

Laska, M. (1996). Manual laterality in spider monkeys: Solving visually and tactually guided food-reaching tasks. *Cortex, 32*, 717-726.

McCasland, J. S. and Konishi, M. (1981). Interaction between auditory and motor activities in an avian song control nucleus. *Proceedings of the National Academy of Sciences, U.S.A., 78,* 7815-7819.

McGrew, W. C., & Marchant, L. F. (1993). Primate ethology: A perspective on human and nonhuman handedness. In P.K. Bock (ed.), *Handbook of psychological anthropology.* New York : Greenwood Press.

McKenzie, R., Andrew, R. J., & Jones, R. B. (1998). Lateralization in chicks and hens: New evidence for control of response by the right eye system. *Neuropsychologia, 36*, 51-58.

MacNeilage, P. F., Studdert-Kennedy, M. J., & Lindblom, B. (1987). Primate handedness reconsidered. *The Behavioural and Brain Sciences, 10*, 247-303.

Marchant, L. F., & McGrew, W. C. (1996). Laterality of limb function in wild chimpanzees of Gombe National Park: Comprehensive study of spontaneous activities. *Journal of Human Evolution, 30*, 427-443.

Mathews, L. J. (1978). *The natural history of the whale.* London: Weidenfeld and Nicholson.

Mench, J. A., & Andrew, R. J. (1986). Lateralization of a food search task in the domestic chick. *Behavioral and Neural Biology, 46*, 107-114.

Miklósi, A., Andrew, R. J., & Dharmaretnam, M. (1996). Auditory lateralization: Shifts in ear use during attachment in the domestic chick. *Laterality, 1*, 215-224.

Miklósi, A., Andrew, R. J., & Savage, H. (1998). Behavioural lateralization of the tetrapod type in the zebrafish (*Brachydanio rerio*). *Physiology and Behavior*, *63*, 127-135.

Naitoh, T., & Wassersug, R. (1996). Why are toads right handed? *Nature, 380*, 353.

Neveu, P. J. (1988). Cerebral neocortex modulation of immune functions. *Life Sciences, 42*, 1917-1923.

Neveu, P. K., Barnéoud, P., Vitiello, S., Betancur, C., & LeMoal, M. (1988). Brain modulation of the immune system: Association between lymphocyte responsiveness and paw preference in mice. *Brain Research, 157*, 392-394.

Nottebohm, F. (1971). Neural lateralization of vocal control in a Passerine bird. I. Song. *The Journal of Experimental Zoology,177*, 229-261.

Nottebohm, F. (1977). Asymmetries in neural control of vocalization in the canary. In S. Harnad, R. W. Doty, L. Goldstein, J. Jaynes, & G. Krauthamer, (eds.), *Lateralization in the nervous system* (pp. 23-44). New York: Academic Press.

Nottebohm, F. (1980). Brain pathways for vocal learning in birds: A review of the first 10 years. In J. M. Sprague, & A. N. Epstein (eds.), *Progress in psychobiology and physiological psychology* (pp. 85-124). New York: Academic Press.

Nottebohm, F., Stokes, T. M., & Leonard, C. M. (1976). Central control of song in the canary, *Serinus canarius*. *Journal of Comparative Neurology, 165*, 457.

Petersen, M., Beecher, M., Zoloth, S., Moody, D., & Stebbins, W. (1978). Neural lateralization of species-specific vocalisations by Japanese macaques (*Macaca fuscata*). *Science, 202,* 324-327.

Rashid, N., & Andrew, R. J. (1989). Right hemisphere advantage for topographic orientation in the domestic chick. *Neuropsychologia, 27,* 937-948.

Ridgway, S.H. (1986). Physiological observation on dolphins' brains. In R.J. Schusterman, J.A. Thomas, & F.J. Wood (eds.), *Dolphin cognition and behavior : A comparitive approach* (pp. 31-60). Hillsdale, N.J : Erlbaum.

Robins, A., Lippolis, G., Bisazza, A., Vallortigara, G., & Rogers, L. J. (1998). Lateralized agonistic responses and hindlimb use in toads. *Animal Behaviour, 56,* 875-881.

Rogers, L. J. (1980). Lateralization in the avian brain. *Bird Behaviour, 2,* 1-12.

Rogers, L. J. (1981). Enviromental influences on brain lateralization. *Behavioural and Brain Sciences, 4,* 35-36.

Rogers, L. J. (1989). Laterality in animals. *International Journal of Comparative Psychology, 3,* 5-25.

Rogers, L. J. (1990). Light input and the reversal of functional lateralization in the chicken brain. *Behavioural Brain Research, 38,* 211-221.

Rogers, L. J. (1991). Development of lateralization. In R.J. Andrew (ed.), *Neural and behavioural plasticity : The use of the domestic chick as a model* (pp. 507-535). Oxford : Oxford University Press.

Rogers, L. J. (1995). *The development of brain and behaviour in the chicken.* Oxon : CAB International.

Rogers, L. J. (1996). Behavioural, structural and neurochemical asymmetries in the avian brain: A model system for studying visual development and processing. *Neuroscience and Biobehavioral Reviews, 20,* 487-503.

Rogers, L. J. (1997). Early experiential effects on laterality. In J. Fagot, L.J. Rogers, J.P. Ward, B. Bulman-Fleming, & W. Hopkins (eds.), *Hemispheric specialisation in animals and humans* (pp. 199-220). Hove : Psychology Press.

Rogers, L. J. (1998). Light experience and hormone levels in chick embryo affect posthatching behaviour. In N.J. Adams & R.H. Slotow (eds.), *Making rain for African ornithology: Proceedings of the 22nd International Ornithological Congress 16-22 August 1998, Durban.* Johannesburg: Birdlife South Africa, S46.2.

Rogers, L. J. (2000). Evolution of hemispheric specialisation; advantages and disadvantages. *Brain and Language,* in press.

Rogers, L. J., & Anson, J. M. (1979). Lateralization of function in the chicken forebrain. *Pharmacology, Biochemistry and Behaviour, 10,* 679-686.

Rogers, L. J., & Bradshaw, J. L. (1996). Motor asymmetries in birds and nonprimate mammals. In D. Elliott & E.A. Roy (eds.), *Manual asymmetries in motor performance* (pp. 3-31). Boca Raton : CRC Press.

Rogers, L. J., & Kaplan, G. (1996). Hand preferences and other lateral biases in rehabilitated orang-utans (*Pongo pygmaeus pygmaeus*). *Animal Behaviour, 51,* 13-25.

Rogers, L. J., & Workman, L. (1993). Footedness in birds. *Animal Behaviour, 45,* 409-411.

Rogers, L. J., & Workman, L. (1989). Light exposure during incubation affects competitive in domestic chicks. *Applied Animal Behaviour Science, 23,* 187-198.

Rogers, L. J., Ward, J. P., & Stafford, D. (1994). Eye dominance in the small-eared bushbaby, *Otolemur garnettii. Neuropsychologia, 32,* 257-264.

Rogers, L. J., Andrew, R. J., & Burne, T. H. J. (1998). Light exposure of the embryo and the development of behavioural lateralization in chicks., I: Olfactory responses. *Behavioural Brain Research, 97,* 195-200.

Rogers, L. J., Zappia, J. V., & Bullock, S. P. (1985). Testosterone and eye-brain asymmetry for copulation in chickens. *Experientia, 1,* 1447-1449.

Sherman, G. F., Garbanati, J. A., Rosen, G. D., Yutzey, D. A., & Denenberg, V. H. (1980). Brain and behavioral asymmetries for spatial preference in rats. *Brain Research, 192,* 61-67.

Sobel, N., Supin, A. Y., & Mislobodoski, M. S. (1994). Rotational swimming tendencies in the dolphin (*Tursiops truncatus*). *Behavioural Brain Research, 65,* 41-45.

Suthers, R. A. (1990). Contributions to birdsong from the left and right sides of the syrinx. *Nature, 347,* 473.

Tommasi, L., & Vallortigara, G. (1999). Footedness in binocular and monocular chicks. *Laterality, 4,* 89-95.

Vallortigara, G., Rogers, L. J., Bisazza, A., Lippolis, G., & Robins, A. (1998). Complementary right and left hemifield use for predatory and agonistic behavior. *Neuroreport, 9,* 3341-3344.

Vallortigara, G., Zandforlin, M., & Cailotto, M. (1988). Right-left asymmetry in position learning of male chicks. *Behavioural Brain Research, 27,* 189-191.

Vicario, D. S., & Yohay, K. H. (1994). Song-selective auditory input to a forebrain vocal control nucleus in the zebra finch. *Journal of Neurobiology, 24,* 288.

Walker, S. F. (1980). Lateralization of function in the vertebrate brain: A review. *British Journal of Psychology, 71,* 329-367.

Ward, J. P., & Hopkins, W. D. (1993). *Primate laterality: Current behavioral evidence of primate asymmetries.* New York : Springer-Verlag.

Ward, J. P., Milliken, G. W., & Stafford, D. K. (1993). Patterns of lateralized behavior in prosimians. In J.P. Ward, & W.D. Hopkins (Eds.), *Primate laterality: Current behavioral evidence of primate asymmetries* (pp. 43-75). New York : Springer-Verlag.

Wiechmann, A. F., & Wirsig-Wiechmann, C. R. (1992). Asymmetric distribution of melatonin receptors in the brain of the lizard *Anolis carolinensis*. *Brain Research, 593,* 281-286.

Williams, H. (1990). Bird song. In R.R. Kesner & D.S. Olton (eds.), *Neurobiology of comparative cognition* (pp. 77-126). Hillsdale : Erlbaum.

Williams, H., Crane, L.A., Hale, T.K., & Espositeo, M.A. (1992). Right-side dominance of song control in the zebra finch. *Journal of Neurobiology, 23,* 1006-1020.

Zappia, J. V., & Rogers, L. J. (1987). Sex differences and reversal of brain asymmetry by testosterone in chickens. *Behavioural Brain Research, 23*, 261-267.

Chapter 2

Genetic, Intrauterine, and Cultural Origins of Human Handedness

Jan W. Van Strien
Vrije Universiteit, The Netherlands

1. MODELS OF HANDEDNESS

Numerous models have been proposed to explain why about 90% of humans are right-handed and about 10% of humans are left handed. In this chapter the various models will be discussed.

Table 1 presents the most influential models. Some models stress the influence of pathological factors on handedness. Factors such as prenatal problems, perinatal problems, or complications soon after birth may influence normal development of the brain. These pathological influences result in a reorganization of brain anatomy leading to left handedness instead of the normal development leading to right handedness. Intrauterine models propose that prenatal forces are responsible for the distribution of left and right handedness. The prenatal influences are not necessarily pathological. Hormonal mechanisms and intrauterine position may influence the development of handedness in rather subtle ways.

M.K. Mandal, M.B. Bulman-Fleming and G. Tiwari (eds.), Side Bias: A Neuropsychological Perspective, 41-61.

Table 1. Theories of human handedness

Model	Sample references
Pathological influences	
Birth-stress	Bakan, Dibb, & Reed, 1975
Pathological left handedness	Satz, Orsini, Saslow, & Henry, 1985
Intrauterine influences	
Testosterone hypothesis	Geschwind & Galaburda, 1987
Position *in utero*	Previc, 1991
Genetic influences	
Single-gene models	Annett, 1985
Two-gene models	Levy & Nagylaki, 1972
Polygenic models	Gangestad & Yeo, 1994
Cultural influences	Collins, 1975
Hybrid models	
Gene-cultural	Laland, Kumm, Van Horn, & Feldman, 1995
Gene-intrauterine	Orlebeke, Knol, Koopmans, Boomsma, Bleker, 1996

Genetic models explain the variations in handedness by postulating genes that code for left- or right handedness or chance. Cultural models, on the other hand, state that handedness is transmitted by the interaction of the individual and his/her environment. Right-handed parents will teach their offspring to use their right hand, while the physical environment also forces a child to be right-handed. Most of the models acknowledge that there is more than one specific origin of handedness. Some models however, explicitly combine genetic and non-genetic influences. In these hybrid models, an interaction of genetic factors and either cultural or intrauterine factors is explicitly stated.

2. DEFINING HANDEDNESS

In daily life, someone is considered left-handed if he or she uses the left hand for writing. For research purposes, handedness can be defined by either preference or skill. Hand preference is determined by asking people (by means of a questionnaire) which hand they use for a number of activities. Questions concern actions like writing, grasping a tennis racket, cutting with a knife, and throwing a ball. A number of hand preference inventories have been published, with the Edinburgh Handedness Inventory

(Oldfield, 1971) being the best known. In general, this type of questionnaire is both reliable and valid (Bryden, 1987): repeated testing of a subject results in the same score and the agreement between the indicated hand preference on a certain item and the actual execution of an activity is very high.

The skill of each hand can be measured by means of a motor test, such as finger tapping or the pegboard task (Annett, 1985). Preference measures result in a J-shaped distribution (with a small peak of extremely left-handed participants , ambidexter participants in the middle, and a large peak of extremely right-handed participants). Skill measures result in a unimodal distribution with only a few extremely left-handed or right-handed participants .

The models of handedness may be based on either hand skill (e.g., Annett's right-shift model) or hand preference (e.g., the McManus model). McManus and Bryden (1992) have concluded that there is evidence that preference may be prior to skill asymmetry, that is, preference causes greater practice and thus better skill.

3. PATHOLOGICAL INFLUENCES

3.1 Birth-stress

Bakan, Dibb, and Reed (1973) argued that left handedness is a consequence of cerebral anoxia due to birth stress. The anoxia causes left-hemisphere motor damage, thus leading to a shift in handedness. Bakan et al. found that left-handed students indicated one or more birth-stress conditions on a 8-item questionnaire twice as often (41%) as did right-handed students (22%). In our own research (Van Strien, Bouma, & Bakker, 1987), we too found that left-handed students reported twice as much birth-stress conditions than did right-handed students. However, these conditions were not related to cerebral anoxia at birth but rather indicated intrauterine complications (high blood pressure in mother, low birth weight, very short labour, jaundice of newborn). Geschwind and Galaburda (1987) have suggested that complications during birth are a consequence of the same factors that influence fetal brain development (see below) rather than the cause of left handedness.

Because of obstretic complications, left handedness is thought to be more common among the first- or late- (fourth and higher) born. In addition, left handedness is thought to be more common in those born to very young or older mothers. Bakan has found a higher prevalence of left handedness among the firstborn infants of older mothers (Bakan et al., 1973) and among first- and lateborn (Bakan, 1977). However, most studies, including our own (Van Strien et al., 1987), have failed to confirm the hypothesis that birth rank and maternal age are related to handedness.

Contrary to other nongenetic theories of handedness, such as intrauterine theories, the birth-stress hypothesis maintains that all left handedness represents a pathological condition. Schwartz (1990) has called this the hard pathological position. The hard pathological position is not supported by empirical findings. The soft pathological position therefore seems more tenable. This position argues that in the majority of left-handers, handedness is defined by genetic and intrauterine influences, whereas in a subgroup of left-handers, handedness is a consequence of early pathology to the left hemisphere.

4. PATHOLOGICAL LEFT HANDEDNESS

The pathological left handedness model is an example of a soft pathological position. Satz, Orsini, Saslow, and Henry (1985) distinguished between natural left handedness and pathological left handedness (PLH). Natural left handedness is primarily determined by genetic influences, and is found in the normal population. PLH is the result of early left-hemisphere pathology and is found in clinical populations, such as patients with epilepsy or with mental retardation. By PLH is not meant the forced use of the left hand due to damage or palsy of the right hand. PLH denotes a clinical syndrome that comes about in children who are natural right-handers and in whom a predominantly left-sided lesion in the frontotemporal/frontoparietal cortex occurs before the age of six. The PLH syndrome has been described by Satz et al. (1985) and is characterized by impaired visuospatial abilities, relatively intact verbal abilities, right hemihypoplasia, and an altered pattern of speech lateralization with speech mediated by the right hemisphere or by both hemispheres. Satz et al. cite the Rasmussen and Milner (1977) study in which it was established that in left-handers of whom it was certain that they had left-hemisphere damage at an early age, 53% were right-brained for speech. In a non-clinical sample of left-handers only 15% were right-brained.

5. INTRAUTERINE INFLUENCES

Several theories postulate intrauterine influences on handedness. In contrast to the pathology hypotheses, intrauterine hypotheses do not necessarily imply (minor) brain damage that changes the standard pattern of cerebral dominance. Rather, the intrauterine models explain variation in handedness as the consequence of subtle developmental mechanisms that affect lateralization. There are two theories about the role of the intrauterine environment in the origin of handedness, the testosterone hypothesis and the intrauterine-position hypothesis.

5.1 The testosterone hypothesis

The testosterone hypothesis was put forth by Geschwind and his colleagues (e.g., Geschwind & Galaburda, 1987). According to Geschwind, functional asymmetries are rooted in anatomical asymmetries of the cerebral cortex. The most important anatomical asymmetry thought to be related to language specialization is the asymmetry of the planum temporale (PT). In 65% of the population the PT is larger on the left side (Geschwind & Levitsky, 1968). Geschwind hypothesized that, because the right hemisphere develops earlier, intrauterine factors are more likely to affect the left hemisphere, which is vulnerable over a longer period. In particular, elevated levels of fetal testosterone retard the growth of posterior locations in the left hemisphere. This slowing down of the development of the left hemisphere leads to a compensatory growth of homologous locations in the right hemisphere, so that anatomical brain asymmetries will be reduced.

According to Geschwind, individuals with a symmetric brain have an equal chance of being right-handed or left-handed. These individuals comprise the *anomalous dominance* group, which also includes individuals with a less strong left-hemisphere dominance for language. Individuals with an asymmetric brain in favour of the left side comprise the *standard dominance group*. They are mainly strong right-handers with a strong left-hemisphere dominance for language.

The key issue of Geschwind's theory is the role of elevated testosterone levels in the development of a symmetric brain. Elevated testosterone levels also affect the development of the thymus, leading to a higher incidence of immune disorders. For this reason, Geschwind's model predicts a relation between handedness and autoimmune diseases. Geschwind and Behan (1982) reported raised frequencies of autoimune diseases in left-handers and their families, a finding that could not be replicated in our own research (Van Strien et al., 1987). Bryden, McManus, and Bulman-Fleming (1994) have done a meta-analysis of the relation between handedness and immune disorders, reviewing the results for over 56,000 individuals. The conclusive outcome was that left handedness is only very slightly (1.003 times) more frequent in people with immune disorders than in controls.

Another problem with Geschwind's model is that it postulates that testosterone delays left-hemisphere development, whereas there is now evidence that it is the right hemisphere that is more sensitive to intrauterine factors. Galaburda, Corsiglia, Rosen, and Sherman (1987), reanalyzing Geschwind and Levitsky's (1968) brain data, found that leftward PT asymmetry was associated with a smaller right PT rather than a larger left PT. In symmetrical brains, the area of the right PT appeared to be increased without a significant reduction of the left PT area. Galaburda et al. concluded that the intrauterine factor that produced symmetry must act on the right hemisphere. Reviewing Galaburda et al.'s finding on PT asymmetry, Habib, Touze, and Galaburda (1990) have suggested that testosterone has a trophic influence on the right PT rather than a slowing influence on the left PT.

Although the influence of testosterone on brain asymmetry is debated, there is some support for Geschwind's suggestion that a more symmetric PT leads to anomalous dominance. Brain imaging studies have found evidence for a relation between anatomical (PT) asymmetries and handedness (e.g., Steinmetz, Volkman, Jäncke, and Freund, 1991; Foundas, Leonard, & Heilman, 1995; see Beaton, 1997, for a review).

5.2 Position in utero

Previc (1991) has traced the origins of human cerebral lateralization to asymmetries in prenatal development of the ear and vestibular apparatus. Previc argued that handedness is determined by the position of the fetus in the uterus. During the final trimester of pregnancy, about two thirds of fetuses lie head-down, with the right ear facing toward the mother's front. This position, in combination with maternal locomotory patterns, stimulates the development of the left vestibular apparatus in particular. Part of vestibular apparatus is the utricle, which responds to linear acceleration by means of hair-cell shearing. During maternal walking, the backward inertial force results in a more effective stimulation of the left fetal utricle, because the right side of the fetus faces outward and the inertial force acting upon the fetus thus is directed leftward. The left vestibular advantage has major consequences for motoric lateralization. According to Previc, the left side of the body will be better able to control balance and posture due to its stronger vestibular reflexes, whereas the right side of the body will be used for voluntary motor behaviour. The left-side advantage for controlling balance and posture emerges before the right-side advantage for voluntary motor functions. The left vestibular advantage also underlies the right-hemispheric specialization for most visuospatial functions.

Previc asserted that the emergence of motoric lateralization in humans is mainly caused by the switch to an upright position. The asymmetric stimulation of the fetal vestibular organ is a consequence of human bipedal locomotion. With quadrupedal locomotion in animals, the inertial force is directed toward the fetus' head (i.e. upward) due to the orientation of the fetus relative to forward acceleration. In this case, there is more or less symmetrical shearing of the left and right utricles. Previc also suggested that auditory lateralization develops independently of motoric lateralization. In his view, craniofacial asymmetries underlie the establishment of auditory lateralization. In most individuals, the cranial bones are larger on the right side, whereas the facial region itself is larger on the left side. The smaller right fetal craniofacial bones may cause enhanced middle-ear conduction on the right side, especially in the 1,000 to 6,000 Hz range. This right-ear advantage in the speech range contributes to the left-hemispheric advantage in speech perception. Craniofacial asymmetry emerges in early fetal development whereas vestibular asymmetry emerges in later fetal development.

Previc's theory has not been tested directly. The 2:1 ratio for right- to left-sided motoric dominance most probably cannot account for the 8:1 ratio of right- to left handedness (see Corballis, 1997), although Previc proposed that parental prompting and other sociocultural pressures increase the percentage of right handedness up to 90%.

6. GENETIC INFLUENCES

The prevalence of left handedness in children of left-handed parents is much higher than in children of right-handed parents. McManus and Bryden (1992) summarized data from 25 sets of parent-child data, with a total of 72,600 offspring. They found that if both parents are right-handed, 9.5% of the children are left-handed. If one parent is left-handed, the prevalence rises to 19.5%, and if both parents are left-handed, to 26.1%. Simple Mendelian models fail to fit these data. For instance, if one assumes that there is a dominant allele (R) for right handedness and a recessive allele (l) for left handedness, all left-handers would be homozygote ll individuals. Then, if both parents are left-handed, all offspring should be left-handed (instead of only 26.1%).

Another empirical finding that must be accounted for by a genetic model is the high proportion of monozygotic twins that are discordant for handedness (McManus & Bryden, 1992). More recent genetic models can account more or less for these findings.

The genetic models can be divided into single-gene models, which propose that one gene determines handedness, two-gene models, which propose that one gene determines handedness and a second gene determines speech lateralization, and polygenic models, which propose that no single gene is responsible for handedness.

6.1 Single-gene models

6.1.1 The Annett models

Annett was the first to introduce the role of chance in the determination of handedness. In her 1972 paper, she proposed that the basic characteristic of laterality is a normal distribution of differences between the sides. In addition, a factor, which she called the right-shift (RS) factor, biases this distribution toward the right. Handedness thus depends on two factors, an accidental and congenital but non-genetic factor, and a second, in her view possibly genetic, RS factor. The important leap in Annett's model was that handedness was conceived as the outcome of a probabilistic process. In the absence of the RS+ factor, the chance of left handedness equals the chance of right handedness ($P_l = P_r = .50$). The chance variation in handedness implies that there is no specific cause of left handedness. Or as Annett put it

"The suggestion that a genetic factor may be involved in the shift toward dextrality but not in the origin of the basic bell-shaped distribution which underlies all lateral asymmetry has the paradoxical implication that right handedness may be inherited while left handedness is not". (Annett, 1972, p. 355).

The model resolved the issue that handedness did not "breed true" and influenced other models, both genetic (McManus, 1985) and intrauterine (Geschwind & Galaburda, 1987).

Annett's early formulation of the model was based on her definition of handedness as a difference in skill between the hands. She proposed that in the group of individuals lacking the RS factor, the distribution of intermanual differences in skill is centred on zero. Annett (1974) examined differential hand skill in children of two left-handed parents, and found, as expected, an equal division between children more skillful with the right hand and those more skillful with the left hand. Annett's finding of a prevalence of about 50% left-handers in children with two left-handed parents is at variance with the 26% found in the accumulated data of McManus and Bryden (1992). Note however, that Annett (1974) based her handedness data on skill rather than preference, and that she screened out parents in whom left handedness might have been of pathological origin (see Corballis, 1997).

More recently, Annett (1985, 1995) has reformulated her model. She now explicitly proposes a RS+ allele that codes for speech representation in the left cerebral hemisphere. Right handedness is a consequence of this standard pattern of cerebral dominance. When the RS+ allele is absent (RS- allele), both the pattern of cerebral dominance and handedness are random.

In addition, Annett (1985; Annett & Kilshaw, 1983) has suggested that the RS+ allele exhibits *semi-dominance* or *additivity*. This means that in homozygote (RS++) genotypes the RS bias is expressed more strongly than in heterozygote (RS+-) genotypes. Her earlier work suggested a dominant RS+ allele and a recessive RS- allele, which implied that the RS bias was expressed equally strong in homozygote and heterozygote genotypes.

6.1.2 The McManus model

McManus (1985) asserted that handedness is determined by an autosomal locus, at which there are two alleles: D (dextral) and C (chance). Genotype DD individuals are all right-handed, whereas genotype CC individuals are right-handed (50%) or left-handed (50%) by chance. For heterozygote (DC) genotypes the model proposed *additivity*, that is, there is a 25% chance of a heterozygote individual being left-handed and a 75% chance of him or her being right-handed.

Interestingly, McManus extends his model to language dominance. The core supposition is that the C allele codes for both chance handedness and chance language dominance, and that these chances are independent of each other, whereas the D allele codes for right handedness and left-hemisphere dominance for language. In the heterozygote DC individuals there is a 25% chance of being left-handed and, independently, a 25% chance of being right-hemisphere dominant for language. So, out of sixteen DC individuals, one will be left-handed and right language dominant, nine will be right-handed and left language dominant, three will left-handed and left language dominant, and three will be right-handed and right language dominant. In homozygotic CC individuals the independent chances for handedness and language dominance are 50%, resulting in four equally probable handedness/language dominance groups. Homozygotic DD individuals are all right-handed and left language dominant. From these figures, McManus predicts that 5.98% of right-handers and 28.87% of left-handers will exhibit right-hemisphere dominance for language.

Note that according to McManus both handedness and cerebral dominance are influenced by a single gene. Annett has a similar view regarding the association of handedness and cerebral dominance, the main difference with McManus being that Annett does not assume that all homozygote RS++ individuals by definition are right-handed. Other authors have suggested that handedness and cerebral dominance are determined by two different genes instead of one single gene.

6.2 Two-gene models

Levy and Nagylaki (1972) proposed that cerebral dominance and handedness are determined by two diallelic loci. One gene locus determines the dominant language hemisphere (L or l), the other gene locus determines whether the hand ipsi- or contralateral to the dominant hemisphere will be the preferred hand (c or C, respectively). The alleles for left-hemisphere dominance (L) and contralateral hand control (C) are dominant, whereas the genes for right-hemisphere dominance (l) and ipsilateral hand control (c)are recessive. Left-handers either possess at least one L-allelle in the absence of a C-allele (left-hemisphere dominance, ipsilateral hand control) or possess at least one C allele in the absence of a L-allele (right-hemisphere dominance, contralateral hand control). Right-handers either possess at least one L-allele and one C-allele (left-hemisphere dominance, contralateral hand control) or possess no dominant alleles (right-hemisphere dominance, ipsilateral control).

In the Levy-Nagylaki model, the relation between language dominance and handedness is under genetic control, whereas in the single-gene models a random association between both phenomena is postulated. As to the degree of lateral specialization, Levy and Nagylaki hypothesized that full expression of the L-allele only occurs when a dominant C-allele is present. In other words, strong left-hemisphere language dominance occurs in the presence of contralateral hand control.

Although the Levy-Nagylaki model was in agreement with Rife's (1940) family data, the goodness of fit to other data sets was less than satisfactory. This model, too, was not able to account for the relatively low MZ and DZ twin concordances, and is no longer debated in the literature.

6.3 Polygenic models

Gangestad and Yeo (1994) have suggested a near-universal developmental design that tends to result in moderate right handedness. In their view, the precision with which this design is expressed may be affected by genetic factors that predispose for so-called developmental instability. In humans, developmental instability is reflected by minor physical anomalies (e.g., wide-spaced eyes, malformed ears) and fluctuating asymmetries (e.g., foot-breadth asymmetry). Developmental instability most probably has a highly polygenic basis and is thought to be associated with polygenic homozygosity, disadvantageous combinations of genes, and genetic imbalance. Developmental instability disrupts the developmental design of moderate right handedness and leads to either left handedness or extreme right handedness. Yeo, Gangestad and Daniel (1993) demonstrated that scores on a composite measure of developmental instability were significantly correlated with scores for hand preference and relative hand skill. The developmental-instability hypothesis predicts that left-handed parents, when compared to right-handed parents, not only produce more left-handed children but also more extremely right-handed children. Gangestad and Yeo (1994) measured hand skill (peg moving) in students and asked them to report the handedness of both parents (left vs. right). The authors found a curvilinear relation between students' relative hand performance and parental handedness. Both participants with extreme left handedness scores and participants with extreme right handedness scores reported more left-handed parents than did participants with moderate scores.

7. CULTURAL INFLUENCES

A pure cultural theory would assert that left handedness is a consequence of social conditioning and practice. For instance, the left-handed mother would teach the child to use the left hand. There is no support for this theory. The most important argument against a social-conditioning theory is the fact that children reared by left-handed foster-parents display no increase in the use of the left hand (Carter-Saltzman, 1980). Furthermore, social conditioning in the past suppressed rather than evoked left handedness. No doubt there still are strong cultural influences on handedness. There are many societies in which left-handed children are forced to use their right hand for writing (Harris, 1990). The most well-known example of social pressure is the report about Chinese school children living in Taiwan, of whom only 0.7% used their left hand for writing (Teng, Lee, Yang, & Chang, 1976). This social pressure for right-handed writing however, showed no influence on hand use in other activities.

Furthermore, left-handers live in a right-handers' world, with knives, scissors, dispensers, and equipment designed for right-handed use. According to Collins (1975), environmental biases resulting from asymmetric worlds can strongly modify lateral preferences. Collins tested a large sample of inbred mice for paw preference and found that in an unbiased environment (feeding tube positioned in the middle of the front wall of a testing cubicle), most mice were either strongly right-pawed or strongly left-pawed, resulting in a symmetrical U-shaped distribution. Because the inbred mice possessed almost no genetic variance, yet exhibited maximum phenotypic variation in pawedness, this result demonstrated that the direction of pawedness is not under genetic control. Collins therefore proposed that native pawedness is the outcome a seemingly random process.

When mice were tested in a biased environment (feeding tube positioned to either the left or the right side of the front wall) approximately 90% exhibited pawedness consistent with the environmental bias. Mice that were first exposed twice to an environmental bias in one direction, and then to an environmental bias in the opposite direction, either gravitated toward the anti-bias (group A) or resisted the anti-bias (group B). Collins assumed that group A consisted of native left-pawed mice initially tested in the right-handed environment and native right-pawed mice initially tested in the left-handed environment. Group B consisted of native right-pawed mice initially tested in a right-handed environment and native left-pawed mice originally tested in a left-handed environment. If the data of the initial tests were partitioned, mice of group B indeed appeared to have adapted more readily to the biased environment (which was compatible with their native pawedness) than did mice of group A. This outcome led Collins to conclude that the environmental bias is superimposed upon an already laterally dichotomized population. Neither a social-conditioning model nor an environmental-bias model can explain why in all human societies right handedness came to be the standard.

8. HYBRID MODELS OF HANDEDNESS

Hybrid models of handedness are models that combine aspects of different models. Two such models, the gene-cultural model and the gene-intrauterine model, will be discussed here. The gene-cultural model deviates from the pure cultural model in that it assumes a genetic factor that biases handedness to the right, while the variation in handedness is thought to be under cultural control rather than under genetic control. The gene-intrauterine model makes the important point that maternal genes may influence the intrauterine environment of the fetus, without necessarily being transmitted to the offspring.

8.1 The gene-cultural model

The gene-cultural model of Laland, Kumm, van Horn, and Feldman (1995) maintained that left- and right-handers have the same genotype, that is, no genetic variation underlies variation in handedness. Natural selection has increased the probability of right handedness from chance to a probability, estimated by Laland et al., of approximately 0.78. They hypothesized that cultural factors play a substantial role, and they considered parental handedness the most important cultural factor. The model stated that when both parents are right-handed, the probability of right handedness in their offspring increases by 0.14 (i.e., a probability of .92), whereas when both parents are left-handed, the probability of right handedness decreases by a similar amount (i.e., a probability of .64). When parents differ in their handedness, there is no parental influence and the probability of right handedness in their offspring will be the genotypic 0.78. Laland et al. explained that the parental influence on the individual's phenotype does not imply that children voluntarily copy the handedness of their parent or that parents willfully teach handedness to their children. Rather, the parental influence should be viewed only as a parameter that changes the probability of a child becoming right-handed.

Laland et al. found that the estimates based on the gene-culture model provided a good fit to 16 out of 17 data sets of family handedness, whereas the McManus (1985) model was a good fit to 12 out of 16. Despite the good fit, the theoretical rationale remains debatable. For the gene-culture model it makes no difference whether biological parents or foster-parents transmit hand preferences to their offspring. Although Laland et al. dissociate themselves from a social-conditioning theory, their model cannot account for the fact that children reared by left-handed foster-parents display no increased incidence of left handedness (Carter-Saltzman, 1980).

8.2 The gene-intrauterine model

Orlebeke, Knol, Koopmans, Boomsma, and Bleker (1996) examined hand preference in 1700 adolescent twin pairs and their parents. They found a significantly higher prevalence of left handedness in first-born twins than in second-born co-twins. Only first-born twins showed an association between low birth weight and an increased probability of left handedness. More specifically, the larger the intra-pair birth weight difference in the direction of lower weight of the first-born twin, the higher the prevalence of left handedness in the first-born twin. In addition, Orlebeke et al. found that left-handed fathers increased the probability of left handedness in their sons but not in their daughters, whereas left-handed mothers increased the probability of left handedness in both sons and daughters. Orlebeke et al. hypothesized that a maternal gene codes for the production of a hormone (possibly testosterone) in the mother herself, thus influencing the prenatal environment, whereas the paternal (Y-chromosomal) gene codes for testosterone production in the male fetus. According to Orlebeke et al., low-birth-weight and high-birth-stress children in particular are vulnerable to these hormonal influences.

9. HANDEDNESS IN TWINS

In a large meta-analysis of twins and singletons, Sicotte, Woods, and Mazziotta (1999) found that the prevalence of left handedness was significantly higher in twins than in singletons. They also found that monozygotic twins were more likely to be concordant for handedness than dizygotic twins. This outcome is in agreement with McManus and Bryden (1992), who summarized the data from 14 twin studies. McManus and Bryden concluded that 21.7% of monozygotic pairs and 22.6% of dizygotic pairs were discordant for handedness. These figures indicate that the proportion of monozygotic and dizygotic twins that are discordant for handedness is high, and that the difference in concordance rate between monozygotics and dizygotics is relatively small. McManus and Bryden calculated the expected number of discordant pairs under a binomial distribution. In dizygotic twins, discordance did not differ from binomial expectations (observed/expected ratio = .993), whereas in monozygotic twins more discordant pairs were found than would be expected (observed/expected ratio = .901). According to McManus and Bryden, these data undoubtedly suggest that there is a genetic influence on handedness.

The low concordance rate in twins must be explained by any genetic model. A conventional Mendelian model, such as that of Levy and Nagylaki (1972) fails to account for the twin data. If genes code for handedness in an absolute fashion, then twins with the same genotype must have identical handedness. As we saw, more recent genetic models of handedness draw on chance factors that contribute to the determination of handedness, and can account for the observed concordance rates in twins. For instance, the McManus (1985) model could be fitted to family and twin studies, without requiring different parameters for twins and singletons.

It has been proposed that the excess of left handedness and the low concordance in monozygotic twins may be the consequence of 'mirror imaging'. Using functional magnetic resonance imaging, Sommer, Ramsey, Bouma, and Kahn (1999) clearly demonstrated that mirror-imaging for cerebral functions can occur in healthy monozygotic twins of discordant handedness. The meta-analysis of Sicotte et al. (1999) however, revealed no difference in the prevalence of left handedness among monozygotic vs. dizygotic pairs. The similar frequencies of left handedness in monozygotic and dizygotic pairs eliminate mirror-imaging as an explanation of excess left handedness and frequent discordance in monozygotic twins.

The raised prevalence of left handedness among twins can be explained by pathological influences, such as premature birth, low birth weight, or other perinatal complications. Orlebeke et al.'s (1996) finding of a higher frequency of left handedness in first-born twins than in second-born co-twins, in connection with low birth weight provides evidence in favour of pathological mechanisms underlying left handedness in twins.

10. SEX DIFFERENCES

There is consistent evidence that the prevalence of left handedness is higher in men than in women (Harris, 1990). A Dutch survey among 9000 participants revealed that that 11.8% of men and 9.6% of women were left-handed (Dutch Central Bureau of Statistics, 1986). Various models have offered an explanation for the sex difference in handedness. The cultural model asserts that women are more apt than men to give in to social pressure against left handedness (see Harris, 1990). The testosterone theory maintains that the hormonal influences that cause left handedness are more prominent in male than in female fetuses. The birth-stress hypothesis explains the sex differences by the fact that newborn boys are larger than newborn girls, and hence have been more vulnerable to birth stress. McManus and Bryden (1992) have theorized that a recessive modifier gene (m) on the X-chromosome will inhibit the directional asymmetry of DD and DC genotypes. The modifier gene results in chance asymmetry in m males and mm females (but not in Mm females). Because m males will be [$1/P_m$ times, where P_m is the frequency of the m modifier allele] more common than mm females, there is a higher prevalence of left handedness in males. As McManus and Bryden (1992) have noted, this relatively rare sex-linked modifier gene could have interesting implications for understanding problems such as stuttering or dyslexia, which are characterized by an excess of males and an excess of left-handers. It could be hypothesized that the modifier gene not only acts on the D alleles, but also on autosomal genes involved in speaking or reading.

11. CONCLUSIONS

From this chapter it will be clear that many ideas exist about the origins of handedness. Most probably, no single model explains all aspects of the origins of human handedness. It therefore is tempting to try to integrate several views. When we focus on the similarities rather than on the differences of the various models, common properties can be found. The Annett model and the McManus model share the idea of a genetically determined bias to the right side and random sidedness in the absence of such a side bias, be it for cerebral dominance or for handedness. Annett's idea of chance handedness profoundly influenced Geschwind's thinking about anomalous dominance (Geschwind & Galaburda, 1987, p.69). In Geschwind's view, a symmetric development of the brain (due to hormonal influences), rather than the absence of a side-bias gene, will cause chance handedness. The suggestion of Orlebeke et al. (1996) that maternal genes can influence the intrauterine environment of the fetus may constitute a bridge between genetic models and the testosterone hypothesis.

An important issue that needs to be resolved however, is whether or not there is an innate bias to right handedness. In contrast to Annett's (RS+) and McManus' (D) proposal of chance asymmetry in the absence of directional alleles, other authors have suggested a natural bias to right handedness. Previc (1991) supposed that two-thirds of individuals will exhibit right motoric dominance due to the position of the fetus in the uterus. Gangestad and Yeo (1994) suggested a developmental design that results in moderate right handedness, whereas Laland et al. (1995) asserted that natural selection has increased the probability of right handedness. Likewise, Geschwind and Galaburda (1987) postulated that there is an innate bias toward standard dominance. Recent animal studies support the innate-bias hypothesis. Vallortigara, Rogers, and Bisazza (1999) review novel evidence from comparative neurosciences that shows that functional and structural lateralization of the brain is widespread among vertebrates.

In humans, the morphological asymmetries of the PT and the pars triangularis (PTr) are important in relation to functional lateralization (e.g., Foundas, Leonard, Gilmore, Fennell, & Heilman, 1994). Foundas et al. (1995) measured PT and PTr asymmetry in eight healthy right-handers and eight healthy left-handers. Twelve participants (seven right-handers, five left-handers) had a leftward asymmetry and four participants (one right-hander, three left-handers) exhibited no asymmetry or a rightward asymmetry. Left-handers as a group did not show a significant leftward asymmetry, which could be viewed as support for Geschwind's hypothesis of chance handedness caused by a symmetric PT. However, the high frequency of left-handed individuals with a leftward asymmetry, that is, a standard anatomical asymmetry, rules out this conclusion.

Anterior rather than posterior regions may be closely connected to handedness. Examining neuronal dipole generators with magnetoencephalography, Volkman, Schnitzler, Witte, and Freund (1998) found a correlation of -.76 between the asymmetry of the hand-area size in the primary motor cortex and the asymmetry of hand performance on a standardized handedness test. Amunts, Jäncke, Mohlberg, Steinmetz, & Zilles (2000) used magnetic resonance morphometry to analyse the asymmetry in depth of the central sulcus (CS) in the hand region in healthy participants. In males, the asymmetry scores decreased linearly from consistent right-handers (deeper central sulcus on the left) over mixed-handers to consistent left-handers (deeper central sulcus on the right). In females, the asymmetry scores were reduced and did not vary with handedness.

In sum, handedness appears to be determined by genetic, intrauterine and perinatal factors, the only important cultural factor being the social suppression of left handedness. Current neural imaging techniques can shed light on the influence of these factors on the biological correlates of handedness (see also Chapter 8, this volume). For instance, it would be laborious but highly informative to conduct a twin study with morphometric and functional imaging to investigate the variation in PT, PTr, and CS asymmetry. These structural and functional-imaging data, combined with the twin's handedness data, could reveal the extent of the genetic influences on these morphological asymmetries and elucidate the relations between these structures and handedness.

12. REFERENCES

Amunts, K., Jäncke, L., Mohlberg, H., Steinmetz, H., & Zilles, K. (2000). Interhemispheric asymmetry of the human motor cortex related to handedness and gender. *Neuropsychologia, 38*, 304-312.

Annett, M. (1972) The distribution of manual asymmetry. *British Journal of Psychology, 63*, 343-358.

Annett, M. (1974). Handedness in the children of two left handed parents. *British Journal of Psychology, 65*, 129-131.

Annett, M. (1985). *Left, right, hand and brain: The right shift theory.* London: Erlbaum.

Annett, M. (1995). The right shift theory of a genetic balanced polymorphism for cerebral dominance and cognitive processing. *Cahiers de Psychologie, 14*, 427-480.

Annett, M., & Kilshaw, D. (1983). Right- and left hand skill: II. Estimating the parameters of the distribution of L-R differences in males and females. *British Journal of Psychology, 74*, 269-283.

Bakan, P. (1977). Left handedness and birth order revisited. *Neuropsychologia, 15*, 837-839.

Bakan, P., Dibb, G., & Reed, P. (1973). Handedness and birth stress. *Neuropsychologia, 11*, 363-366.

Beaton, A. A. (1997). The relation of planum temporale asymmetry and morphology of the corpus callosum to handedness, gender, and dyslexia: A review of the evidence. *Brain and Language, 60,* 255-322.

Bryden, M. P. (1987). Handedness and cerebral organization: data from clinical and normal populations. In D. Ottoson (Ed.), *Duality and unity of the brain,* (pp. 55-70). Houndmills: Macmillan Press.

Bryden, M. P., McManus, I. C., & Bulman-Fleming, M. B. (1994). Evaluating the empirical support for the Geschwind-Behan-Galaburda model of cerebral lateralization. *Brain & Cognition, 26,* 103-167.

Carter-Saltzman, L. (1980). Biological and sociocultural effects on handedness: Comparison between biological and adoptive families. *Science, 209,* 1263-1265.

Collins, R.L. (1975). When left handed mice live in a right handed world. *Science, 187,* 181-184.

Corballis, M.C. (1997). The genetics and evolution of handedness. *Psychological Review, 104,* 714-727.

Dutch Central Bureau of Statistics. (1986). Left handedness. *Maandbericht gezondheidsstatistiek, 5,* 5-10.

Foundas, A. L., Leonard, C. M., Gilmore, R., Fennell, E., & Heilman, K. M. (1994). Planum temporale asymmetry and language dominance. *Neuropsychologia, 32,* 1225-1231.

Foundas, A. L., Leonard, C. M., & Heilman, K. M. (1995). Morphologic cerebral asymmetries and handedness: The pars triangularis and planum temporale. *Archives of Neurology, 52,* 501-508.

Galaburda, A. M., Corsiglia, J., Rosen, G. D., & Sherman, G. F. (1987). Planum temporale asymmetry: Reappraisal since Geschwind and Levitsky, *Neuropsychologia, 25,* 853-868.

Gangestad, S. W., & Yeo, R. A. (1994). Parental handedness and relative hand skill: A test of the developmental instability hypothesis, *Neuropsychology, 8,* 572-578.

Geschwind, N., & Behan, P. (1982). Left handedness: Association with immune disease, migraine, and developmental learning disorder. *Proceedings of the National Academy of Sciences U.S.A., 79,* 5097-5100.

Geschwind, N., & Levitsky, W.(1968). Human brain: Left-right asymmetries in temporal speech region. *Science, 161,* 186-187.

Geschwind, N., & Galaburda, A.M. (1987). *Cerebral lateralization: Biological mechanisms, associations, and pathology.* Cambridge MA: MIT press.

Habib, M., Touze, F., & Galaburda, A.M. (1990). Intrauterine factors in sinistrality: A review. In S. Coren (Ed.), *Left handedness: behavioural implications and anomalies.* Amsterdam, Netherlands: North-Holland.

Harris, L. J. (1990). Cultural influences on handedness: Historical and contemporary theory and evidence. In S. Coren (Ed.), *Left handedness: behavioural implications and anomalies.* Amsterdam, Netherlands: North-Holland.

Laland, K.N., Kumm, J., Van Horn, J. D., & Feldman, M.W. (1995). A gene-culture model of human handedness. *Behavior Genetics, 25*, 433-445.

Levy, J., & Nagylaki, T. (1972). A model for the genetics of handedness. *Genetics, 72*, 117-128.

McManus, I. C. (1985) Handedness, language dominance and aphasia: A genetic model. *Psychological Medicine, Monograph Supplement no. 8,* 1-40.

McManus, I. C., & Bryden, M. P. (1992). The genetics of handedness, cerebral dominance, and lateralization. In: Rapin, I.,& Segalowitz, S. J.(Eds.), *Handbook of neuropsychology, Vol. 6: Child neuropsychology.* Amsterdam, Netherlands: Elsevier.

Orlebeke, J. F., Knol, D. L., Koopmans, J. R., Boomsma, D. I., & Bleker, O. P. (1996). Left handedness in twins: Genes or environment*? Cortex, 32,* 479-490.

Oldfield, R. C. (1971). The assessment and analyses of handedness: The Edinburgh inventory. *Neuropsychologia, 9*, 97-113.

Previc, F. H. (1991). A general theory concerning the prenatal origins of cerebral lateralization in humans. *Psychological Review, 98,* 299-334.

Rasmussen, T., & Milner, B. (1977). The role of early left-brain injury in determining lateralization of cerebral speech functions. *Annals of the New York Academy of Sciences, 299,* 355-369.

Rife, D. C. (1940). Handedness with special reference to twins. *Genetics, 25,* 178-186.

Satz, P., Orsini, D.L., Saslow, E., Henry, R. (1985). The pathological left handedness syndrome. *Brain & Cognition, 4,* 27-46.

Schwartz, M. (1990). Left handedness and prenatal complications. In S. Coren (ed.), *Left handedness: Behavioral implications and anomalies* (pp. 75-97). Amsterdam: North-Holland.

Sicotte, N. L., Woods, R. P., & Mazziotta, J. C. (1999). Handedness in twins: A meta-analysis. *Laterality, 4*, 265-286.

Sommer, I., Ramsey, N., Bouma, A., & Kahn, R. (1999). Cerebral mirror-imaging in a monozygotic twin. *Lancet, 354*, 1445-1446.

Steinmetz, H., Volkmann, J., Jancke, L., & Freund, H. J. (1991). Anatomical left-right asymmetry of language-related temporal cortex is different in left- and right-handers. *Annals of Neurology, 29*, 315-319.

Teng, E. L., Lee, P., Yang, K., & Chang, P. C. (1976). Handedness in a Chinese population: Biological, social, and pathological factors. *Science, 193,* 1148-1150.

Vallortigara, G., Rogers, L. J., & Bisazza, A. (1999). Possible evolutionary origins of cognitive brain lateralization. *Brain Research Reviews, 30,* 164-175.

Van Strien, J. W., Bouma, A., & Bakker, D. J. (1987). Birth stress, autoimmune diseases, and handedness. *Journal of Clinical and Experimental Neuropsychology, 9,* 775-780.

Van Strien, J.W. (1995). Levels of analysis, gene proportions, left hand weakness, and genetic determinants of cerebral asymmetry. *Cahiers de Psychologie, 14,* 615-622.

Volkmann, J., Schnitzler, A., Witte, O.W., & Freund H. (1998). Handedness and asymmetry of hand representation in human motor cortex. *Journal of Neurophysiology, 79,* 2149-2154.

Yeo, R. A., Gangestad, S. W., & Daniel, W. F. (1993). Hand preference and developmental instability. *Psychobiology, 21,* 161-168.

Chapter 3

Grasp-reflex in Human Neonates: Distribution, Sex Difference, Familial Sinistrality, and Testosterone

Üner Tan
Blacksea Technical University, Turkey

1. GRASP REFLEX

The palmar grasp-reflex is one of the primitive reflexes observed in neonates. It emerges at around 11 weeks *in utero* and is inhibited or suppressed at about 2-4 months after birth. The palmar grasp-reflex is due to an inborn coordination of movements, which were described as an instinctive motion allowing a baby to practice grasping and letting go of objects (Lorenz, 1937, 1943). Stirnimann (1941) has argued that the palmar grasp-reflex might be a first expression of a social instinctive behaviour. This reflex is a cutaneo-muscular, polysynaptic reflex, which is elicited by touching the ulnar part of the palm. Characteristically, the thumb comes over and locks the object, then the baby's other fingers lock the object very tightly. One can, in fact, easily visualize the importance of this reflex for grasping mother's hands, arms, and, of course, nipples for nutrition. The palmar grasp-reflex also shares one of the most important functions of the polysynaptic reflexes, i.e., protective action.

M.K. Mandal, M.B. Bulman-Fleming and G. Tiwari (eds.), Side Bias: A Neuropsychological Perspective, 63-82.

The palmar grasp-reflex should not be present after the first year of the postnatal period. If it persists, there may be problems with writing and speaking. These children cannot juxtapose the fingers and thumb for rapid alternate movements, indicating an immaturity of the cerebellum, and poor speech/language is a likely consequence. There is no crossing in cerebellar control. Therefore, a rightsided persistence in the palmar grasp-reflex will indicate a rightsided cerebellar lesion. A palmar grasp-reflex can be enhanced by sucking, indicating a connection with the neuro-muscular system of the mouth muscles, which are of considerable importance for speaking. Accordingly, Polack (1960) reported that sucking increases activity in the hands, and active sucking and active grasping are closely associated. So, there is a coordination between the hand and mouth in the early months of life, as a tool for exploration and expression. In fact, the residual reflexes in these areas can affect speech and articulation and fine motor control in later life. Thus, it is conceivable that the mouth and hand muscles are co-activated in newborns, as is evident much later in life. There is indeed an association between fine motor control of the mouth and hand muscles from birth to death. A retained palmar grasping reflex can result in poor manual dexterity, indicating an association of this primitive neonatal reflex with handedness.

Considering the above-mentioned co-activation of the palmar grasping reflex in neonates with the exploring mouth reflexes, it can be hypothesized that the grasp-reflex observed in human neonates may be an essential element for the development of speech and handedness in human beings. That is, the grasp-reflex may be a basic element for the development of fine motor activities. If so, we should be able to see some features of the adult manual and speech lateralizations in the grasp-reflex of the human neonates. It was also hypothesized that speech and handedness may develop in parallel under genetic and environmental influences. These hypotheses were tested in the present article.

It is well known that babies use their mouth muscles to explore objects during early development; this action is then modified and replaced by speech action, a more elaborate and a more skilled kind of motor activity. At the same time, the grasp-reflex is also modified and replaced by a more elaborate and a more skilled action of the hands. During this developmental stage, the pure grasping action of the neonate is not inhibited by the cerebral cortical areas, it is rather modified by the developing cortex, to take over the actions that are more involved in fine motor activity in order to exert more accurate movements to reach a goal.

The present chapter will consider (i) the distribution of the right minus left grasp-reflex strengths in neonates, to examine the similarities with the distribution of handedness in adults; (ii) the effects of familial sinistrality on the grasp-reflex asymmetries, to examine genetic influences; (iii) the sex-related differences in the grasp-reflex asymmetries; (iv) and the association of testosterone with the grasp-reflex strengths, to reveal the effects of one of the most important environmental factors acting during perinatal development.

To analyse the above mentioned issues, the palmar grasp-reflex was quantitatively measured in 160 female and 167 male neonates (N = 327). The grasp reflex was elicited by a small balloon brought into contact with the ulnar part of the palmar surface of hand. This balloon was connected to a pressure transducer to record the grasp-reflex strength on a polygraph. The grasp-reflex strength was measured as peak-to-peak amplitude of the polygraph deflections, which were expressed as arbitrary units. Ten reflexes were measured from the right and left hands alternately in fully awake babies. The parents were asked about the prevalence of left handedness in the family. If there were one or more left-handers in the family (mother, father, siblings), the baby was considered as positive for familial sinistrality (FS+), otherwise the baby was taken as negative for familial sinistrality (FS). To measure the free and total testosterone levels, blood samples were taken from the umbilical artery just after birth. The serum testosterone concentration was then quantitatively measured using a solid-phase, radioimmunoassay technique (Coat-A-Count), which is commercially available (Diagnostic Products Corporation, USA).

2. THE DISTRIBUTION OF HAND PREFERENCE (THE RIGHT MINUS LEFT GRASP REFLEX)

For females, 42 (26.3%) were right handed (significantly stronger right hand), 15 (9.4%) were left handed (significantly stronger left hand), and 103 (64.4%) were mixed handed (no significant difference between hands). For males, 42 (25.1%) were right handed, 12 (7.2%) were left handed, and 113 (67.7%) were mixed handed. These results were recently reported by Tan & Tan (1999) elsewhere. Sex was not a significant factor for these distributions, χ^2 (2) = 0.70, $p > .10$. Our further unpublished observations showed that the number of the mixed-handed males (N = 113) significantly exceeded the number of the mixed-handed females (N = 103). The number of males with asymmetric reflexes (N = 54) was significantly less than the number of females with asymmetric reflexes (N = 57), χ^2 (1) = 5.32, p < .05.

The ratios of reflex strengths to body weight were taken as relative reflex strengths, since body weight significantly correlated with the grasp-reflex strengths from the right and left hands (right: r = .22, t (323) = 4.00, p < .001; left: r = .24, t (323) = 4.47, p < .001). Table 1 shows that the mean right-left (R-L) reflex strength was significantly greater than zero [raw score: t (323) = 6.65, p < .001, and relative score: t (323) = 7.17, p < .001].

Table 1. Right minus left grasp-reflex strengths in the male and female neonates

Participants	N	Mean (raw)	Mean(/weight)
Total	324		
Females	158	0.61 ± 1.44	0.21 ± 0.50
Males	166	0.48 ± 1.50	0.17 ± 0.46
Total (R-L>0)	211		
Females	99	1.21 ± 1.51	0.41 ± 0.54
Males	112	1.01 ± 1.43	0.33 ± 0.45
Total (R-L<0)	113		
Females	59	-0.40 ± 0.44	-0.12 ± 0.13
Males	54	-0.63 ± 0.94	-0.18 ± 0.26

The mean R-L reflex strength tended to be more right shifted in females than in males, but the difference was not statistically significant [raw score : F (1, 322) = 0.66, p > .10]. For the neonates with R-L > 0, the mean grasp-reflex strength was greater in females than males, but sex was not a significant factor [raw score : F (1, 209) = 0.94, p > .10, and relative score : F (1, 209) = 1.56, p > .10].

For participants with R-L < 0, the mean R-L grasp-reflex strength was smaller in males than females, that is, males tended to be more left handed than females. However, the difference did not reach the traditional level of significance, F (1, 111) = 3.07, p = .08). The sex effect was not significant for the relative R-L reflex strengths, F (1, 111) = 1.09, p > .10). The number of participants with R-L > 0 (N = 211) significantly exceeded that with R-L < 0 (N = 113, χ^2 = 19.36, p < .001).

The above presented observations indicated that the difference between the relative numbers for males and females including right-, left-, and mixed-handed participants was not significant, although females tended to be more right- and left handed and less ambidextrous than males. Considering only the mixed-handed participants , males were significantly more ambidextrous than females. Furthermore, female neonates tended to be more asymmetrical than males. Considering the hand preference in adults, females were also found to be more right- and left handed than males, but males were more mixed handed than females (Tan, 1988). Thus the grasp-reflex strength may be taken as an index for the future development of human handedness.

Interestingly, the percentages for the right-, left-, and mixed-handed participants (26.3%, 9.4%, and 64.4%, respectively) were very close to those found in most of the studies concerning the morphological and physiological asymmetries. For instance, Annett (1972) has reported that 66.8% of her participants exhibited consistent right handedness, 3.7% consistent left handedness, and 29.5% were mixed-handers. Tan (1988) found similar proportions for a Turkish sample: 66.1% consistent right-handers, 3.4% consistent left-handers, and 30.5% mixed-handers. From a morphological standpoint, Geschwind and Levitsky (1968) studied 100 brains after death and found the planum temporale larger on the left in 65 cases, on the right in 11, and not clearly different in 24 cases. These results were confirmed for fetal brains (Chi, Dooling, & Gilles, 1977; Wada, Clarke, & Hamm, 1975; Witelson & Pallie, 1973). Similarly, about two-thirds of humans possess a larger left facial region (Burke, 1971; Keles, Diyarbakirli, Tan, & Tan, 1997; Lundstrom, 1961; Vig & Hewitt, 1975; Woo, 1931), which presumably originates in early fetal life (Trenouth, 1985).

The percentages reported by Tan & Tan (1999), 64.4, 9.4, and 26,3% (see above) are close to those reported by Geschwind and Levitsky (65, 11, and 24%). Tan & Tan s percentages seem to follow a "*2/3 principle*" in cerebral laterality. Accordingly, it has been shown that the right ear is more sensitive than the left ear in approximately two-thirds of the adult population (Ward, 1957). The summated potentials recorded from the right cochlea also follow the same proportion (Chatrian, Wirch, Edwards, Turella, Kaufman, & Snyder, 1985). It was frequently reported that the "natural" ratio of right to leftsided motor laterality is in the vicinity of 2/3 (e.g., Azemar, 1970; Grapin & Perpere, 1968).

It is remarkable that the proportion for left handedness is almost the same as reported for handedness (5-10%, see Annett, 1985) and planum temporale asymmetries (11.0%, see Geschwind & Levitsky, 1968). Tan & Tan (1999) have reported that the percentages for left handedness were 7.2 and 9.4 for the female and male neonates, respectively. Interestingly enough, Coren &

Porac (1977) examined the distribution of handedness dating back some 5000 years and found that the proportion of left-handers has been remarkably constant at about 10% over the entire period. Thus here is a parallelism between the prevalences of left- handedness in adults over a time period of 5000 years and left handedness in human neonates. This clearly implies that left handedness is not chance; it must be largely genetic, operating throughout human history. Accordingly, Tan and Tan (1997) have analysed the distribution of the L-R peg-moving times and found that left-handers comprised a single normal distribution on the left side of zero difference, which did not match with a chance event.

Comparing males and females, the relative number for the males with symmetric grasp-reflex strengths (mixed-handers) significantly exceeded that for the female neonates with symmetric grasp-reflex strengths. This implied that females were more left and/or right handed than males, but males were more ambidextrous than females for the human neonates, indicating that females were more lateralized than males. Gur et al. (1982) have suggested that females may be more asymmetrical than males in the cerebral representation of functions. Borod, Caron, and Koff (1984) have reported that females had larger dominance ratios than males for preference measures, indicating that females were more lateralized than males. Tan (1988) has reported that women are more right- or left handed than men for hand preference. Moreover, Keles et al. (1997) have found that a lack of facial asymmetry was encountered more frequently in left-handed men than in left-handed women. Thus there is supporting evidence for a more asymmetric organization of the female than the male brain. There are, however, also studies showing reverse findings, i.e., males are more lateralized than females (see for instance Bryden, 1979; McGlone, 1978, 1986; Levy, 1972, 1976; Waber, 1976; Wada, Clarke, & Hamm, 1975).

One of the results of my unpublished observations (see above) was that the mean R-L grasp-reflex strength was significantly greater than zero; the number of participants with R-L > 0 (right dominant: 65.1%) significantly exceeded the number of participants with R-L < 0 (left dominant: 34.9%). This indicates that the R-L grasp-reflex strength exhibited a right hand dominance for the total sample (see also Tan, Ors, Kurkcuoglu, Kutlu, & Cankaya, 1992a). However, a great majority of the neonates, about 65.0%, did not exhibit an asymmetric grasp reflex, if the mean values from right and left were compared statistically; only about 25.0% were found to be right handed. This implies that the grasp reflex is not lateralized in most of the human neonates, but it tends towards a right shift slightly. Accordingly, Pollack (1960) could not detect an asymmetry between the right and left hands in response latency, strength, and endurance of grasp reflex in neonates. Roberts and Smart (1981) have reported no difference between sides in grasp endurance. McGraw (1940) has investigated the gripping or clinging (proprioceptive) phase of the grasp reflex and reported no difference between the two hands. On the contrary, Halverson (1936, 1937a, 1937b, 1937c), identifying two aspects of the grasp reflex (finger closure and gripping or clinging response to a pull on the finger tendons), found a slight overall superiority of the left hand compared with the right hand. Yu-Yan, Cun-Ren, and Over (1983) have also found that the grasp duration for holding the rattle was significantly longer for the left hand than for the right hand.

There are also reports indicating a right dominance for the grasp reflex. For instance, Caplan and Kinsbourne (1976) have recorded the time of holding a rattle in one or both hands and found a significantly longer holding time for the right hand than for the left hand when the hands were tested separately, but not when tested together. Petrie and Peters (1980) have reported similar results for their 2-month-old infants (see also Hawn & Harris, 1983).

A consideration of the above studies does not reveal any clear-cut evidence for an asymmetric grasp reflex in human neonates. The present work showed a symmetric grasp reflex in about two thirds of the human neonates. The grasp-reflex strength exhibited a slight tendency toward right handedness. For this reason, it can be concluded that the grasp reflex tends to be right dominant.

As pointed out above (see also Table 1), the mean R-L grasp-reflex strength was greater in females than males for the total sample and for the total right-handers. Thus females seemed to be more right handed than males, but the difference did not reach the traditional level of statistical significance. On the other hand, males tended to be more left handed than females although the difference was not statistically significant. However, Tan, Ors, Kurkcuoglu, Kutlu, and Cankaya (1992c) have reported that there was no significant difference between the mean grasp-reflex strengths from the right and left hands of the male neonates, although the the mean grasp-reflex strength from the right hand was significantly greater than that from the left hand for females. This indicated stronger lateralization in females than males towards right handedness.

The nonsignificant sex differences may be due to small sample sizes. Accordingly, Annett (1985, p. 77) has argued that "when the small absolute size of the differences between the sexes .... is noted, it will be readily understood that in smaller samples (hundreds rather than thousands) the sex difference may be statistically insignificant, absent, or even reversed. However, there can be no reasonable doubt that males are more likely to be left handed than females, but by a small margin". On the other hand, Annett (1972) has studied hand skill (peg-moving task) in school children and in undergraduates and found that females were considerably more asymmetrical in favour of the right hand than males. The small sex difference for the grasp reflex in neonates and the large sex difference for the hand skill in adults can be explained by the development of the cortical motor system. Namely, there is no cortical control for the grasp reflex in neonates, but hand skill is under a strong cortical motor control, to create fine motor skills in adults. It is therefore conceivable that sex differences in fine motoric skill may be due to differential development of the motor cortex in males and females.

3. FAMILIAL SINISTRALITY

Table 2 shows the mean R-L grasp-reflex strengths in FS- and FS+ neonates for the total sample, right-handers (R-L > 0) and left-handers (R-L < 0). The mean R-L grasp reflex was found to be significantly less in FS+ participants than in FS- participants for the total sample, females, and right-handed participants . For the left-handed participants , the mean R-L grasp reflex was found to be significantly smaller in FS+ than FS- participants . There was an insignificant difference between the mean R-L reflex strengths for FS- and FS+ females; FS+ males showed significantly more left handedness than FS- males.

The mean R-L grasp-reflex strengths were found to be 2.1±1.99 for the FS- right-handers (N = 63), 1.16±1.01 for the FS+ right-handers (N = 21), - 0.34±1.12 for the FS- left-handers (N = 20), and - 1.19±1.18 for the FS+ left-handers (N = 7). ANOVA yielded the following results: $\underline{F} = 4.63$, $\underline{p} < .05$ for the main effects (combined), $\underline{F} = 1.46$, $\underline{p} > .10$ for sex, $\underline{F} = 7.53$, $\underline{p} < .005$ for FS, and $\underline{F} = 1.36$, $\underline{p} > .10$ for the Sex * FS interaction.

FS was not a significant factor for the grasp-reflex strength from the right hand [raw score: $\underline{F}\ (1, 321) = 1.14$, $\underline{p} > .10$, and relative score : $\underline{F}\ (1, 321) = 0.37$, $\underline{p} > .10$]. However, FS was found to be a significant factor influencing the grasp-reflex strength of the left hand [raw score : $\underline{F}\ (1, 321) = 6.98$, $\underline{p} < .01$, and relative score : $\underline{F}\ (1, 321) = 3.76$, $\underline{p} = .05$]. That is, the mean grasp-reflex strength of the left hand was significantly greater in FS+ participants (8.06±2.61) than in FS- participants (7.06±3.09).

Table 2 The mean R-L grasp-reflex strengths for FS- and FS+ participants

Participants	N	Mean	SD	t	df	P
Total(R-L)	324					
FS-	236	0.70	1.05			
FS+	88	0.12	0.94	3.22	322	< .005
FS- fem.	117	0.68	0.66			
FS+ fem.	41	0.43	0.09	2.85	156	=.005
FS- males	118	0.73	0.98			
FS+males	47	-0.15	0.48	5.88	163	< .001
Total (RH)	210					
FS-	157	1.25	1.19			
FS+	53	0.69	0.62	3.28	208	< .001
FS- fem.	71	1.36	1.15			
FS+ fem.	28	0.83	0.56	2.32	97	< .05
FS- males	86	1.16	1.25			
FS+males	25	0.53	0.49	2.38	109	< .05

R-L: right minus left reflex strength; RH:R-L > 0; LH: R-L < 0

The above results (see also Table 2) indicated that the R-L grasp-reflex strength was significantly smaller in FS+ than in FS- participants . That is, FS was a significant factor for the grasp-reflex asymmetry, which was shifted to the left under the influence of familial sinistrality (see also Tan et al., 1992b). Considering males and females separately, the R-L grasp reflex was found to be smaller in FS+ females than FS- females, but the difference was statistically insignificant. The reflex asymmetry was also shifted to left, and even became negative (left handedness) under the influence of FS for the male participants . Such a reversed asymmetry was not established for visual asymmetries (see Annett, 1985, p. 136).

McKeever and VanDeventer (1977) reviewed studies of perceptual asymmetries considering sex and FS and reported that there are many inconsistencies among studies, but these authors concluded that both FS and sex may moderate the degree of cerebral dominance for language processing. This conclusion is consistent with the above results for the grasp reflex in neonates. Annett (1985, p. 146) has argued that " .... if FS+ reduces the presence of the typical pattern of cerebral specialization, groups should be ordered as follows: The strongest asymmetries should be found in FS- right-handers, followed by FS+ right-handers, then FS- left-handers, and FS+ left-handers". The same order for the mean R-L grasp-reflex strengths was established in the present study (see above). So, FS+ may be taken as an important factor clearly reducing the grasp-reflex asymmetry in human neonates. The influence of FS on the grasp-reflex asymmetry suggests that FS+ participants may have a genetic predisposition to left handedness. Hopkins, Bales, and Bennett (1994) have reported that offspring of chimpanzees had the same hand preference as their biological parents significantly more often than would be predicted by chance alone. These results strongly suggested a heritability component to the expression of hand preference in these animals.

According to Annett (1985, p. 246), the classical model assumed that right-handers were left-brained and left-handers right-brained. If handedness was genetically determined, then a gene for right handedness would be a gene for left-brainedness and a gene for left handedness would be a gene for right-brainedness. Heterozygosity could be associated with variability of brainedness; patients with FS could have a better chance of recovery from dysphasia; FS+ right-handers might be more likely to be heterozygotes (see Annett, 1985, p. 246). So, the outcome of some cerebral diseases might be predicted by the assessment of the R-L grasp-reflex strength with FS. Annett (1985, p. 387) argued, however, that ".. the presence of sinistral relatives should not have strong implications for individual laterality". There are some other concerns about using FS in laterality research (see Peters, 1995; Bradshaw, 1989; Bishop, 1990). Some authors have not found FS to be important (e.g., Bryden, 1975; McKeever &VanDeventer, 1977; Newcombe & Ratcliff, 1973). However, there are also a considerable number of articles reporting significant differences between FS- and FS+ participants (e.g., Bradshaw, Nettleton, & Taylor, 1981; Searleman, Hermann, & Coventry, 1984; Pipe, 1987). There is, in fact, evidence for a familial influence on handedness, since two left-handed parents are more likely to have left-handed offspring than are two right-handed parents. When only one parent is left handed, the probability of left-handed offspring becomes intermediate (Annett, 1974; Chamberlain, 1928; Rife, 1940). Hécaen and Sauguet (1971) have reported that FS- left-handers were essentially indistinguishable from right-handers in dyslexia. By contrast, the FS+ left-handers showed evidence of a higher prevalence of bilateral representation of language, with more diffuse representation both within and between cerebral hemispheres (see also Andrews, 1977; Satz, Fennel, & Jones, 1969; Zurif & Bryden, 1969). Moreover, Hécaen, De Agostini, and Monzon-Montes (1981) have found that FS+ left-handers showed a higher incidence of ambilateral representation of language, whereas FS- left-handers had language representation in the left cerebral hemisphere as did the right-handed participants . Hardyck and Petrinovic (1977) have suggested that perceptual asymmetry is attenuated in FS+ participants (see also Bryden, 1965; Kraft, 1981; Satz et al., 1969; Zurif & Bryden, 1969). A diminished laterality in FS+ participants was more consistent for dextrals and the majority of these studies reported less perceptual asymmetry in FS+ dextrals than in FS- dextrals (Hines & Satz, 1971; Kraft, 1981; McKeever & VanDeventer, 1977; Snyder, 1978; Springer & Searleman, 1980; Varney & Benton, 1975). There were, however, the opposite results, too (Briggs & Nebes, 1976; Hines & Satz, 1974; Snyder, 1979). Kee, Bathurst, and Hellige (1983) have studied the effects of concurrent verbal tasks on repetitive finger tapping from the right and left hands. These authors have found that concurrent verbal tasks interfered with right-hand tapping than with left-hand tapping. This pattern was more pronounced for FS- participants than FS+ participants , who exhibited a left-hand interference. The results of the present work also showed that FS+ increased the grasp-reflex strength from the left hand and caused a decrease in the R-L grasp-reflex strength. So, the latter studies are consistent with those reported in the present work (see my unpublished observations). That is, FS reduces the grasp-reflex asymmetry, and even shifts it towards left handedness by a selective action on the grasp reflex from the left hand.

4. TESTOSTERONE AND GRASP REFLEX

Total sample Table 3 presents the mean testosterone levels for the right-, left-, and mixed-handed participants for the grasp reflex in human neonates (unpublished observations). For the free testosterone levels, the difference between groups did not reach a traditional level of significance, F (2, 55) =2.56, $p < .10$. The left-handers had the lowest mean testosterone level, which was significantly less than that of the right-handers, t (35) = 2.23, $p < .05$, and mixed-handers, t (31) = 2,01, $p < .05$. For the total testosterone levels, the difference between groups was also statistically significant, F (2, 55) = 3,34, $p < .05$. The right-handers had the highest testosterone level followed by the mixed-handers and the left-handers.

Table 3. The mean testosterone levels in right- (RH), left- (LH), and mixed-handed (MH) neonates

Participants	**N**	**free testos. (Ng/dL)**	**SD**	**tot. testos. (Ng/dL)**	**SD**
RHs	25	7.81	4.21	127.1	62.7
LHs	12	4.88	2.44	77.3	38.2
MHs	21	8.44	5.55	124.0	60.4
RH male	17	8.26	4.42	127.8	57.1
LH male	4	6.79	3.56	107.2	50.9
MH male	11	9.87	5.47	153.3	57.0
RH fem.	8	6.78	3.27	124.0	67.4
LH fem.	8	3.93	1.38	62.3	18.5
MH fem.	10	6.67	5.43	90.9	59.3

4.1 Males and females

The mean free testosterone levels were found to be 5.9±3.6 ng/dL and 7.8±4.2 ng/dL for the female (N = 35) and male (N = 47) neonates, respectively. The difference between males and females was statistically significant, $t\ (80) = 2.15$, $p < .05$. The mean total testosterone levels were found to be 112.5±58.9 ng/dL and 142.2±69.2 ng/dL for the female (N = 35) and male (N = 47) neonates, respectively. The difference was statistically significant, $t\ (80) = 2.05$, $p < .05$. In females, the mean total testosterone level was significantly greater in right-handers than left-handers, $t\ (14) = 2.27$, $p < .05$. There was, however, no significant difference between the mean free testosterone levels of the left- and mixed-handed participants , $t\ (16) = 1.38$, $p > .10$. An identical result was also found for the mean total testosterone level, that is, the mean testosterone concentration was significantly greater in right- than in left-handers, $t\ (14) = 2.5$, $p < .05$. The difference between the mean testosterone levels for the left- and mixed-handers was not significant, $t\ (16) = 1.31$, $p > .10$. In males, the difference between the left- and right-handed participants was insignificant both for the mean free testosterone levels, $F\ (2, 29) = 0.75$, $p > .10$, and for the mean total testosterone level, $F\ (2, 29) = 1.20$, $p > .10$.

The total testosterone concentrations were 34.9±41.2 and 74.5±76.2 ng/dL for the FS- and FS+ females, respectively. The difference was statistically significant, $\underline{t}$ (44) = 2.26, $\underline{p} < .05$. The free testosterone levels were 4.2±2.8 and 7.7±3.3 ng/dL for the FS- and FS+ participants , respectively. The difference was statistically significant, $\underline{t}$ (24) = 2.93, $\underline{p} < .01$. There was not a significant difference between the total and free testosterone levels for the FS- (total: 95.9±93.6 ng/dL) and FS+ males (total: 96.7±89.5 ng/dL).

The above presented results indicated that the mean neonatal testosterone level was significantly higher in males than females. This is consistent with most of the previous reports (see Jacklin, Maccoby, Doering, & King, 1983). Interestingly enough, the free and total testosterone concentrations were found to be significantly lower in left-handers than right- and mixed-handers. This is inconsistent with Geschwind`s testosterone theory of cerebral lateralization (Geschwind & Behan, 1982; Geschwind & Galaburda, 1987), associating sinistrality to prenatal testosterone, which purportedly slows down the normal development of the left cerebral hemisphere with a subsequent compensatory growth of the right cerebral hemisphere. If this theory were correct, the neonatal testosterone level should be highest in left-handers. By contrast, the above presented unpublished observations indicated that left-handers had the lowest testosterone levels in human neonates (see Table 3) with higher testosterone levels in the right- and mixed-handers. Drea, Wallen, Akinbami, and Mann (1995) have examined hand use in 1-year-old rhesus monkeys that experienced different neonatal hormone environments. These authors did not find any relation between exogenous neonatal hormone treatments and left hand use, but elevated neonatal testosterone levels strengthened the degree to which monkeys showed a hand preference, i.e., neonatal exposure to elevated testosterone increased sinistrality in some monkeys, but promoted right handedness in others. In the present study, testosterone was lowest in left-handers and highest in right-handers. Both studies, showing some similarities, did not support Geshwind`s testosterone theory of cerebral lateralization. However, my results are consistent with those reported by Moffat and Hampson (1996), who found that salivary testosterone levels were significantly lower in left-handed adults than in their right-handed counterparts of both sexes. Thus, it is conceivable to conclude that prenatal and postnatal testosterone may be involved in cerebral lateralization but not in line with Geschwind`s theory. However, there may be other factors playing a role in neonatal cerebral laterality. For instance, Churchill, Igna, and Snef (1962) have reported a significant correlation between adult handedness and position at birth. Other studies indicated that handedness might be associated with fetal position (Gesel & Ames, 1950; Michel & Goodwin, 1979; Moss, 1929). Previc (1991) offered another explanation for the prenatal origins of cerebral lateralization. Accordingly, Tan (1994c) has reported that the grasp-reflex asymmetry may, at least partly, depend upon prenatal position, which may influence the later developing handedness in humans (see also Tan & Zor, 1994). Hormones other than testosterone were also reported to be associated with grasp-reflex asymmetries (see for instance Tan, 1994a,b; Tan & Zor, 1993, 1994).

5. CONCLUSIONS

It was suggested that the percentages for the handedness groups assessed by grasp reflex in human neonates are consistent with percentages for the morphological and physiological asymmetries. This was tentatively called a "2:3 principle" of cerebral lateralization. About 66.0% ambidexterity suggests a basic symmetry in grasp reflex in human neonates which will be modified later by the development of the cerebral cortex. That is, the asymmetric development of the cerebral cortex seems to be essential for the development of future handedness. Interestingly, the percentage of the left-handed neonates was very close to that observed in adults, suggesting a prenatal development of left handedness under genetic influences. The overall tendency was towards right handedness. Males were more mixed handed than females; females were more right handed than males. These sex differences were, however, not large enough to be conclusive. The genetic origin of the grasp-reflex asymmetry was clear from the analysis of familial sinistrality. Namely, familial sinistrality caused a left shift, which was created by a special increase in the grasp-reflex strength from the left hand. Geschwind`s theory of cerebral lateralization was not supported by the above presented unpublished findings, since left handedness was found to be associated with the lowest testosterone levels in neonates; right-handers had the highest testosterone levels. There are, however, many other environmental factors affecting the laterality of the grasp reflex. The present unpublished results suggest that left handedness may be inborn but the emergence of right and mixed handedness seems to be a longlasting developmental event awaiting especially the asymmetric or symmetric development of the cerebral cortex in humans.

6. REFERENCES

Andrews, R. J. (1977). Aspects of language lateralization correlated with familial handedness. *Neuropsychologia, 15,* 769-778.

Annett, M. (1972). The distribution of manual asymmetry. *British Journal of Psychology, 63,* 343-358.

Annett, M. (1974). Handedness in the children of two left handed parents. *British Journal of Psychology, 65,* 129-131.

Annett, M. (1985). *Left, right, hand and brain: The right shift theory.* Hillsdale : Lawrence Erlbaum.

Azemar, G. (1970). *Sport et laterality. Les asymmetries fonctionelles (Sports and laterality. Functional asymmetries).* Paris: Editions Universitaires.

Bishop, D. V. M. (1990). *Handedness and developmental disorder.* Hillsdale : Lawrence Erlbaum.

Borod, J. C., Caron, H. S., & Koff, E. (1984). Left handers and right handers performance and preference measures of lateral dominance. *British Journal of Psychology, 75,* 177-186.

Bradshaw, J. L. (1989). *Hemispheric specialization and psychological function. Chichester* : John Wiley & Sons.

Bradshaw, J. L., Nettleton, N. C., & Taylor, M. J. (1981). Right hemisphere language and cognitive deficit in sinistrals? *Neuropsychologia, 19,* 113-132.

Briggs, G. G., & Nebes, R. D. (1976). The effects of handedness, family history, and sex on the performance of a dichotic listening task. *Neuropsychologia, 14,* 129-133.

Bryden, M. P. (1965). Tachistoscopic recognition, handedness, and cerebral dominance. *Neuropsychologia, 3,* 1-8.

Bryden, M. P. (1975). Speech lateralization in families: A preliminary study using dichotic listening. *Brain & Language, 2,* 201-211.

Bryden, M. P. (1979). Evidence for sex-related differences in cerebral organization. In M. A. Whitting & A. Peterson (Eds.), *Sex-related differences in cognitive functioning* (pp. 121-143). New York: Academic Press.

Burke, P. H. (1971). Stereophotogrammatic measurement of normal facial asymmetry in children. *Human Biology, 43,* 536-548.

Caplan, P. J., & Kinsbourne, M. (1976). Baby drops the rattle: Asymmetry of duration of grasp by infants. *Child Development, 47,* 532-534.

Chamberlain, H.D. (1928). Inheritance of left handedness. *Journal of Heredity, 19,* 557-559.

Chatrian, G. E., Wirch, A. L., Edwards, K. H., Turella, G. S., Kaufman, M. A., & Snyder, J. M. (1985). Cochlear summating potential to broadband clicks from the human external auditory meatus. A study of subjects with normal hearing for age. *Ear and Hearing, 6,* 130-138.

Chi, J. G., Dooling, E. C., & Gilles, F. H. (1977). Gyral development of the human brain. *Annals of Neurology, 1,* 86-93.

Churchill, J. A., Igna, E., & Senf, R. (1962). The association of position at birth and handedness. *Pediatrics, 29,* 307-309.

Coren, S., & Porac, C. (1977). Fifty centuries of right handedness : The historical record. *Science, 198,* 631-632.

Drea, C. M., Wallen, K., Akinbami, M. A., & Mann, D. R. (1995). Neonatal testosterone and handedness in yearling rhesus monkeys (macaca mulatta). *Physiology & Behavior, 58,* 1257-1262.

Geschwind, N., & Levitsky, W. (1968). Human brain: Left right asymmetries in temporal speech region. *Science, 161,* 186-187.

Geschwind, N., & Behan, P. (1982). Left handedness: Association with immune disease, migraine, and developmental learning disorder. *Proceedings of the National Academy of Sciences, 79,* 5097-5100.

Geschwind, N., & Galaburda, A. M. (1987). *Cerebral lateralization.* Cambridge, MA : MIT Press.

Gesel, A., & Ames, L. B. (1950). Tonic-neck reflex and symmetro-tonic behaviour. *Journal of Pediatrics, 36,* 165-176.

Grapin, P., & Perpere, C. (1968). Symétrie et lateralisation du nourrison (Lateralization of infants). In R. Kourilsky & P. Grapin (Eds.), *Main Droite et main gauche. Norme et lateralité* (pp. 83-100). Paris: Presses Universitaires de France.

Gur, R. C., Gur, R. E., Obrist, W. D., Hungerbuhler, J. P., Younkin, D., Rosen, A. D., Skolnick, B.E., & Reivich, M. (1982). Sex and handedness differences in cerebral blood flow during rest and cognitive activity. *Science, 217,* 659-661.

Halverson, H.M. (1936). Complications of the early grasping reactions. *Psychologica Monographs, 47,* 47-63.

Halverson, H. M. (1937a). Studies of the grasping responses of early infancy. *Journal of Genetic Psychology, 51,* 371-424.

Halverson, H. M. (1937b). Studies of the grasping responses of early infancy. *Journal of Genetic Psychology, 51,* 393-424.

Halverson, H. M. (1937c). Studies of the grasping responses of early infancy. *Journal of Genetic Psychology, 51,* 425-449.

Hardyck, C., & Petrinovich, L.F. (1977). Left handedness. *Psychological Bulletin, 84,* 385-404.

Hawn, P. R., & Harris, L.J. (1983). Hand differences in grasp duration and reaching in two and five month old infants. In G. Young, S.J.

Segalowitz, C.M. Corter, & S.E. Trehub (Eds.), Manual specialization and the developing brain (pp. 331-348). New York: Academic Press.

Hécaen, H., De Agostini, M., & Monzon-Montes, A. (1981). Cerebral organization in left handers. *Brain & Language, 12*, 261-284.

Hécaen, H., & Sauguet, J. (1971). Cerebral dominance in left handed subjects. Cortex, 7, 19-48.

Hines, D., & Satz, P. (1971). Superiority of right visual half fields in right handers for recall of digits presented at varying rates. *Neuropsychologia, 9*, 21-25.

Hines, D., & Satz, P. (1974). Cross-modal asymmetries in perception related to asymmetry in cerebral function. *Neuropsychologia, 12*, 239-247.

Hopkins, W. D., Bales, S. A., & Bennett, A. J. (1994). Heritability of hand preference in chimpanzees. *International Journal of Neuroscience, 74*, 17-26.

Jacklin, C. N., Maccoby, E. E., Doering, C. H., & King, D. R. (1983). Neonatal sex steroid hormones and muscular strength of boys and girls in the first three years. *Developmental Psychobiology, 17*, 301-310.

Kee, D. W., Bathurst, K., & Hellige, J. B. (1983). Lateralized interference of repetitive finger tapping: Influence of familial handedness, cognitive load and verbal production. *Neuropsychologia, 21*, 617-624.

Keles, P., Diyarbakirli, S., Tan, M., & Tan, Ü. (1997). Facial asymmetry in right and left handed men and women. *International Journal of Neuroscience, 91*, 147-160.

Kraft, R. H. (1981). The relationship between right handed children's assessed and familial handedness and lateral specialization. *Neuropsychologia, 18*, 696-705.

Levy, J. (1972). Lateral specialization of the human brain. Behavioural manifestations and possible evolutionary basis. In J. A. Kiger (Ed.), *The biology of behavior*. Corvallis: Oregon State University.

Levy, J. (1976). Cerebral lateralization and spatial ability. *Behavior Genetics, 6*,171-188.

Lorenz, K. (1937). Die bildung des instinktbegriffes. *Naturwissenschaften, 25*, 289-300.

Lorenz, K. (1943). Angeborene formen moeglicher erfahrung. *Z. Tierpsychol., 5*, 235-409.

Lundstrom, A. (1961). Some asymmetries of the dental arches, jaws, and skull, and their etiological significance. *American Journal of Orthodontics, 47*, 81-106.

McGlone, J. (1978). Sex differences in functional brain asymmetry. *Cortex, 14*,122-128.

McGlone, J. (1986). The neuropsychology of sex differences in human brain organization. In G. Goldstein and R. E. Tarter (eds.), *Advances in Clinical Neuropsychology,* Vol. 3, pp. 1-30. New York: Plenum Press.

McGraw, M. B. (1940). Suspension grasp behavior of the human infant. *American Journal of Diseases of Children, 60,* 799-811.

McKeever, W. F., & Van Deventer, A. D. (1977). Visual and auditory language processing asymmetries: Influences of handedness, familial sinistrality and sex. *Cortex, 13,* 225-241.

Michel, G. F., & Goodwin, R. (1979). Intra-uterine birth position predicts newborn spine head position preference. *Infant Behavior and Development, 2,* 29.

Moffat, S. D., & Hampson, E. (1996). Salivary testosterone levels in left and right handed adults. *Neuropsychologia, 34,* 225-233.

Moss, F. A. (1929). *Applications of psychology.* Cambridge, MA: Harvard University Press.

Newcombe, F., & Ratcliff, G. (1973). Handedness, speech lateralization, and ability. *Neuropsychologia, 11,* 399-407.

Peters, M. (1995). Handedness and its relation to other indices of cerebral lateralization. In R.J. Davidson & K. Hughdahl (Eds.), Brain asymmetry, Cambridge: MIT Press.

Petrie, B. F., & Peters, M. (1980). Handedness: Left/right differences in intensity of grasp response and duration of rattle holding in infants. *Infant Behavior and Development, 3,* 215-221.

Pipe, M. E. (1987). Pathological left handedness: Is it familial? *Neuropsychologia, 21,* 91-98.

Polack, S. (1960). The grasp response of the neonate: Its characteristics and interaction with the tonic neck reflex. *Achives of Neurology, 3,* 574-581.

Previc, F. H. (1991). A general theory concerning the prenatal origins of cerebral lateralization in humans. *Psychological Review, 98,* 299-334.

Roberts, A. M., & Smart, J. L. (1981). An investigation of handedness and headedness in newborn babies. *Behavioral Brain Research, 2,* 275-276.

Rife, D. C. (1940). Handedness, with special reference to twins. *Genetics, 25,* 178-186.

Satz, P., Fennel, E., & Jones, M. B. (1969). Comments on: A model of inheritance of handedness and cerebral dominance. *Neuropsychologia, 7,* 101-103.

Searleman, A., Herrmann, D. J., Coventry, A. K. (1984). Cognitive abilities and left handedness: An interaction between family sinistrality and strength of handedness. *Intelligence, 8,* 295-304.

Snyder, T. J. (1978). Handedness, speech Lateralization, and hand-foot speech interactions. Unpublished Masters Thesis, Virginia Commonwealth University.

Snyder, T. J. (1979). Familial handedness, handedness, and sequential arm tapping performance: Implications for differences in cerebral organization. Unpublished Doctoral Thesis, Virginia Commonwealth University.

Springer, S. P., & Searleman, A. (1980). Left handedness in twins: implications for the mechanisms underlying cerebral asymmetry function. In J. Herron (Ed.), *Neuropsychology of left handedness*, pp. 139-158. New York: Academic Press.

Stirnimann, F. (1941). Greifversuche mit der hand neugeborener. *Ann. paediat., 157*,17-27.

Tan, Ü. (1988). The distribution of hand preference in normal men and women. *International Journal of Neuroscience, 41*, 35-55.

Tan, Ü. (1994a). Human growth hormone may differentially influence the grasp-reflex in human neonates on the basis of genetically predetermined neural pattern of brain organization in utero. *International Journal of Neuroscience, 74*, 87-93.

Tan, Ü. (1994b). Correlations between grasp-reflex strengths and serum thyroid-hormone levels depending upon sex and familial sinistrality in human neonates: importance of genetically predetermined cerebral organization. *International Journal of Neuroscience, 75*, 31-43.

Tan, Ü. (1994c). Role of prenatal position in grasp-reflex asymmetry in human neonates. *Perceptual and Motor skills, 78*, 287-290.

Tan, Ü., Ors, R., Kurkcuoglu, M., & Kutlu, N. (1992a). The lateralization of the grasp-reflex in human newborns. *International Journal of Neuroscience, 62*, 1-8.

Tan, Ü., Ors, R., Kurkcuoglu, M., Kutlu, N., & Cankaya, A. (1992b). There is a relatively leftbiased grasp-reflex asymmetry in human newborns with familial sinistrality compared to those without familial sinistrality. *International Journal of Neuroscience, 62*, 9-16.

Tan, Ü., Ors, R., Kurkcuoglu, M., Kutlu, N., & Cankaya, A. (1992c). Lateralization of the grasp-reflex in male and female human newborns. *International Journal of Neuroscience, 62*, 155-163.

Tan, Ü., & Tan, M. (1997). The mixture distribution of left minus right hand skill in men and women. *International Journal of Neuroscience, 92*, 01-08.

Tan, Ü., & Tan, M. (1999). Incidences of asymmetries for the palmar grasp reflex in neonates and hand preference in adults. *NeuroReport, 10*, 3253-3256.

Tan, Ü., & Zor, N. (1993). Sex-dependent relations of grasp-reflex strengths from right and left hands to serum gonadotropin (FSH and LH) levels in human neonates with regard to cerebral lateralization. *International Journal of Neuroscience, 73*, 21-226.

Tan, Ü., & Zor, N. (1994). Grasp-reflex strength from right and left hands in relation to serum cortisol level and fetal position in human neonates. *International Journal of Neuroscience, 74*, 27-32.

Tan, Ü., & Zor, N. (1994). Relation of serum free-testosterone level to grasp-reflex strength in human neonates with rightear and leftear facing out in utero positions. *International Journal of Neuroscience, 75*, 9-18.

Trenouth, M. J. (1985). Asymmetry of the human skull during fetal growth. *The Anatomical Record, 211*, 205-212.

Varney, N. R., & Benton, A. L. (1975). Tactile perception of direction in relation to handedness and familial handedness. *Neuropsychologia, 11*, 423-428.

Vig, P. S., & Hewitt, A. B. (1975). Asymmetry of the human facial skeleton. *The Angle Orthodontist, 45*, 125-129.

Waber, D. P. (1976). Sex differences in cognition: A function of matuıation rate? *Science, 192*, 572-574.

Wada, J. A., Clarke, R., & Hamm, A. (1975). Cerebral hemispheric asymmetry in humans. *Achives of Neurology, 32*, 239-246.

Ward, W. D. (1957). Hearing of naval aircraft maintenance personnel. *The Journal of the Acoustical Society of America, 29*, 1289-1301.

Witelson, S. F., & Pallie, W. (1973). Left hemisphere specialization for language in the newborn: Neuroanatomical evidence of asymmetry. *Brain, 96*, 641-646.

Woo, T. L. (1931). On the asymmetry of the human skull. *Biometrika, 22*, 324-352.

Yu-Yan, M., Cun-Ren, F., & Over, R. (1983). Lateral asymmetry in duration of grasp by infants. *Australian Journal of Psychology, 35*, 81-84.

Zurif, E.B., & Bryden, M.P. (1969). Familial handedness and left-right differences in auditory and visual perception. *Neuropsychologia, 7*, 179-187.

Chapter 4

Age and Generation Trends in Handedness: An Eastern Perspective

Syoichi Iwasaki
Fukushima Medical University, Japan

Of the many documented lateralized cerebral functions exhibited by human beings, unimanual control of tools and objects is the most obvious and easy-to-identify behaviour. Therefore it has been attracting the attention of both lay persons and researchers. Still little consensus is achieved as to why there is a small minority (about 10 % according to Hardyck & Petrinovich, 1977) who are left handed, whereas the majority are right handed. Both genetic and environmental theories of right handedness have been proposed. As recent examples, Provins (1997), in his article that appeared in Psychological Review, maintains that right handedness is a result of adaptation to the right-handed world, whereas Corballis (1997), whose article appeared in the same journal, argues for genetic determination of right handedness. (See also Chapter 2 of this volume.) As an example of a genetic theory, Annett (1985) proposed that handedness is a byproduct of left-hemisphere lateralization of speech caused by a right-shift gene, which creates an underlying bias of manual skill to the right side thus causing right handedness. The unique point of her right-shift theory is the idea that the left handers are produced simply by dividing a normally distributed underlying manual skill variation with an arbitrary criterion. Therefore, there may be left handers even among those who have a right-shift gene and thus develop language in the left hemisphere. If they lack the gene, the proportions of left handers and right handers are equal with a substantial

M.K. Mandal, M.B. Bulman-Fleming and G. Tiwari (eds.), Side Bias: A Neuropsychological Perspective, 83-100.

number of ambidextrous persons in between. Thus, according to this theory, left handedness is a chance result that does not require any specific explanations.

In contrast to Annett's right-shift theory, which admits existence of left handers even among those who have a right-shift gene, there are other theories that propose environmental causes that lead to left handedness, while assuming that we are all genetically right handed (Coren, 1995a). Left handers are created because an unfortunate minority of us are converted to become so by environmental interference. Thus, Bakan (1971) maintains that left handedness is caused by prenatal damage to the left hemisphere, which controls the right side of our body. Another environmental theory of left handedness was proposed by Geschwind and Galaburda (1987) who pointed out that sex hormones (especially testosterone) cause delayed development of the left hemisphere leading to more symmetrical cerebral functions. Left handedness is a chance result of this symmetrical brain. These researchers argued that this is why there are more left-handed men than left handed-women. A more detailed description of these theories of the development of human handedness can be found in Chapter 2 of this volume.

A satisfactory theory of handedness must explain several facts concerning human hand preference: 1) There are, by a great margin, more right handers than left handers. 2) Proportions of these different handedness groups differ among different countries and among different age groups. 3) There are more male left handers than female left handers. Any existing theories do not seem to be able to explain fully all of these facts without resorting to factors other than their proposed cause of handedness. Thus, both genetic and prenatal explanations must resort to other environmental factors that would affect the developing brain to promote or inhibit a particular inclination of hand preference if they want to explain the cultural and age differences in handedness (that is, an age trend that is not a reflection of a generation trend of a relaxed attitude against left handedness; see the following section).

Although signs of the right hand preference are found early in infancy (Michel & Harkins, 1986; Thompson & Smart, 1993), even in the uterus (Hepper, Shahidullah, & White, 1991), it gradually develops with age and becomes fully stable by approximately 10 years of age (Gesell & Ames, 1947). This early development of right handedness may be a genetically programmed maturational process. In contrast to this evidence for early establishment of handedness, cross-sectional studies all over the world including those conducted in Japan (Iwasaki, Kaiho, & Iseki, 1995; Maehara et al., 1988) have repeatedly found a much prolonged developmental course of increasing numbers of right handers with age. Some western researchers (especially when writing hand is used to classify handedness sub-categories, e.g., Beukelaar & Kroonenberg, 1986; Brackenridge, 1981) have argued that the age trend of increased right handedness beyond early childhood is not really an age trend, but actually is a reflection of a generation trend of progressively relaxing social censorship against left handedness.

In this article, I will argue that age and generation trends are caused by different factors, by using the datasets on handedness that have been accumulated in Japan. If there is no evidence for a relaxation of social pressures or the process has already reached its limit (i.e., there is no more room for relaxation) decades before, as was suggested by Gilbert and Wysocki (1992) for the age trend of writing hand in the USA, then in the absence of other factors influencing handedness throughout life, one would predict a flat age trend beyond age 10.

Secondly, I will attempt to estimate a "true" prevalence of left handedness. Conventional estimates of the prevalence of left handers are 10% of the population, which according to some western studies (e.g., Spiegler & Yeni-Komshian, 1983 - they found 13.8 % of young adults were left handed) appear to be too low. (See, however, McManus (1995), in which he stated that "a true incidence of left handedness is 7.75%".) In contrast to these western estimates, Asian studies have reported much lower frequencies of left handers on the order of several % (5 to 8 %) of the total population. Thus, the Oriental prevalence is closer to the McManus estimate. And even lower values, of less than 1 %, were reported in some rural communities (Bryden, Ardila, & Ardila, 1993; Verhaegen & Ntumba, 1964). These different prevalences of left handedness are certainly partly accounted for by different social stances against left hand use. However, many researchers (e.g., Teng, Lee, Yang, & Chang, 1976) believe that the effect of correction is specific to the targeted acts (which usually are eating and writing) so that other acts are less likely to be affected by the intervention. Consequently, if multiple acts besides writing and eating are checked, obtained frequencies of different handedness sub-categories would show limited sensitivity to the different degrees of social intervention. Thus, it is important to know whether and to what extent the prevalence of left handedness obtained by a survey that checks multiple acts is influenced by the severity of the social pressures.

1. THE AGE TREND IS NOT A REFLECTION OF A GENERATION TREND

Many western researchers ascribed the increase in the prevalence of right handers with age to decreased social pressures against left hand use (i.e., a generation shift of the social pressures). For this to be true, it must be shown that there is some indication of the relaxation of such pressures. Without such evidence, factors directly related to age should be sought as possible causes of the trend. Conversely, without evidence for such a relaxation, a flat age trend would be expected, as previously stated.

There are two ways to infer such a relaxation; it may be seen as an increase in the prevalence of left handedness when recent studies are compared with older ones (see Spiegler & Yeni-Komshian, 1983, for such a generation shift). In making comparisons of this sort, one must be careful in equating the age of participants of different studies, because this factor may affect the prevalence independently of the generation effect. A second method is to compare the prevalence of correction of left hand use across different age cohorts. If people become less eager to correct their children's and pupil's left hand use, this must be reflected in the reduced frequency of the experience of correction among younger generations.

Relying on these criteria, there is little evidence for an easing of the social pressures in Japan. First, as may be seen in Table 1, different studies spanning more than half a century have found a relatively stable trend of the prevalence of left handedness, which was 4 to 6 % for boys and 2 to 4 % for girls. Second, we (Iwasaki et al., 1995) found no evidence that the practice of left hand-use correction has declined in recent years. On the contrary, there was a significant tendency of older people reporting less frequently on their experiences of left hand correction than younger people. On the other hand, for evidence of a decline of the social pressures in Japan, see Hatta & Kawakami (1995) , who found some increase in the number of female left handers as compared with a previous study (Hatta & Nakatsuka, 1976).

If the age trend is a reflection of the generation trend of relaxed social pressure generally, then in Japan in particular there should be found a flat age trend, because there is no marked change in the prevalence of the correction. Both Maehara et al. (1988) and our own study (Iwasaki et al., 1995) have shown that there is a clear increase in the prevalence of right handedness up to the age of the thirties, which is much more conspicuous for men than for women. Thus, the age trend found in Japan does not appear to have resulted from declining social pressures against left hand use.

Table 1. Juvenile prevalence of left handedness in Japan reported in four studies spanning more than half a century

Researchers	**Year**	**N**	**Age range**	**Male**	**Female**	**Sex Diff.**	**Criterion of LH**
Komai & Fukuoka	1934	2046	13 -15	6.3	4.1	Unknown	mean of 4 acts
Sasaki	1965	308	13 -15	4.0	3.9	Unknown	LQ*
Shimizu & Endo	1983	4282	16 - 18	4.0	2.4	Significant	LQ
Iwasaki	1995	544	13 - 19	6.1	2.4	Significant	LQ

*LQ : Right – Left / Right + Left

As an additional check on the independence of the age trend from the generation change in the social pressures, the age trend of "true" right handers is calculated based on the data reported in Iwasaki et al. (1995). The "true" right handers are defined as those who use their right hand for eating and writing and have overall right hand preference for other acts as well (i.e., Laterality Quotient > .90). Furthermore, to be classified as "true" right handers they should have no experience of the correction of their left hand use. If one shows no tendency to use one's left hand during development, even a strict society that is watchful of its members against deviation from the norm would have no influence on one's hand use. On the assumption that the age trend is a reflection of decreasing social pressures against left hand use, the age trend of these "true" right handers should turn out to be flat, because any amount of fluctuation of the pressures against left hand use is of no concern for them and would not affect their natural hand preference. As shown in Figure 1, however, the prevalence of "true" right handers does increase with age just like the overall prevalence of the general category of right handers (the group that includes right handers who *did* report having been corrected for using their left hand).

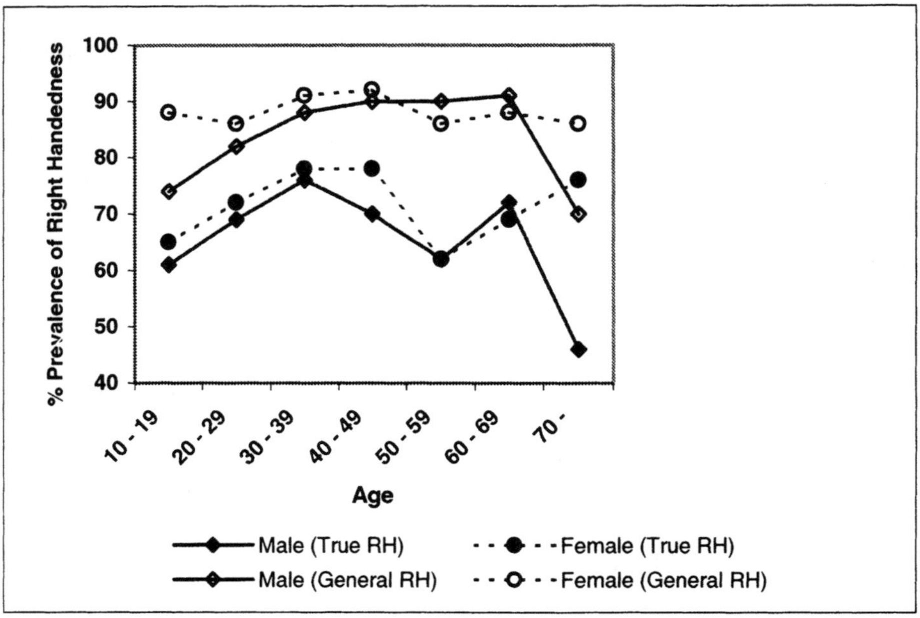

Figure 1. Age trends for true right handers (filled diamonds for men and filled circles for women) who are supposedly born to be right handed as compared with overall age trends that include both 'true' and converted right handers.

For men in their thirties, the two curves run parallel to each other. After that they begin to diverge. Surprisingly, for women the age trend of the "true" right handers deviates from that of the much flatter original female trend and approaches that of the male "true" right handers. The original age trends for the two sexes were widely separate for the youngest age groups but the discrepancy between them gradually narrowed with age. Thus there were more significant sex differences in the frequencies of handedness of a variety of measures among younger groups than among older groups (see table 4 of Iwasaki et al., 1995). However, these sex differences in younger age groups are almost gone when the frequencies of "true" right handers are compared between sexes. Although a sex difference in the prevalence of left handedness has been found in many parts of the world, this analysis suggests that it is really a result of the fact that women conform more readily to social norms than men, rather than a result of some other true differences between sexes such as prenatal hormonal environment. These analyses suggest that the age trend of an increasing prevalence of right handedness (at least up to the thirties) found in Japan is a true age effect, which is independent of recent relaxation in the social attitude against left hand use.

2. WHAT DETERMINES TREND?

As argued in the previous section, if this age trend is really independent of the social changes in the attitude against left handedness, then what factors are responsible for it? The factors that have been proposed to explain the age trend are: (1) adaptation to the right-hand world in which many tools are made for right handers, thus compelling left handers to use them with their right hand, which would facilitate their conversion to right handedness (Coren, 1992); (2) selective elimination of the left handers due to their shorter life span (Coren & Halpern, 1991); (3) a practice effect, in which the consistency of handedness increases with age as people become better at using one hand with practice (Porac, 1993); (4) different criteria used by different age cohorts when they judge each item of a handedness questionnaire. Although little attention is paid to this last possibility, it can make a subtle difference in the frequencies of each handedness sub-category if a multiple-item questionnaire is used to assess handedness. For example, if older people are more decisive in their judgment of their hand preferences and thus tend to mark more extreme categories or choose categories more consistently across different acts, then this would produce an apparent age trend of increasing right handedness.

Of these possibilities, adaptation to the right-hand world does not appear to be strong enough to affect one's actual hand use. Besides, it might be possible to argue that people would become ambidextrous as they grow older, since they would encounter many occasions during the life-span when they are forced to use their non-preferred hand at least temporarily owing, for instance, to injury to the preferred hand (Dellatolas, Moreau, Jallon, & Lellouch, 1993).

The elimination hypothesis is also not very likely, because from the age of about 10 to their 30's people are most vigorous and healthy and unlikely to become ill due to minor physical anomalies, which are postulated to cause left handedness and exert adverse effects on physical fitness. A much more plausible cause of death of the young left handers is accidents (Coren, 1992), which are the number one cause of deaths among the young people in Japan [the age covered the range from 1 to 29 years of age for males and from 1 to 24 years for females according to the National Public Health Statistics of Japan (Health and Welfare Statistics Association, 1997)]. However, to my knowledge there has been no study in Japan that investigated the relation between accident-proneness and handedness. Therefore, this possibility, although a viable one, is not very likely because a difference in death tolls between different handedness groups, even if it really exists, would not likely to be large enough to produce the increase of more than 10 % in the prevalence of right handers.

A practice effect is also inconsistent with the evidence that the prevalence of left handers does not increase with age, although that of right handers does (Coren, 1995b). Furthermore, as argued above, it seems possible to maintain that the probability is greater for people to become mixed handers with experiences of temporary inability of using the preferred hand as a result of injury, than to become strong left or right handers (Dellatolas et al., 1993).

The last possibility of the age-dependent criterion shift, although apparently not a very attractive one for the researchers of handedness, has not received due attention in the literature. It argues that people may form more and more firm opinions on their own behaviours as they grow older, contrary to their actual behaviour of increased ambidextrality. This increase of confidence in one's own behaviours may continue up to one's thirties, which may explain the age trend found for the Japanese right handers. One problem with this explanation is, as for the practice hypothesis, lack of evidence of a comparable increase in the prevalence of left handers. There seems to be no plausible reason that left handers behave differently in this respect than do right handers. The fact is, however, that the prevalence of left handedness decreases with age (Ellis, Ellis, Marshall, Windridge, & Jones, 1998; Plato, Fox, & Garruto, 1984). The decrement might be accountable by left handers' reluctance in expressing their actual preference in a straightforward way, because as they grow older they surely learn the fact that they are members of a minority group who share the feeling that they might be regarded as different by the majority.

Thus, at present all the available explanations, if considered alone, are not satisfactory. Either there are other unknown causes of the age trend or all these factors could contribute synergistically to modify the hand-use acts toward apparent increase in the prevalence of right handedness as people grow older.

3. WHAT IS THE "TRUE" PREVALENCE OF LEFT HANDEDNESS?

Although the figure of 10 % is often mentioned as an estimated prevalence of left handedness, according to Hécaen and Ajuriaguerra (1964), the reported figures vary from as high as 30 % to less than 1 %. Of these widely different estimates, smaller values are surely due to stricter cultural censorship in some societies, whereas larger ones may be explainable either as an over-representation of left handers in a sample or as less reliable estimates (Hardyck & Petrinovich, 1977). Such an over-representation may occur when participants are recruited on a voluntary basis. Since handedness is by far a greater concern for the left handers who belong to a minority group and would have received unpleasant attention from others who are mostly right handers, they tend to be eager to participate in a study of handedness-related behaviours. A large sample size is also important to obtain a reliable estimate of the prevalence of left handedness if there is an imbalance in the motivation of participants according to their handedness, because one or two more left handers would inflate the prevalence of left handedness considerably when the overall number of participants is small. To avoid such a sampling bias, an all-inclusive study in which data are collected from all members of a society would produce more representative figures of each handedness subcategory. Thus, large-scale studies conducted in western societies, which were published in the 90's, found the prevalence of left handedness among young people to be over 10 %, ranging from 13 % in the US (Gilbert & Wysocki, 1992; total N = 1177507) to 11.2 % in England (Ellis et al., 1998; total N = 6097). In contrast, as may be seen in table 1, the Japanese figures are much lower than these western ones. For example, our own study (Iwasaki et al., 1995, N = 544) found the prevalence of left handedness to be 6 % for men and 2.4 % for women among the age group of 13 to 19 years, which is comparable to the earlier figures of 4.03 % for men and 2.36 % for women (Shimizu & Endo, 1983, N = 4282, all of whom were high-school students). In the largest sample study ever conducted in Japan, Maehara et al. (1988; total N = 8693) found that about 5% of the young people (aged 14 to 15 years) were left handed. [In their original report, they only listed the prevalence of right handers. The cited figure is found in Maehara (1989).] Thus, the Japanese prevalence of young left handers is about half the figure reported in western societies. In other Asian countries a similarly low prevalence of left handedness has been reported. For example, in Hong Kong, 8.2% of young men and 2.7 % of young women (mean age of 19.9) are reported to be left handed (Hoosain, 1990).

The discrepancy between the eastern and western frequencies of left handedness is certainly at least partly due to the cross-cultural differences in the general tolerance of left handedness. To obtain a "culture-free" estimate of the prevalence of left handedness, a regression analysis was conducted on an international database that registers both the prevalence of left-handed writers and the severity of social pressures. To be included in the analysis the dataset must contain both the prevalence of the left hand use for writing, and, as a measure of severity of the social pressure, the reported frequency of correction of left-hand writing. The main part of the data came from a cross-cultural study of Perelle and Ehrman (1994), in which the values of these two items from 13 different countries were listed. Sources of additional data were two African countries (De Agostini, Khamis, Ahui, & Dellatolas, 1997), Hong Kong (Hoosain, 1990), Taiwan (Teng et al., 1976), and Japan (Iwasaki et al., 1995). Writing hand was chosen as an index of handedness because this act is one of the two main targets of the cultural intervention (the other is eating) and has been reported to be one of the most reliable items in the assessment of handedness (Raczkowski, Kalat, & Nebes, 1974; Roszkowski & Snelbecker, 1982). Therefore, it was expected that the prevalence of left-handed writers would be sensitive to changes in the cultural attitude against left hand use. As shown in figure 2, the analysis revealed that the two variables were negatively correlated with the regression coefficient of -.391, the probability of which was .109 (two-tailed test).

Although it does not reach a conventional level of significance partly because of the small number of samples (N = 18), this is too good to be simply dismissed as non-significant, because the relation between these two variables is a logically expected one. Thus, on the assumption that the obtained relation between the strength of correction and the prevalence of left-hand writers is reliable, it may be possible to estimate the prevalence of left hand use for writing in the ideal case, when there is no social intervention, by resorting to the linear equation shown in figure 2. By extrapolating the prevalence of correction to zero, we can obtain the estimated prevalence of left-hand writing of 9.55 %, which is in good agreement with the widely held figure of 10 %.

Although western societies are much more liberal in the use of the left hand for writing, there seems to be some residue of the old practice. Thus, even in the US, the reported prevalence of correction is 6% (Perelle & Ehrman, 1994). Substituting this figure for X in the equation for the regression line shown in figure 2, one obtains an estimated prevalence of left hand use for writing of 7.4 %. The actual prevalence of 13 % (Gilbert & Wysocki, 1992) for the younger generation is nearly double this estimate. Although this discrepancy may be due to the unreliability of the regression equation, there may be some thus far unsuspected factors that are at work to increase the prevalence of left handedness in recent generations (Hugdahl, Satz, Mitrushina, & Miller, 1993).

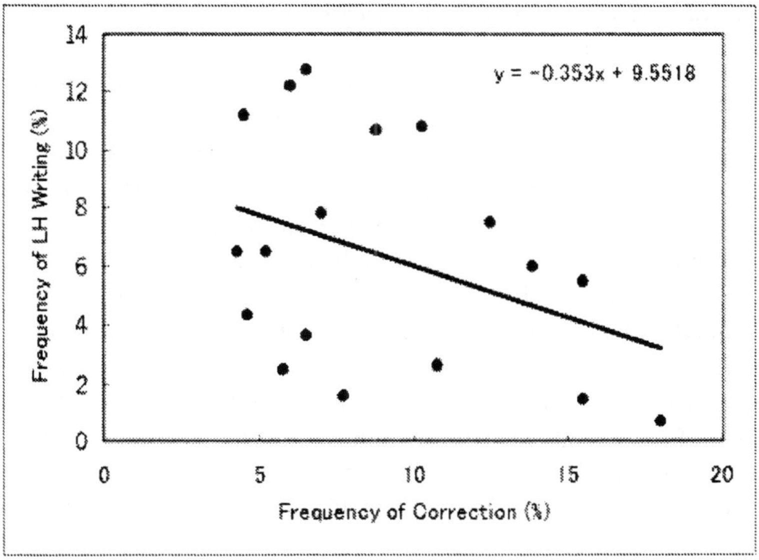

Figure 2. Cross-cultural relation between reported frequency of correction of left-hand writing and prevalence of left-hand writing. The fitted linear equation is also depicted in the graph.

4. TO WHAT EXTENT DOES CULTURAL INTERVENTION AFFECT THE PREVALENCE OF LEFT HANDEDNESS?

The major findings of handedness have been repeatedly confirmed in many countries; that is, the age trend of increasing prevalence of right handers, the sex difference of more left-handed males than females, and the familial sinistrality effect of increased left handers in the family with other left-handed members, especially a left-handed mother. In Japan too, all of these variables have been reported to affect handedness. Thus, besides the age trend mentioned above, both the sex difference and the familial sinistrality effect have been reported in Japan (Maehara, 1989; Shimizu & Endo, 1983). Furthermore, Maehara (1989) found a stronger maternal influence than paternal one on the prevalence of left handedness. Thus, except for the actual frequency of left handedness, these findings are also true for the Japanese population, suggesting cross-cultural consistency of the phenomena related to handedness. The universality of handedness-related facts implies a genetic background for the phenomenon, shared by different peoples of vastly different cultural backgrounds.

In spite of the cross-cultural consistency of these handedness-related phenomena, the prevalence of left handedness differs widely among different nations. Can this be attributed to different strengths of the social pressures against left hand use? Many researchers expressed their opinion that social intervention can only change the hand used for the targeted act (e.g., Teng et al., 1976) or affect only modestly an overall score derived from multiple items (Ellis et al., 1998; Leiber & Axelrod, 1981). In Japan, Komai and Fukuoka (1934) traced developmental trends for individual acts when the social pressures against left-handed writing were enforced (i.e., starting from the period when the participants began to attend primary school). Figure 3 depicts these trends, which illustrate that although the number of children who used their left hand for writing declined steadily with age and was almost zero by age 15, the prevalence of left hand use for other acts stabilized after age 10 and remained relatively high compared to that of left-handed writing. Thus, the obtained prevalence of left handedness should be relatively insensitive to the level of the social pressures if it is estimated by an overall measure like a laterality quotient. Being relatively immune to the different levels of social pressures, cross-cultural comparisons based on such an overall score should find similar frequencies of left handedness even among countries of vastly different cultures. Contrary to this expectation, the studies that used a multiple-item handedness inventory found that the prevalence of left handedness for young people ranged from 0 % in native Amazonians (Bryden et al., 1993) to more than 10 % in western countries (e.g., Ellis, et al., 1998) with the prevalence in many countries falling between these two extremities.

One reason for this discrepancy may be found in the age of the social intervention. In most countries, the main targets of correction are eating and writing. Of these, as writing is usually taught in school, the actual parental and teacher's interventions would start at around the time when children begin to attend a primary school at the age of 6 to 7 (see Hugdahl et al., 1993, for actual comments made by some old switched writers who experienced such interventions). As mentioned above, the effect of these interventions can also be seen in Japan as a rapid termination of left-handed writing among primary school children (Komai & Fukuoka, 1934). In contrast, if eating is the main target, the intervention may start much earlier, just after weaning. The earlier intervention and more strict method of correction such as tying down the left hand (as was once done even in western societies, see Coren, 1992, p. 55) would convert natural left handers

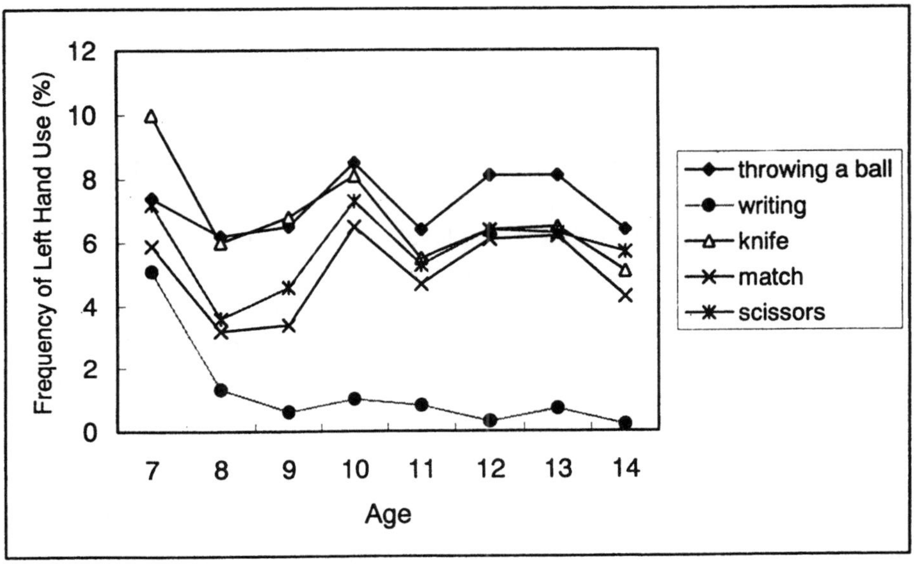

Figure 3. Developmental trends of the prevalence of left hand used for individual acts during the very time when children were under attack from both parents and teachers who were eager to correct left-hand writing (based on table 2 of Komai and Fukuoka, 1934)

into perfect right handers, whereas later intervention would affect mainly the acts being modified. Some hint for this possibility is found in the effects of upper-limb injuries on the later development of handedness reported by Dellatolas and his associates (Dellatolas et al., 1993). They found an increased frequency of injuries among weak right handers as compared with strong right handers. However, if the age of injury was limited to before 7 years of age, the injuries were associated with an increased frequency of left handedness. Thus, an earlier temporary incapacity of the upper limb might induce conversion of handedness, whereas a later one would simply weaken the original preference but would not totally reverse it. Similarly, forced restriction of left hand use, if it occurs before 7 years, may well convert left-handed children into complete right handers. In this connection, the effect of early head injury on later development of lateralization of language may be of particular importance (Rasmussen & Milner, 1977). These researchers obtained evidence for such a sensitive period in the lateralization of language when they probed the hemisphere responsible for language functions by selectively anaesthetizing each hemisphere with sodium amobarbitol. They concluded that only if the lesion of the left-hemisphere language areas occurs before the age of 5 years can it cause a shift of language functions to the right hemisphere.

5. CONCLUSIONS

I started this research project with the expectation that handedness is genetically determined and that the "true" prevalence of the left handedness would be well below 10 %, that is, the level reported in most Asian countries (Maehara, 1989, expressed a similar opinion). However, the analysis of the relation between writing hand and the reported prevalence of correction of left hand use suggested that approximately 10 % of us are left handed in an ideal world where there are no such social pressures, which is in good accordance with the widely held view of western investigators. Although many researchers, including myself, believe that other manual acts are relatively immune to the corrective intervention directed toward the targeted acts (i.e., eating and writing), I reconsidered the possibility that early intervention by adults (mostly parents) could exert greater and more pervasive influences on the children's overall handedness, such as full conversion to right handedness. This would happen if eating rather than writing were the main target of correction, because practice of an adult form of behaviour for the former act would start much earlier than would that for the latter one.

Considering the cross-cultural consistency of major findings on handedness, the handedness of our species is undoubtedly a genetically predisposed trait. However, we are very adaptive and malleable especially in the first several years whereas the genetic predisposition toward right hand preference constitutes a relatively weak bias. Thus the phenotype of handedness develops gradually under the influence of many environmental factors, strength of social pressures being only one of them. Hormonal aberration and prenatal pathological events can also affect it. There might be some other, thus-far unrecognized, factors such as maternal stress during pregnancy and stressful early experience of infants, which may affect the development of hand preference of newborn babies. For example, Alonso, Castellano, and Rodriguez (1991) have found that prenatal stress affected behavioural lateralization in a sex-dependent manner in rats, weakening absolute lateralization (strength of lateral preference in a T-maze, ignoring its direction) in male offspring, but augmenting it in female offspring.

One remaining puzzle is the age trend of increasing right handedness. It is unlikely that the actual hand used for each act continues to be influenced by environmental factors after one has learned to use one hand for a particular act (this would occur by the age of 10) unless one intentionally chooses to change one's handedness. This may be the case for those who change their throwing hand to be a "southpaw" pitcher, because a southpaw is, simply by his or her rarity, superior to the majority of pitchers who throw a ball with the right hand. Such a conversion is sometimes seen in young Japanese baseball players. Selective elimination of the left handers remains a possibility, although this factor alone is unlikely to explain the more than 10 % increase of right handers from teenagers to the thirties as found in Japan. One relatively neglected factor that might influence the outcome of a multiple-item questionnaire survey is the age-dependent shift in the consistency of choosing alternatives of each question. It may be that one simply becomes more confident in one's judgment of the hand used for a particular act as one grows older. Thus, with age one's choice is determined more by one's belief of one's own handedness, rather than by actual behaviours.

Although it is relatively easy to study the prevalence of different handedness groups, especially with a questionnaire, it is much harder to find out how the prevalence thus found is realized. Careful titration of the factors responsible for the findings (both universal and culture-sensitive) related to handedness including the age trend might help us to solve this intricate puzzle of our species' distinctive characteristics.

6. REFERENCES

Alonso, J., Castellano, M.A., & Rodriguez, M. (1991). Behavioral lateralization in rats: prenatal stress effects on sex differences. *Brain Research, 539*, 45-50.

Annett, M. (1985). *Left, right, hand and brain: The right shift theory.* London : Lawrence Earlbaum.

Bakan, P. (1971). Handedness and birth order. *Nature, 229*, 195.

Beukelaar, L.J., & Kroonenberg, P.M. (1986). Changes over time in the relationship between hand preference and writing hand among left handers. *Neuropsychologia, 24*, 301-303.

Brackenridge, C.J. (1981). Secular variation in handedness over ninety years. *Neuropsychologia, 19*, 459-462.

Bryden, M.P., Ardila, A., & Ardila, O. (1993). Handedness in native Amazonians. *Neuropsychologia, 31*, 301-308.

Corballis, M.C. (1997). The genetics and evolution of handedness. *Psychological Review, 104*, 714-727.

Coren, S. (1992). *The left hander syndrome. The causes & consequences of left handedness.* New York : The Free Press.

Coren, S. (1994). The diminished number of older left handers: Differential mortality or social-historical trend?. *International Journal of Neuroscience, 75*, 1-8.

Coren, S. (1995a). Family patterns in handedness: Evidence for indirect inheritance mediated by birth stress. *Behavior Genetics, 25*, 517-524.

Coren, S. (1995b). Age and handedness: Patterns of change in the population and sex differences become visible with increased statistical power. *Canadian Journal of Experimental Psychology, 49,* 376-386.

Coren, S., & Halpern, D.F. (1991). Left handedness: A marker for decreased survival fitness. *Psychological Bulletin, 109*, 90-106.

De Agostini, M., Khamis, A.H., Ahui, A.M., & Dellatolas, G. (1997). Environmental influences in hand preference: An African point of view. *Brain and Cognition, 35,* 151-167.

Dellatolas, G., Moreau, T., Jallon, P., & Lellouch, J. (1993). Upper limb injuries and handedness plasticity. *British Journal of Psychology, 84*, 201-205.

Ellis, S.J., Ellis, P.J., Marshall, E., Windridge, C., & Jones, S. (1998). Is forced dextrality an explanation for the fall in the prevalence of sinistrality with age? A study in northern England. *Journal of Epidemiology and Community Health*, 52, 41-44.

Geschwind, N., & Galaburda, A. M. (1987). *Cerebral lateralization. Biological mechanisms, associations, and pathology.* Cambridge, Massachusetts : The MIT Press:

Gesell, A., & Ames, L.B. (1947). The development of handedness. *Journal of Genetic Psychology, 70*, 155-175.

Gilbert, A.N., & Wysocki, C.J. (1992). Hand preference and age in the United States. *Neuropsychologia, 30*, 601-608.

Hardyck, C. & Petrinovich, L.F. (1977). Left handedness. *Psychological Bulletin, 84*, 385-404.

Health and Welfare Association (1997). *The National Public Health Statistics, 44*, 428-431. (in Japanese).

Hécaen, H., & Ajuriaguerra, J. (1964). *Left handedness: Manual superiority and cerebral dominance.* Grune & Stratton: New York.

Hepper, P.G., Shahidullah, S., & White, R. (1991). Handedness in the human fetus. *Neuropsychologia, 29*, 1107-1111.

Hoosain, R. (1990). Left handedness and handedness switch amongst the Chinese. *Cortex, 26*, 451-454.

Hugdahl, K., Satz, P., Mitrushina, M., & Miller, E.N. (1993). Left handedness and old age: Do left handers die earlier? *Neuropsychologia, 31*, 325-333.

Hatta, T., & Kawakami, A. (1995). Patterns of handedness in modern Japanese: A cohort effect shown by re-administration of the H.N. Handedness Inventory after 20 years. *Canadian Journal of Experimental Psychology, 49*, 505-512.

Hatta, T., & Nakatsuka, Z. (1976). Note on hand preference of Japanese people. *Perceptual and Motor Skills, 42*, 530.

Iwasaki, S., Kaiho, T., & Iseki, K. (1995). Handedness trends across age groups in a Japanese sample of 2316. *Perceptual and Motor Skills, 80*, 979-994.

Komai, T., & Fukuoka, G. (1934). A study on the frequency of left handedness and left-footedness among Japanese school children. *Human Biology, 6*, 33-42.

Leiber, L., & Axelrod, S. (1981). Intra-familial learning is only a minor factor in manifest handedness. *Neuropsychologia, 19*, 273-288.

Maehara, K. (1989). *Migikikihidarikiki no kagaku (Science of right handedness and left handedness)*, Kohdansha: Tokyo (in Japanese).

Maehara, K., Negishi, N., Tsai, A., Otuki, N., Suzuki, S. Takahashi,T., & Sumiyoshi, Y. (1988). Handedness in the Japanese. *Developmental Neuropsychology, 4*, 117-127.

McManus, I.C. (1995). Familial sinistrality: The utility of calculating exact genotype probabilities for individuals. *Cortex, 31*, 3-24.

Michel, G.F., & Harkins, D.A. (1986). Postural and lateral asymmetries in the ontogeny of handedness during infancy. *Developmental Psychobiology, 19*, 247-258.

Perelle, I.B., & Ehrman, L. (1994). An international study of human handedness: the data. *Behavior Genetics, 24*, 217-227.

Plato, C.C., Fox, K.M., & Garruto, R.M. (1984). Measures of lateral functional dominance: Hand dominance. *Human Biology, 56*, 259-275.

Porac, C. (1993). Are age trends in adult hand preference best explained by developmental shifts or generational differences? *Canadian Journal of Experimental Psychology, 47*, 697-713.

Provins, K.A. (1997). Handedness and speech: A critical reappraisal of the role of genetic and environmental factors in the cerebral lateralization of function. *Psychological Review, 104*, 554-571.

Raczkowski, D., Kalat, J.W., & Nebes, R. (1974). Reliability and validity of some handedness questionnaire items. *Neuropsychologia, 12*, 43-47.

Rasmussen, T., & Milner, B. (1977). The role of early left-brain injury in determining lateralization of cerebral speech function. *Annals of New York Academy of Sciences, 299*, 355-369.

Roszkowski, M.J., & Snelbecker, G.E. (1982). Temporal stability and predictive validity of self-assessed hand preference with first and second graders. *Brain and Cognition, 1*, 405-409.

Sasaki, A. (1965). Studies on the lateral dominance (Part 2) Difference in percentages among sexes and age groups. *Kaseigaku Zasshi, 16*, 153-157 (in Japanese).

Shimizu, A., & Endo, M. (1983). Handedness and familial sinistrality in a Japanese student population. *Cortex, 19*, 265-272.

Spiegler, B.J., & Yeni-Komshian, G.H. (1983). Incidence of left handed writing in a college population with reference to family patterns of hand preference. *Neuropsychologia, 21*, 651-659.

Teng, E.L., Lee, P-H., Yang, K-S., & Chang, P.C. (1976). Handedness in a Chinese population: Biological, social, and pathological factors. *Science, 193*, 1148-1150.

Thompson, A.M., & Smart, J.L. (1993). A prospective study of the development of laterality: Neonatal laterality in relation to perinatal factors and maternal behaviour. *Cortex, 29*, 649-659.

Verhaegen, P., & Ntumba, A. (1964). Note on the frequency of left handedness in African children. *Journal of Educational Psychology, 55*, 89-90.

Chapter 5

Lateral Asymmetries and Interhemispheric Transfer in Aging: A Review and Some New Data

Alan A. Beaton[1], Kenneth Hugdahl[2] and Philip Ray[3]
[1]University of Wales, UK : [2] University of Bergen, Norway : [3]University of Wales, UK

The emergence of bipedalism in hominid evolution allowed the hands to be used for purposes other than locomotion, as in food gathering or tool manufacture, and this shaped the early social and cultural development of our species. It has been proposed that specialization of the right hand, either for gestures or tool-making, led to the development of vocal language and to its lateralization in the left hemisphere (see Hewes, 1973; Corballis, 1989, 1991; Bradshaw & Rogers, 1993; Davidson & Noble, 1993; Noble & Davidson, 1996).

The human propensity to use the right hand more than the left hand for skilled motor activities is found in all cultures that have been studied (Hardyck & Petronovich, 1977; Harris, 1980; 1990; Peters, 1995). Over a century ago, Ireland (1880) wrote: "It is...difficult to understand how all nations and tribes, without exception, have in all times of which we know anything given the preference to the right hand" (p. 207). This preference for the right hand occurs throughout recorded history (Coren & Porac, 1977) and may have existed for well over one million years (Toth, 1985).

M.K. Mandal, M.B. Bulman-Fleming and G.Tiwari (eds.), Side Bias: A Neuropsychological Perspective, 101-152. © 2000 Kluwer Academic Publishers.

It is commonly accepted that handedness, regarded either as a preference for one hand or greater skill of one hand in comparison with the other (see Bishop, 1989), derives from some advantage of the contralateral cerebral hemisphere. In the words of Jones (1944) "What we are admiring in the multitude of actions of the useful human hand is the human cerebral perfection, not the bones, muscles, and joints that carry out the complex volitions" (p. 301).

Recent theories of individual differences in human handedness range from those in which genetic mechanisms play an important (but not exclusive) role (Annett, 1985; 1995; Corballis, Lee, McManus, & Crow, 1996; McManus, 1985; McManus & Bryden, 1992;) to those that emphasize learning and experience (Perelle, Ehrman, & Manowitz, 1981; Provins, 1967, 1997) or early biological events (Yeo, Gangestad, & Daniel, 1993).

1. BIMANUAL PERFORMANCE

An emphasis on unilateral manual preference ignores the fact that many, if not most, manual activities involve the co-ordinated use of two hands rather than one. Even the act of writing with one hand is associated with movements involving the other hand whereas throwing a ball entails compensatory postural adjustments (Guiard, 1987). Many everyday actions involving the combined activity of two hands, such as tying a shoelace or playing a musical instrument – even unscrewing a lid – become so over-learned that we forget how difficult they are for young children.

Under normal circumstances there is a powerful tendency for the upper limbs to work together in temporal and spatial synchrony (Kelso, Putnam, & Goodman, 1983) although this is not perfect since one limb tends to lead slightly while the other lags (Berlucchi, Aglioti, & Tassinari, 1994; Kelso, Southard, & Goodman, 1979). In the course of daily life, however, the behaviour of the two limbs must be uncoupled. It is well known that when making movements with one limb, young children make unintended mirror movements with the opposite limb. Normal children (and adults) quickly learn to inhibit such mirror movements thus allowing the limbs to be co-ordinated in novel ways. Fagard, Morioka, and Wolff (1985) have suggested that in the early stages of acquiring a bimanual skill "unintended bilateral coactivation occurs at multiple levels of motor organization". This may facilitate symmetrical motor output from the two upper limbs but will interfere when the task calls for asynchronous or asymmetrical output. The term temporal coupling refers to the fact that it is difficult to perform different temporal patterns with the two hands; spatial coupling refers to the fact that it is difficult to perform two different spatial patterns with the left and right hands (such as patting one's head and rubbing one's stomach). As Fagard et al. (1985) put it "the movement routines performed by the two limbs must be uncoupled before they can be re-integrated in a more complex pattern" (p. 535). Recent evidence from a single commissurotomized patient suggests that although temporal coupling between the upper limbs does not require the participation of the corpus callosum, spatial interference between the two sides does (Franz, Eliassen, Ivry, & Gazzaniga, 1996).

2. THE ROLE OF THE CORPUS CALLOSUM IN BIMANUAL PERFORMANCE

Given that each hand is controlled primarily (but not exclusively) by the opposite hemisphere (Gazzaniga, Bogen & Sperry (1967); Zaidel & Sperry, 1977), the co-operative activity of left and right hands must at some stage involve interhemispheric integration, especially for tasks that are not highly overlearned. Studies of patients with total and partial forebrain commissurotomy have demonstrated that efficient bimanual performance of certain unfamiliar tasks, especially if carried out in the absence of visual feedback, requires the integrity of the corpus callosum and especially of its anterior portion (Preilowski, 1972, 1975; Zaidel & Sperry, 1977).

There is evidence (in the monkey at least) for both homotopic and heterotopic callosal connections between the sensorimotor cortical areas on the two sides of the brain (Jenny, 1979). However, regions representing the "distal segments of the fore- and hind-limbs" (Jones & Powell, 1969) or the "portions of the hand area corresponding physiologically to the finger area" (Jenny, 1979) (but not the thumb) appear to lack direct commissural projections. Nonetheless, the two sides of the brain are richly connected by about 100 million large- and small-diameter fibres running through the corpus callosum. There is a complex relation between fibre thickness and brain size (Schütz & Preißl, 1996) but it is not known whether fibres of different thickness have different functions (Berlucchi, Aglioti, Marzi, & Tassinari, 1995).

The results of investigations of bimanual co-ordination in people with congenital absence of the corpus callosum (for general reviews see Chiarello, 1980; Jeeves, 1979, 1990, 1994; Lassonde, 1994; Milner &

Jeeves, 1979) are largely in agreement with those carried out on patients who have had the callosum surgically sectioned (although the possible contribution of associated cerebral damage must be kept in mind in interpreting both the split-brain and acallosal data). For example, Jeeves, Silver, and Jacobson (1988) studied one partial commissurotomy patient, three acallosal participants and control participants on Preilowski's task (which involves learning to co-ordinate the two hands to move a single cursor) and concluded that "for fast, co-ordinated highly skilled bimanual performance an intact functional corpus callosum is necessary" (p. 849). The same group has demonstrated that 6-year-old children's performance on this task is similar to that of acallosal participants . It was argued that the relatively poor level of bimanual skill could be attributed to "reduced efficiency of information interchange between the hemispheres due to an immature corpus callosum at age 6" (p. 322). Rauch and Jinkins (1994) have shown that in the first decade of life the corpus callosum is smaller relative to brain area, as well as in absolute size, compared with adults.

Experiments with mature acallosal participants led Ferris and Dorsen (1975) to conclude that the corpus callosum contributes inter alia to precision of movement. Jeeves and Silver (1988) reported that a single adult acallosal patient did not show normal prehension movements in reaching to pick up a briefly illuminated object (a saucer) in the dark but maintained an open hand until contact with the object was made (see also Jakobson, Servos, Goodale & Lassonde, 1994; Silver & Jeeves, 1994). A 12-year-old acallosal patient was reported by Reynolds and Jeeves (1977) to be slower than controls on the Minnesota Formboard test whether she used one hand or both hands. The 5-year-old acallosal patient studied by Jeeves et al. (1988) showed "early signs of dyspraxic difficulties accounting for clumsiness and poor co-ordination". At least one other patient with congenital absence of the callosum has been said to be relatively clumsy in using the fingers of her two hands, especially on the right (Meerwaldt, 1983). In this context it is interesting to note that following callosotomy, ipsilateral cerebral control of the fingers of the left hand is better than ipsilateral control of the fingers of the right hand (Trope, Fishman, Gur, Sussman, & Gur, 1987).

There is evidence, then, that the callosum is involved in motor functions. In cases of congenital absence of the callosum there may be continued (Lassonde, Sauerwein, Geoffrey, & Décarie, 1986) or increased reliance on ipsilateral pathways which are thought to be in competition with the contralateral pathways and not inhibited as would be the case if the callosum were present (Reynolds & Jeeves, 1977; Jeeves, 1990; Silver & Jeeves, 1994). Although there is considerable individual variability in the extent to which acallosals can compensate for the disconnection effects so typical of adult patients with total forebrain commissuromy, longitudinal studies show that certain manual deficits shown by acallosal patients may persist for many years (Jeeves, 1979).

It has been claimed that the role of the callosum in normal efficient, bimanual motor output is inhibitory (Chiarello & Maxfield, 1996; Dennis, 1976; Jeeves et al., 1988), a proposal for which there is some direct evidence. After application of a magnetic stimulus to one side of the head (over the motor cortex of one hemisphere) the evoked potential over the motor cortex of the opposite hemisphere can be recorded. This stimulus is known as a conditioning stimulus. Muscle responses can also be elicited on the contralateral side. If a second magnetic stimulus (the test stimulus) is applied to the previously unstimulated hemisphere at intervals within a period of around 5-6 milliseconds following application of the conditioning stimulus, the amplitude of the muscle response is reduced. That is, the conditioning stimulus has inhibited the response. The inhibitory effect is thought to be mediated at a cortical rather than spinal level via a transcallosal mechanism and is therefore referred to as interhemispheric inhibition (Ferbert, Prior, Rothwell, Day, Colebatch, & Marsden, 1992).

Magnetic stimulation to one hemisphere is associated with a period of electrical inactivity in the tonically activated muscles of the opposite side - the so-called silent period. The silent period can be reduced by application of a second magnetic stimulus (the conditioning stimulus) to the opposite hemisphere and this too has been attributed to a transcallosally mediated mechanism partly on the grounds that the effect was not observed in a patient with callosal agenesis (Schnitzler, Kessler, & Benecke, 1996).

Callosally mediated effects of magnetic stimulation are not always inhibitory. Evidence for both excitatory and inhibitory transcallosally mediated activity (see Cook, 1984) has been provided by Meyer, Röricht, Grafin, Kruggel, and Weindl (1995) and Schnitzler et al. (1996). Because it is believed that movements of the distal muscles are first generated bilaterally, transcallosal inhibition might ensure production of purely unilateral movement. Excitatory effects might facilitate finely co-ordinated bimanual activities.

3. BIMANUAL CO-ORDINATION, UNIMANUAL ASYMMETRY, AND AGING

Although many investigators have examined age-related changes in hand preference (see below and Chapter 4 of this volume), relatively few appear to have examined changes in asymmetry of motor skill with age. Elderly persons are reportedly slower than younger participants in both bimanual and unimanual tasks. In an experiment by Stelmach, Amrhein, and Goggin (1988) an elderly group of participants (mean age 69.8 years, range 67-75 years) exhibited twice the asynchrony between the left and right hands in initiating bimanual movements as did a younger group (mean age 22.4 years, range 21-25 years). Stelmach et al. attributed the locus of the age-related decrement in bimanual co-ordination to the stage of movement execution rather than to a preparatory stage.

Meudell and Greenhalgh (1987), using a unimanual peg-moving task, reported that the difference between left and right hand performance was relatively greater for older participants (mean age 72 years, range 63-82 years) than for younger participants (mean age 15 years, range 14.2 - 15 years). The older people took relatively longer with their left hand. The interaction between age and manual performance asymmetry was interpreted in terms of a faster age-related decline in abilities subserved by the right compared with the left hemisphere. Somewhat similar results to those of Meudell and Greenhalgh were reported by Weller and Latimer-Sayer (1985), who used a peg-moving task in a cross-sectional investigation. Speed of unimanual peg moving declined with age for both hands but to a greater extent for the left hand.

Further evidence of disproportionate slowing of the left hand in aging was found by Mitrushina, Fogel, D'Elia, Uchiyama, and Satz (1995) on a task that required participants to push a pin through a series of holes. There was a significant correlation between age and an index of performance asymmetry between the hands due to relatively greater slowing of the left hand with increasing age. According to the authors, this task imposed demands on the "highly specialized praxic processes governed by the left hemisphere" (p. 363). In fact, it is debatable whether the performance of the left hand on this task required the participation of the left hemisphere. Nonetheless, Mitrushina et al. (1995) proposed that "those functions which are controlled via callosal relay are predominantly affected by aging" (p. 363). Thus both bimanual and unimanual aging effects have been attributed to callosal factors.

4. AGING AND THE CORPUS CALLOSUM

The possibility that beyond a certain age the corpus callosum becomes increasingly less efficient at relaying information between the two sides of the brain is suggested by neuroanatomical findings. There have been a number of studies reporting that the size of the corpus callosum decreases with age although this effect has sometimes been found to interact with gender. Using post mortem material, Witelson (1989,1991) reported a decrease in total callosal area with age only in men. The age effect was not replicated by Aboitiz, Scheibel, and Zaidel (1992) but their specimens came from people who had died relatively young (in their forties) and any reduction in callosal size may not yet have been sufficiently advanced to show up.

Magnetic resonance imaging techniques have also been employed in studies of morphological changes with age in the corpus callosum. Bleier, Houston, and Byne (1986) refer to unpublished findings from their laboratory of "an age-associated decrease in anterior posterior distance" but no details are provided. It seems likely that the data were incorporated in the subsequent paper by Byne, Bleier, and Houston (1988) in which it is reported that there was a significant effect for age in anterior-posterior distance in which age was defined as above and below 40 years. There was also an interaction with gender, there being no difference between males and females in the younger group but a smaller callosal length in the anterior 4/5ths in men over 40 compared with women. Holloway and de Lacoste (1986) found no correlation between overall callosal area and age. On the other hand, Allen, Richey, Chai, and Gorski (1991) reported a significant decrease in total callosal area with advancing age and in its anterior components considered separately as did Weis, Kimbacher, Wenger, & Neuhold (1993). Doraiswamy, Figiel, Husain, McDonald, Shah, Boyko, Ellinwood, and Krishnan (1991) found total callosal area to be negatively correlated with age after covarying for gender.

Woodruff, McManus, and David (1995) carried out a meta-analysis of 11 MR studies of callosal size in schizophrenic patients and controls. They reported that if one particular study (in which callosal size was small and the participants were older than in other studies) was excluded there was no effect of age on callosal area in either patients or controls. However, in the largest single study to date, Burke & Yeo (1994) obtained a significant negative correlation between age (56-90 years) and total as well as anterior callosal area in 38 men but not 59 women.

Although there is striking individual variability in callosal morphology (Bleier, Houston, & Byne, 1986) and not all studies controlled for overall brain size, there seems to be some agreement that callosal thickness (e.g., Hayakawa, Konishi, Matsuda, Kuriyama, & Konishi, 1989) and/or cross-sectional area (Doraiswamy et al., 1991; Weis et al., 1993; Witelson, 1991) declines with age, at least in men if not women (Byne et al., 1988; Doraiswamy et al., 1991; Witelson, 1989, 1991), and especially in the anterior portion of the callosum (Allen et al., 1991; Weis et al.,1993; Burke & Yeo, 1994). Holloway and de Lacoste (1986) found a hint of an age effect in relation to the posterior (splenial) section (corrected for total size of callosum) but concluded "We believe ... that larger samples, with a greater spread of ages ... will be necessary to more accurately assess the interrelationships between age, brain weight and sex in the human corpus callosum" (p. 90). Cowell, Allen, Zallatemo, and Denenberg (1992) reported that maximum callosal width in males was attained at the age of 20 years (declining thereafter) but not until 41-50 years of age in females.

For some purposes it is sufficient to show only that there is a relation between callosal size and age. For others, it would be relevant to know whether the callosum was relatively large or small for a given overall size, weight or volume of brain. Jäncke, Staiger, Schlaug, Huang, and Steinmetz (1997) found that corpus callosum size increases with forebrain volume (though less than proportionally) and thus overall brain size needs to be controlled for if it is not to be a potentially confounding factor in studies of gender (see especially Rauch & Jinkins, 1994) or handedness differences in callosal size (for review see Beaton, 1997) and so too for age differences. Similarly, studies looking a regional morphology of the callosum in relation to gender, handedness or age need to take account of overall callosal size.

In the studies referred to above some investigators considered the relation between size of the callosum and overall size of the brain. Holloway and de Lacoste (1986) reported that this did not influence the age effect. Others have reported that there there is no significant relation between callosal size and either overall cortical area (Cowell et al., 1992), brain (Witelson, 1991) or body size (Doraiswamy et al., 1991). Witelson (1989) reported a significant relation between callosal size and brain weight but only reported correlations between age and callosal size uncorrected for brain weight. Burke and Yeo (1994) reported a significant relation between several callosal measures and brain volume and between the latter and age in men but not women. However, as their main interest was in handedness and gender differences they did not analyse the co-variation of callosal size and hemispheric volume with respect to age. Nor do most of the remaining studies provide relevant data with which to evaluate the age effect (Bleier et al., 1986; Hayakawa et al., 1989; Allen et al., 1991; Aboitiz et al., 1992; Weis et al., 1993; Woodruff et al. 1995).

It is possible that a relation between callosal size and age reflects nothing more than cerebral atrophy in some proportion of patients. Rauch and Jinkins (1994), using MR imaging, found no relation between age and either callosal size or an index of proportional callosal size that took account of general cerebral area. However, among a group of patients who were judged from their scans to have cerebral atrophy, the callosum was significantly smaller than in patients without such atrophy.

A reduction in overall callosal size with the passage of years suggests that there could be a corresponding reduction in efficiency of interhemispheric transmission across the callosum. The hypothesis that elderly participants would show relatively poor integration between the hands was tested by Moes, Jeeves, and Cook (1995) using a modification of the task first used by Preilowski (1972, 1975) to study bimanual co-ordination in commissurotomy patients. Patients with total or anterior callosal section were reported by Preilowski to be impaired at learning to co-ordinate their two hands to move a single cursor. Moes et al. reported that elderly volunteers (ranging in age from 60 to 85 years) were also significantly impaired on this task even after one accounted for a general slowing of their performances relative to younger participants. This effect (together with other evidence) was seen as consistent with the idea that with increasing age there is a proportionately greater slowing of interhemispheric than intra-hemispheric processes. To the best of our knowledge, however, there is no evidence of a disproportionate decline in number (or of diameter) of fibres in the callosum as compared with elsewhere in the neocortex.

The time taken for interhemispheric transmission of information has been estimated by comparing response times to a stimulus presented on the same side of space as the responding hand with the time taken to respond to the same stimulus presented on the side opposite the responding hand (Poffenberger, 1912; Bashore, 1981). Although interhemispheric transfer time is not all that is being measured in this paradigm (Berlucchi, et al., 1995), there is sufficient evidence to infer that some aspect of interhemispheric integration is reflected in the difference between crossed and uncrossed reaction times. If it is true that interhemispheric integration is to some extent compromised in elderly individuals, then it should be detectable using this paradigm. Consistent with such a prediction, Jeeves and Moes (1996) found that the crossed minus uncrossed reaction time difference was significantly greater in a group of elderly (60 years plus) than in younger (18-30 years) participants.

5. HEMISPHERE FUNCTION AND AGING

It is commonly accepted that there is some reduction in volume of brain tissue with advancing age. Cerebral atrophy, and specifically dendritic atrophy within the superior temporal gyrus, may begin around the age of 50 years (Anderson & Rutledge, 1996). This atrophy may not be uniform at the two sides thereby allowing the possibility that those functions mediated preferentially by one or other side of the brain decline at different rates. By and large, investigators seem agreed that although visuo-spatial ability declines with age, verbal functions do not. This has been attributed to a specific decline in the functions of the right hemisphere, though Nebes (1990) is justifiably sceptical of both the findings and the supposed explanation. Summaries of work related to hemispheric specialization in relation to aging are provided by Goldstein and Shelley (1981), Kocel (1980), Ellis and Oscar-Berman (1989), and Nebes (1990).

The inference that the relation between the two hemispheres changes with age requires to be tested using direct rather than indirect methods of specialized hemispheric function. The relevant evidence with regard to an unequal decline in hemispheric efficiency is, however, equivocal. On a test of tactile recognition of non-verbal shapes, Riege, Metter, and Williams (1980) found that the left hand showed a greater decline in accuracy with age than did the right hand. However, using lateralized tachistoscopic presentation, Obler, Woodward, and Albert (1984) found no evidence that an age-related decline in verbal or non-verbal matching tasks was more precipitate for one visual hemifield than the other. Nor did Borod and Goodglass (1980) find any interaction between age and dichotic-listening asymmetry for verbal or musical materials. Ellis and Oscar-Berman (1989) concluded that "although the neuropsychological decline associated with aging...affects certain cognitive abilities more than others, this is not directly related to to any lateralized hemispheric dysfunction. Rather, it is likely that both hemispheres are influenced to equal degrees by the functional deterioration associated with aging.." (p. 143).

If it were true that hemispheric asymmetry varies with age, this would be expected to affect a wide range of sensorimotor behaviour, including everyday manual activities. The fact that there are age-related trends in hand preference is therefore of interest in this context.

6. AGE-RELATED TRENDS IN ADULT HANDEDNESS

Trankell (1955) refers to "a decreasing frequency of left handedness with age" without giving any data. Fleminger, Dalton, and Standage (1977), using Annett's (1970) questionnaire (but not her method of classification), noted that an increase with age in right handedness (defined as all items carried out with the right hand) was associated with a decrease in mixed handedness (right hand for writing but left for any other action) and (from the age of 45 years) a decrease in left handedness (defined as left-handed writing). Many investigators since then have observed in cross-sectional studies an increase in right handedness with increasing age of the people sampled (Ashton, 1982; Beukelaar & Kroonenberg, 1986; Brackenridge, 1981; Coren & Halpern, 1991; Davis & Annett, 1994; Dellatolas, Tubert-Bitter, Castresana, Mesbah, Giallonardo, Lazaratou, & Lellouch, 1991; Gilbert & Wysocki, 1992; Halpern & Coren, 1988; Hugdahl, Satz, Mitruchina, & Miller, 1996; Kuhlemeier, 1991; Porac, Coren, & Duncan,1980) even when care is taken (Hugdahl, Zaucha, Satz, Mitrushina, & Miller, 1996) to exclude those individuals who have experienced pressure to change hands. Ashton (1982) pointed out that one explanation for the decline in left handedness with age may be "differential morbidity or mortality of left handers" (p. 142).

An increase in dextrality with age apparently applies more to some actions than to others (Porac, 1993) and possibly varies with gender, being reported for males but not females in a Brazilian study (Brito, Brito, Paumgartten, & Lins, 1989). Even if writing hand is removed from consideration, it appears that there is an age-related decline in right-sidedness as defined by activities other than writing, such as throwing a ball or drawing (Hugdahl et al., 1996). One would expect such activities to be less liable to modification through social pressure.

7. HANDEDNESS, AGING, AND LIFE-EXPECTANCY

The reason for the age related trend is controversial. (See also Chapter 4 of this volume.) The elimination hypothesis, that non-right handers die younger than right handers so none remains alive if one goes sufficiently far back in time (see Ashton, 1982, Coren & Halpern, 1991; Halpern & Coren, 1988, 1990, 1991), has been severely criticized (see Anderson, 1989; Kuhlemeier, 1991; Wood, 1988; commentaries in *The New England Journal of Medicine*, *vol. 325, No. 14,* 1991; Harris,1993; and reply by Halpern & Coren, 1993).

The evidence offered by Coren and colleagues for reduced longevity in left handers was of two sorts. First, the mean age of death of consistently left-handed baseball players was slightly younger than that of left-handed players (Halpern & Coren, 1988). The findings were not replicated by Fudin, Renninger, Lembessis, and Hirshon (1993) nor by Hicks, Johnson, Cuevas, Deharo, and Bautista (1994) in their own analyses of baseball data (see also Lembessis & Fudin, 1994). Supporting findings have, however, been claimed (Rogerson, 1993).

The second type of evidence (Halpern & Coren, 1991) came from a study of the mean age of death of right handers and non-right handers (determined by answers given by relatives of the deceased to three questions) in the state of California. Mean age of death was 9 years younger for non-right handers.

Annett (1993) has drawn attention to a weakness in any attempt to explain a decline in dextrality with increasing age in terms of reduced longevity of left handers. She argues that "The fallacy rests on a failure to distinguish between criteria used to define left handedness in the early and the later studies. In the first half of this century, the pressures against left handed writing were so well-known ...that evidence of left handedness was sought in actions *other than writing*" (original italics).. In a sample of people dying in any one year, the oldest sinistrals would have been shifted to dextrality and counted as right handers, while those recorded as left handers would be on average younger.... It is not that left handers die younger, but that left handed writers are younger than right handed writers in the population" (pp. 296-297). She has since published data confirming this for the U.K. (Davis & Annett, 1994).

An alternative hypothesis, that society has in general become more tolerant of deviations from consistent handedness in recent years (the social-modification hypothesis), has been more widely accepted, not that this is the only alternative possible. Porac (1993a) identified 6 different hypotheses but on the basis of her own data favoured "a combination of two developmental hypotheses, one postulating a trend toward increased consistency of preferred hand use and the second proposing the gradual covert shaping role of a right-based environment" (p. 709). It is perhaps not well appreciated that even at the present time some 11 per cent of (Canadian) individuals (Porac, Coren, & Searleman, 1986) experience overt pressure to switch hand, usually from using the left to using the right. Porac, Rees, and Buller (1990) conclude that "approximately 8% of the within-cultural variability in adult handedness scores can be explained by knowledge of overt environmental pressures. This figure rises to 23.5% when one examines cross-cultural variations in handedness patterns" (p. 285).

It is commonly found in family studies of handedness that there is a greater proportion of non-right handedness among the filial than the parental generation (e.g., Annett, 1979, 1994; Ashton, 1982). This generation effect has been found not only in recent studies but in earlier ones, too. This implies that a relaxation of social presure against left handedness in recent times cannot be the only mechanism underlying a reduction in left handedness with age. However, if the different manifestations of sidedness are indeed biologically related and if the elimination hypothesis has some credence, then one might also expect to see an age-related effect in footedness, eyedness or earedness, which presumably are not subject to social control. It is therefore of interest that Dargent-Paré, De Agostini, Mesbah, and Dellatolas (1992) reported finding an age effect in eyedness and footedness (see also Gabbard & Iteya, 1996; Porac, 1996) in a large scale study (n=5,199) of individuals from different countries. A similar effect was reported by Porac, Coren, and Duncan (1980). It is difficult to see why eyedness, for example, should be subject to social pressure and these data therefore offer a measure of support for the elimination hypothesis, which has not yet been unequivocally disconfirmed (see Hugdahl et al., 1993; 1996).

Because it has been claimed that left handedness is associated with life-threatening conditions such as breast cancer (Kramer, Albrecht, & Miller, 1985), it would not necessarily be surprising to find that left handers die earlier than right handers despite the fallacy highlighted by Annett (1993). Even less dramatic associations between non-right handedness and smoking (Harburg, 1981), alcoholism (Bakan, 1973; London, 1986, 1989; London,

Kibbee, & Holt, 1985), responsiveness to centrally active drugs (Irwin, 1984; London, 1986), auto-immune disease (Geschwind & Behan, 1982; Geschwind & Galaburda, 1987) and risk of accident (Coren, 1989; Halpern & Coren, 1991) might be expected to have some effect.

With regard to auto-immune disease, however, a thorough review of the literature (Bryden, McManus, & Bulman-Fleming, 1994a; see also commentaries and reply by Bryden, McManus, & Bulman-Fleming, 1994b) concluded that although "there seem to be real associations between handedness and some immune disorders ... some of these associations ... follow the pattern hypothesized by Geschwind and Galaburda, while others ... show the reverse pattern" (p.152).

8. ACCIDENTS, HANDEDNESS, AND PLASTICITY OF MANUAL FUNCTION

With regard to accident risk in adults, there have been failures to replicate Coren's (1989) finding of an elevated rate of accidents among non-right handers (Dellatolas, Moreau, Jallon, & Lellouch, 1993; Hemenway, Azrael, Rimm, Feskanich, & Willett, 1994; Merckelbach, Muris, & Kop, 1994; Peters & Perry, 1991) but also some supporting evidence. Graham, Dick, Rickert, and Glenn (1993) reported that left-handed children and adolescents were more likely than right handers to suffer injury, their definitions being based on answers to four questions. Respondents answering "right" to 3 or 4 of the questions were designated right handers; those giving 2, 3 or 4 "left" responses were regarded as left handers.

Daniel and Yeo (1991) re-analysed Coren's data distinguishing between left- and mixed-handers and found that those of mixed handedness, rather than left handers, were at greater risk than fully right-handed individuals (Daniel & Yeo, 1991). A similar effect was found by Hicks, Pass, Freeman, Bautista, and Johnson (1993). These findings cannot easily be reconciled with the idea that the layout of the environment is inimical to left handers as it is difficult to see why strong left handers escape the hazards that beset their less-sinistral brethren. An alternative explanation of a greater risk of accident to mixed-handers than to consistent left handers is that some proportion of the mixed-handers have suffered trauma at an earlier age and that this had the effect of shifting their handedness from full right handedness towards the sinistral end of the handedness distribution. Ashton (1982) noted that a decrease in left handedness with age was apparently balanced by an increase in ambidexterity. Whatever the precipitating event, it may have had a correlated effect of increasing the risk of accident. Segalowitz and Brown (1991) found that mixed-handed children suffered comparatively more mild head injuries than did consistent handers. Daniel and Yeo (1992) stated that both mixed- and left handers were more prone to head injury than right handers. In a later communication, in which left and mixed-handers were collapsed into a single category, Daniel and Yeo (1994) obtained comparable results.

Porac (1993b) reported no overall difference in frequency of accidents to left- and right handers but rather found that right handers tended to injure their right hand and left handers their left hand. She suggested that the pattern of injury is determined by the pattern of hand use rather than by intrinsic risk factors. There is some support for such a view (Beaton, Williams, & Moseley, 1994). Among right handers, either little or no difference (Absoud & Harrop, 1984; Hollis & Watson, 1993; Wilkes, 1956) or a slightly greater frequency of injury to the so-called dominant hand than to the non-dominant hand among right handers (Clark, Scott, & Anderson, 1985; Packer & Shaheen, 1993; Hill, Riaz, Mozzam, & Brennen, 1998) has been reported, although the nature or place of injury, as well as the patient's handedness, affects the distribution of injuries to left and right limbs (Meals, 1979; Beaton et al., 1994; Hill et al., 1998).

Patients who have had a stroke that paralyses the preferred arm commonly learn to use the other arm to a degree of skill which, to casual observation, is almost equivalent to pre-morbid levels of the hand that is paralysed. Furthermore, early but transient injury to the upper limb may induce a permanent change in degree of hand preference (Dellatolas et al.,1993; De Agostini, Khamis, Ahui, & Dellatolas, 1997). In rhesus monkeys there is very considerable functional re-organisation, so-called representational re-modelling, of the cortical maps representing body sensations and movements after complete amputation or sensory deafferentation of a limb (Florence, Taub, & Kaas, 1998; Jones & Pons, 1998; Merzenich, 1998). In one study (Florence et al., 1998) the extent of re-modelling was as great in one adult macaque, which had suffered from chronic disuse of one hand following a wrist injury many years previously, as in monkeys with complete arm amputations. These findings suggest that the degree of asymmetry of hand use is not fixed and that differences between the hands are part of a plastic (Dellatolas et al., 1993) rather than a static system. This in turn raises questions as to the nature of the decline in left handedness observed with increasing age. Perhaps as people get older they tend, for one reason or another, to rely increasingly on their right hand.

Whatever the mechanisms underlying age-related changes in handedness (and other lateral preferences), the fact remains that there are fewer older people than younger people who are left handed. It does not necessarily follow, however, that individuals become more and more right handed as they grow older, although this is a possibility (McGee & Cozad, 1980; Harris, 1990). Brown and Jaffe (1975) suggested that degree of cerebral lateralization changes with age, a hypothesis that implies a gradual shift in hand use. This view can be contrasted with the idea that there is an unequal decline in performance on tasks mediated by left and right hemispheres, the latter declining at a faster rate than the former once a critical age has been reached. Either mechanism would lead to a change with age in manifestations of manual asymmetry and would be expected to interact with the effects of prolonged differential use of the two hands.

9. THE EFFECT OF PRACTICE ON DIFFERENCES BETWEEN THE HANDS

Provins (1997) has recently elaborated his earlier arguments (Provins, 1956; 1958; 1967) that handedness derives from practice effects. Briefly, he contends (Provins, 1997) that "what is genetically determined is a neural substrate that has significantly increased its functional plasticity in the course of evolution...What is fine-tuned is the relative motor proficiency or skills achieved by the two sides in any given task according to their use and the demands made on them as a result of social pressure, other environmental influences or habit" (p. 556).

An alternative view was expressed by McManus, Kemp, and Grant (1986) who argued that practice is unlikely to be the cause of performance differences between the hands because degree of improvement in a simple tapping task was similar for both preferred and non-preferred hands and for typists and piano players as well as for non-specialist participants. Were practice to have been the original source of the difference between hands, it was argued, further practice would have been expected to have shown a differential effect on preferred and non-preferred hands in those for whom the non-preferred hand had already had more than usual experience in fine finger movement; that is in typists and piano players.

There are other reports (e.g., Perelle, Ehrman & Manowitz, 1981) that practice has a greater effect on the non-preferred hand, thereby reducing the original difference between the hands. Using himself as a subject and a simple finger-tapping task, Peters (1976) reported that the speed of the non-preferred hand came to equal that of the preferred hand. We have confirmed the findings of Peters with a single volunteer who practised tapping with each hand every day for a month (MacDonald, Beaton, & Folkard, 1999). By contrast, the between-hand difference on the same task was reported by McManus et al. (1986) to be relatively constant even after some practice. It is probable that length of practice can explain these differences; less than one hour in the McManus et al. study versus 10 days in the Peters study and 28 days in that of MacDonald et al. Furthermore, the initial difference between hands is a function of the precise tapping movement that is being made (Peters, 1980). It is therefore probable that the effects of practice vary both with duration and with characteristics of the movement.

Although repetitive tapping is a purely experimental task that is unlikely to be practised outside the laboratory, there are certain manual tasks that commonly receive a great deal of explicit practice. People learning to play a stringed musical instrument or keyboard typically practise for hours every day. What effect might this have on hand function?

10. HANDEDNESS AND HAND FUNCTION IN MUSICIANS

Hand preference has been assessed in a number of studies of musicians (e.g., Aggleton, Kentridge, & Good, 1994; Byrne, 1974; Christman, 1993; Hassler & Gupta, 1993; Hering, Catarci & Steiner, 1995; Oldfield, 1969). Such studies have usually been carried out with a view to drawing inferences regarding the role of the right hemisphere in musical function. Despite differences in the methods of assessing hand preference, there are indications that among at least some sub-groups of musicians there are more left- or mixed-handers than would be expected by chance. This has been seen as consistent with the view that the right hemisphere has a special role with regard to certain musical functions (see Gates & Bradshaw, 1977; Messerli, Pegna, & Sordet, 1995; Plenger, Breier, Wheless, Ridley, Papanicolaou, Brookshire, Thomas, Curtis, & Willmore, 1996; Hugdahl, Brønnick, Kyllingsbaek, Law, Gade, & Paulson, 1999; Penhune, Zatorre & Feindel, 1999). However, it is unlikely that only the right hemisphere is involved in cognitive processing of music for at least two reasons. One is that some work points to a special role of the left hemisphere in appreciation of rhythm (Gates & Bradshaw, 1977; Gordon, 1978). A second reason is that music is extended in time in the same way that language is a temporally organized activity. The left hemisphere has often been considered to operate according to a temporal, sequential mode of processing whereas the right hemisphere favours a more holistic, Gestalt mode of processing. Although Beaton (1985) criticized attempts to characterize hemisphere function in terms of certain fundamental dichotomies, mode of processing music may differ between musicians and non-musicians (Bever & Chiarello, 1974; Messerli et al., 1995) or according to task demands. It seems highly likely that musical stimuli are processed by both halves of the brain. This is not to say that emotional reactions to music do not differ between left and right cerebral hemispheres (Beaton, 1979).

Although there have been several studies of hand preference in musicians, only a few investigators have looked at hand skill from a laterality perspective. McManus et al. (1986) found no difference in asymmetry of unimanual tapping performance between experienced typists, piano players or control participants although the small number of participants in each group (n=4) might have been too small for a difference to have emerged.

Jäncke, Schlaug, and Steinmetz (1997) have recently reported that on a finger-tapping task the between-hand difference in professional classical musicians was reduced in comparison with untrained control participants of the same age. Beaton and Coleman (1998) confirmed these findings using Annett's peg-moving task. Further, Jäncke et al. (1997) reported that hand-skill asymmetry was related to age of commencement (but not duration) of musical training. The earlier the musicians began training, the smaller the between-hand asymmetry. However, cross-sectional results of this kind cannot distinguish cause from effect. Do individuals become competent musicians because their hands are fairly equal in skill or do their hands become skilled through musical training?

Peters (1985) studied the ability of 5 piano players to perform two different tapping tasks simultaneously, one with each hand, and observed that, unlike non-pianists, these participants "show considerable precision in the co-ordination of the two hands" (p. 191), which presumably occurred as a consequence of long hours of practice on the keyboard. More recently Elbert, Pantev, Wienbruch, Rockstroh, and Traub (1995) found in a functional magnetic imaging study that, in comparison with a control group of non-musicians, there was increased cortical representation of the digits of the left hand, but not of the right hand, in experienced string players (mean age 24, s.d. 3 years) who had played their instruments for a mean duration of 11.7 years (range 7-17 years). The correlation between the number of years for which the musicians had been playing and the magnitude of change (relative to controls) in the dipole moment (presumed to reflect the total degree of neuronal activity) elicited from the little finger of the left hand was significant. This suggests that changes in sensory input can induce changes in cortical re-organization. This in turn implies that environmental influences throughout the life span may have long-term cortical as well as functional consequences.

10.1 Music, practice, and the corpus callosum

Jäncke et al. (1997) interpreted their finding of reduced asymmetry between the hands in musicians compared with controls in terms of improved performance of the non-preferred hand through early and intensive training. The data are presumably drawn from the same participants for whom they reported an enlargement of the (anterior) region of the corpus callosum in musicians whose musical training had begun before the age of 7 years in comparison with those who began their training after this age (Schlaug, Jäncke, Huang, Staiger, & Steinmetz, 1995). Although this might point to the role of experience in developing the size of the callosum, it is clearly compatible with the evidence mentioned earlier that the anterior callosum is important for efficient bimanual performance.

Experience appears to have a role in determining cortical representation of auditory as well as tactile or motor functions. Pantev, Oostenveld, Engelien, Ross, Roberts, and Hoke (1998) found that cortical representation in response to piano tones, but not equally loud pure tones of similar fundamental frequency, was approximately 25 % greater among musicians than among control participants who had never played an instrument and that "Enlargement was correlated with the age at which musicians began to practise" (p. 811). However, the conclusion that "musical experience during childhood may influence structural development of the auditory cortex" (p. 813) was criticized on statistical and logical grounds by Monaghan, Metcalfe, and Ruxton (1998). These authors point out that the major contribution to the effects reported came from children aged 3-5 years of age and "perhaps only children with a particular type of cortical response to musical sounds are capable of learning an instrument from a very early age". In short, direction of causation cannot be inferred from the data presented by Pantev et al. any more than it can from cross-sectional studies of handedness in musicians. For this, it is necessary to separate the potential effects of aptitude, a so-called ear (or cortical response!) for music, from those resulting from experience or training.

Dichotic-listening studies and other experiments (usually with young people) suggest that certain musical functions, such as perception or recognition of melodies or chords, are mediated preferentially by the right hemisphere (Bryden, 1988; Gates & Bradshaw, 1977; Hugdahl et al., 1999). If a left-ear advantage implies a right-hemisphere superiority for some aspects of music, and if the efficiency of the right hemisphere declines more rapidly than that of the left hemisphere, then among elderly participants a left-ear superiority in processing certain musical stimuli might become a right-ear advantage. This was not found, however, by Borod and Goodglass (1980), who found no interaction between ear asymmetry and age for either melodic or digit stimulus materials. Unfortunately, we know of few other studies of aging in relation to dichotic listening using musical stimuli. In contrast, there are a number of studies (see Sidtis, 1988; Nebes, 1990) using verbal stimuli with participants from different age groups.

11. DICHOTIC LISTENING AND AGING

It has been reported by some authors, but not others (e.g., Borod & Goodglass, 1980; Nebes, Madden, & Berg, 1983), that verbal dichotic-listening performance declines with age and that this decline occurs predominantly for left-ear performance (Clark & Knowles, 1973; Johnson, Cole, Bowers, Foiles, Nokaido, Patrick, Woliver, & Woliver, 1979). Clark and Knowles (1973) specified the ear to be reported, and items were only scored as correct if they were recalled in the correct serial position. This means that the extent to which interpretation of the ear effect should be in terms of memorial as compared with hemispheric factors is not clear. In the study by Johnson et al. (1979), the stimuli consisted of digits, and similar considerations apply. Order of report was unconstrained and if participants adopt a right-ear-first order of report (Inglis, 1965) or bias their attention towards the right ear (Mondor & Bryden, 1991,1992) then performance at the left ear is likely to be reduced for such reasons alone. The youngest participants in this study apparently showed no difference between left and right ears and therefore it can not be assumed that in this particular case the dichotic-listening procedure provided a valid test of hemispheric asymmetry of function.

One of the major models of ear effects in dichotic listening is a structural model (for reviews of dichotic listening see Bryden, 1988; Hugdahl, 1995). According to this model, within a particular cerebral hemisphere those impulses transmitted by means of the crossed (contralateral) auditory pathway from ear to brain inhibit those arriving by way of the uncrossed (ipsilateral) pathway. In order to be reported verbally, information from the left ear (transmitted by the contralateral pathway) has to be relayed from the right to the left hemisphere across the corpus callosum (Hugdahl, 1995; Kimura, 1961). In agreement with this model, extinction of left-ear scores has been reported in some cases of total (see Milner, Taylor, & Sperry, 1968; Sidtis, 1988) or partial (e.g., Geffen, Walsh, Simpson, & Jeeves, 1980; Alexander & Warren, 1988; Sugishita, Otomo, Yamazaki, Shimizu, Yoshioka, & Shinohara, 1995) callosal section.

If the structural model of dichotic-listening effects is valid, then a less severe impediment to interhemispheric transfer than total commissurotomy would also be expected to interfere with left-ear recall. Reinvang, Bakke, Hugdahl, Karlsen, and Sundet (1994) tested this prediction among multiple sclerosis patients for whom there was MRI evidence of callosal thinning. These authors reported findings broadly consistent with the structural model, as have others (Pelletier, Habib, Lyon-Caen, Salamon, Poncet, & Khalil, 1993; Rao, Bernardin, Ellington, Ryan, & Burg, 1989; Rubens, Froehling, Slater, & Anderson, 1985).

Although the structural explanation of ear asymmetry in dichotic-listening scores is widely accepted, some authors (Andersson & Hugdahl,1987; Asbjørnsen & Hugdahl, 1995; Bryden, 1978; Bryden, 1988; Bryden, Munhall & Allard, 1983; Hugdahl, 1995; Hugdahl & Andersson, 1986; Mondor & Bryden 1991, 1992; Sexton & Geffen, 1979) have cautioned that memory (Inglis, 1965) or attentional factors have a role and may even over-ride the effect of structural mechanisms. It is conceivable that changes in attentional (or memory) capacity with age interact with ear differences on a dichotic-listening task. Using the same dichotic-listening test as that used in the present study and treating age as a dichotomous variable (below or above 41 years) Cowell and Hugdahl (2000) found a decrease in number of correct reports with age. It is clear, therefore, that interpretation of any depressed performance at the left ear or right ear may be couched in terms of cognitive factors as well as, or instead of, structural factors such as right-hemisphere dysfunction, and/or callosal impairment.

12. THE PRESENT STUDY

It is clear from the foregoing discussion that there is some inconsistency in the literature with regard to whether there is an unequal hemispheric decline in function with age and more specifically with regard to the pattern of performance in dichotic listening. There is also uncertainty over whether there is a specific reduction in interhemispheric integration in elderly people and over the effect of prolonged differential use of the left and right hands.

If it is true that the right hemisphere as a whole declines at a faster rate than the left, then this would be expected to show up on all tasks mediated preferentially by the right hemisphere. The prediction would be that the effects of age interact both with ear asymmetry on a dichotic-listening task and with the between-hand difference on a manual-performance task. There should be a disproportionate loss of performance at the left ear and by the left hand.

On the other hand, if there is a reduction in efficient callosal transfer with increasing age this would not be expected to have any influence on unilateral peg-moving performance but it might show up as a left-ear impairment in a dichotic-listening task. Thus the callosal and right-hemisphere hypotheses can be distinguished. Although the right-hemisphere hypothesis predicts relatively poor performance of both left ear and left hand, the callosal hypothesis predicts no hand deficit on a simple unimanual task but a deficit at the left ear on a (verbal) dichotic-listening task. Both accounts are based upon the contribution of structural factors to observed behavioural asymmetries. However, a difference in performance between the left and right hands or at the two ears on a dichotic-listening task might not be the result of neurological mechanisms alone but relate also to strategic factors. During manual tasks, for example, it is possible that attention is distributed strategically between the two hands or sides of space (Peters, 1981, 1983, 1985; Swanson, Ledlow, & Kinsbourne, 1978; Verfaellie & Heilman, 1990). Similarly, on a dichotic-listening task a subject might preferentially attend to one ear.

One way of controlling attentional (or report) strategies is to require participants to attend to (or report from) one ear before the other. This "forced attention" paradigm has to our knowledge not been used in studies with elderly people.

We decided to investigate laterality in relation to age by focusing on two general questions. First, is there an asymmetrical decline in performance of left and right hemispheres with increasing age? Second, does interhemispheric integration decline with age? To examine these questions we used the tasks of peg moving, finger localization and dichotic listening, with instructions to focus attention to either the right- or left-ear stimuli. Since performance on a unimanual peg-moving task is related to laterality of early childhood (Annett, 1973; Bishop, 1980) and adult (Costa, Vaughan, Levita & Farber, 1963) brain damage, peg-moving time can be regarded as a simple but sensitive lateralized measure of overall hemispheric efficiency. Weller and Latimer-Sayer (1985) employed a peg-moving task and reported an age-related decline in performance of both hands but especially of the left hand. They asked their volunteers to move 48 pegs from the bottom to the top half of a peg-board turning the pegs over as they did so. It occurred to us that this feature of the task might be affected by neuro-muscular deterioration in elderly people or by difficulties in articulation of the wrist. In an attempt to circumvent this possibility we chose to use the standard Annett peg-board task, which does not require turning of the pegs (Annett, 1985).

With regard to interhemispheric integration, a number of studies using tactile recognition or discrimination tasks show that somesthetic and tactile information between the hands transfers across the middle and posterior portions of the corpus callosum (Bentin, Sahar, & Moscovitch, 1984; Dimond, Scammell, Brouwers & Weeks, 1977; Lutsep, Wessinger, & Gazzaniga, 1995; McKeever, Sullivan, Ferguson, & Rayport, 1981; Risse, Le Doux, Springer, Wilson, & Gazzaniga, 1978) although the extent of transfer from one side of the brain to the other may not be as great as in the opposite direction (Bisiacchi, Marzi, Nicoletti, Carena, Mucignat, & Tomaiuolo, 1994; Geffen et al., 1985; Lutsep et al., 1995; Satomi, Kinoshita, & Hirakawa, 1991).

Tasks that require a participant to cross-match shapes are liable to be contaminated by a tendency to verbally label even shapes that are unfamiliar. As a measure of hemispheric integration, therefore, we chose a different manual task, this time involving finger localization. Volpe, Sidtis, Holtzman, Wilson, and Gazzaniga (1982) reported that two split-brain patients, with section only of 3 cm. of the posterior portion of the callosum (including the splenium), were unable to indicate on the fingers of one hand the position where they had been lightly touched on the corresponding fingers of the opposite hand. The required response was non-verbal and both patients were at ceiling when the task was performed within a single hand, left or right.

The results therefore imply that inter-manual finger localization is carried out via the corpus callosum. Because the same patients showed no significant further deterioration when the callosum was subsequently divided in its entirety, the results implicate the posterior section of callosum in the between-hands version of the task. Additional findings pointing to the importance of the callosum in finger cross-localization were reported for a further 10 commissurotomized patients (six with partial and four with complete callosal section) by Geffen, Nilsson, Quinn, and Teng (1985).

Our third task was a dichotic-listening task. Although it may generally be the case that the dichotic-listening test can be used to assess hemispheric asymmetry of speech or other lateralized verbal processes, the stimulus tapes and procedures used in dichotic-listening experiments need to be individually validated against neurological or neurosurgical data, such as the Wada test of speech lateralization (Strauss, 1988; Zatorre, 1989). This point is often overlooked in studies reported in the literature. The tape we used was validated against the Wada procedure (Hugdahl, Carlsson, Uvebrant, & Lundervold, 1997). The dichotic-listening test showed correct classification with respect to language dominance in 92% of participants .

Our study was carried out with three groups of volunteers. The mean ages (to the nearest whole year) of our groups were 22 years old (hereafter referred to as the twenties group), 40 years (hereafter referred to as the forties group), and 59 years (hereafter referred to as the sixties group). There were 14 participants in each group with 6, 3, and 5 males, respectively, in each group. Only non-right handers were asked to participate. Any volunteer who reported being forced to use a particular hand for a particular purpose was not accepted into the study. Any individuals who reported having arthritis or similar condition or showed any lack of dexterity using their hands were similarly excluded. The experimental procedures were as follows.

On the Annett peg-moving task participants moved a row of ten pegs from one side of a board to the other as fast as possible. The dimensions of the board were as given by Annett (1985). Participants were timed by stop-watch to the nearest one-tenth of a second as they carried out five trials with each hand. Pegs were moved from right to left with the right hand and from left to right with the left hand. A trial was abandoned and re-started if a peg was dropped. Timing began when the participant first touched a peg and stopped when the peg was placed in the correct hole and had been released. Participants were encouraged to work as quickly as possible and were provided with feedback as to their times after each trial.

The second experimental task of finger localization was carried out by asking participants to rest their forearms comfortably on a table with the palm of their hands facing upwards and their eyes closed. The experimenter lightly touched the tip of a single finger on one hand with a pencil and then touched the tip of either the corresponding finger on the opposite hand or the tip of a non-corresponding finger on the opposite hand. The participant's task was to indicate by means of a head nod or a headshake whether the "same" (i.e. corresponding) or a "different" (non-corresponding) finger had been touched as before. The non-verbal response was chosen so as to eliminate any asymmetric contribution of the hemispheres to the response. After ensuring that the participant understood the task and after practice trials the experimental trials were administered. There were 12 "same" trials and 12 "different" trials.

For the verbal dichotic-listening task participants were first screened so as to eliminate any obvious hearing impairment. The headphones were adjusted and calibrated so that the stimuli were presented at 70-db for each ear. The stimuli consisted of 36 pairs of all possible combinations of the consonant-vowel (CV) stimuli formed by adding /a/ to the stop consonants /b/, /d/, /g/, /p/, /k/ and /t/ - i.e. /ba/, /da/, /ga/, /pa/, /ka/ and /ta/. Stimulus pairs were spoken by a male voice and aligned by computer to ensure simultaneous presentation to left and right ears.These stimulus pairs were inter-leaved with trials on which the same item was presented to the two ears. This was to ensure that the stimuli were being correctly perceived and acted as a control for any hearing impairment. Only the data for the dichotic trials are analysed and referred to below because incorrect reporting of the binaural trials would have indicated a problem with hearing (or reporting), which would have led us to drop the relevant participant from the study. In fact, all binaural trials were correctly reported by all participants. There were three conditions in the dichotic-listening part of the study. All conditions involved presenting the stimuli either binaurally or dichotically but the instructions differed for each condition. In the first condition (non-forced or NF condition), participants were asked to report what they heard at each ear (or binaurally). No order of report was specified and participants were free to attend to whichever ear they chose. In the forced-right ear (FRE) condition participants were asked to pay attention to the right ear and in the forced left ear (FLE) condition they were asked to pay attention to the left ear. In both these conditions the participants were asked to report stimuli from the attended ear. The order of presentation of FRE and FLE conditions was reversed for one-half of the participants and the conditions were administered after the NF or control condition.

The three experimental tasks (peg moving, finger localization and dichotic listening) were administered in the same order to each participant with the following results. (All differences between means referred to as significant are at the 5 per cent level or better on a two-tailed test unless stated otherwise).

On the peg-moving task, the first trial with each hand was regarded as a practice trial.The peg-moving times for the last four trials were combined and averaged so as to provide a mean score for each hand and these are shown in Table 1.

Table 1. Mean (and s.d.) peg-moving time in seconds for each hand and age group

Age group	Right hand	Left hand
Twenties	8.56 (1.13)	9.67 (1.24)
Forties	9.54 (0.55)	10.43 (1.11)
Sixties	10.11 (1.06)	11.15 (1.46)

A two-way (group by hand) analysis of variance (with repeated measures over hand) revealed significant main effects of group ($p<0.01$) and hand ($p<0.001$) but no interaction between these factors. Subsequent testing by Fisher's LSD procedure (as hand and group effects were predicted) revealed that all groups differed significantly from each other. Despite the non-significant interaction term, left and right hands were compared within each group. The right hand was significantly faster in each group and each group differed significantly from the other. For the left hand, the differences between the youngest and oldest group differed significantly.The overall conclusion to be drawn from these data is that there was (unsurprisingly) a right hand superiority on this task for each group and, of particular relevance for this study, the size of this superiority did not differ with age despite the gradual slowing in peg-moving time. There are thus no indications here of a hemisphere- (or hand-) specific decline.

The data from the finger-localization task are shown in Table 2. These were analysed by two-way analysis of variance with group and type of response (same vs. different) as factors, the last being a repeated-measures factor. The only significant effect was due to group ($p<0.001$). Subsequent testing (Fisher's LSD) revealed that the sixties group differed significantly from the other two groups, which did not differ significantly from each other.

Table 2. Mean number (and s.d.) of correct responses in each condition of finger cross-localization task by each group

Age group	'Same' condition	'Different' condition
Twenties	12.0 (0)	11.93 (0.27)
Forties	10.5 (1.34)	10.86 (1.23)
Sixties	9.5 (2.47)	8.79 (2.19)

Unfortunately, for a variety of administrative reasons beyond our control, the two oldest groups of participants were no longer available for testing before we were able to collect control data on within-hand finger localization. (We had previously decided to collect the cross-hand and within-hand data on separate occasions to avoid potentially confusing the oldest participants.) However, pilot observations and published data from both neurologically normal young adults (Bentin, Sahar, & Moscovitch, 1984; Geffen et al., 1985) and split-brain patients (Volpe et al., 1982), as well as from a single 64-year-old patient with a callosal lesion (Satomi, Kinoshita, & Hirakawa, 1991), attest to the fact that within-hand finger localization performance is at or very close to ceiling when only a single finger is touched, as it was here. Even 10-year-old children find the task trivially easy (Beaton & Puddifer, 1999; see also Galin, Diamond, & Herron, 1977) despite the fact that myelination of the callosum may not yet be complete at this age (see Salamy, 1978; Yakovlev & Lecours, 1967). If, then, it is accepted that within-hand performance was probably fully intact for all three age groups in our study, one interpretation of the cross-hand data is that interhemispheric transfer was impaired in the oldest group relative to the youngest.

The dichotic-listening task had three conditions. In the non-forced (NF) condition, the participants were asked simply to report whichever syllables they heard regardless of which ear they came from. In the forced-right ear (FRE) condition they were asked to attend to and report stimuli heard at the right ear and in the forced-left ear (FLE) condition they had to attend to and report from the left ear. All participants carried out the NF condition first, followed by the forced-ear conditions. Half the number of participants were administered the FRE condition followed by the FLE condition and the remaining half carried out these conditions in the reverse order. The results are shown for all three groups and all conditions in Figure 1. Scores for the non-attended ear in the conditions of forced attention represent intrusions from the unattended ear.

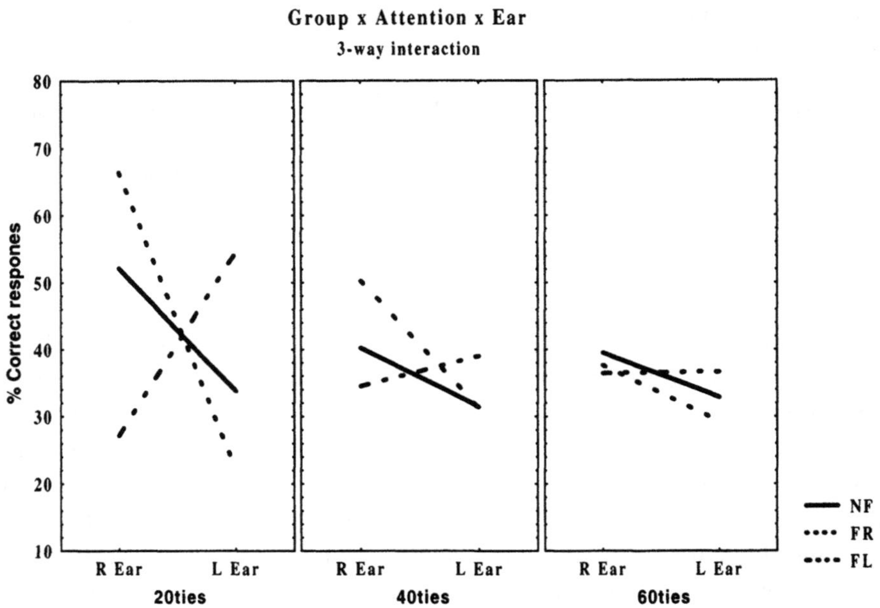

Figure 1. Right and left scores for each group and condition

The data were subjected to three-way analysis of variance with group, ear and condition as factors (repeated measures over ear and condition). The analysis showed significant main effects of group ($p<0.001$) and ear ($p<0.001$) and significant ear-by-condition ($p<0.001$) and condition-by-group ($p<0.001$) interactions.

There was also a significant three-way interaction ($p<0.001$). This interaction was decomposed using Fisher's LSD procedure as we had strong *a priori* hypotheses, namely a significant right-ear advantage (REA) for all groups in the NF and FRE conditions and a significant left-ear advantage (LEA) for each group in the FLE condition. These hypotheses were supported for all groups in the NF and FRE conditions ($p<0.05$ in each case). In the FLE condition, a significant LEA was obtained only for the youngest group ($p<0.001$). For the two older groups there was no significant ear effect in the FLE condition.

It is apparent from Figure 1 that in the NF condition the youngest group is better at the right ear than either of the other two groups ($p<0.05$ in each case), which do not differ between themselves. There is no difference at the left ear among the three groups.

In order to assess the relative magnitude of the ear effect in each group and condition, the data were converted into laterality coefficients according to the formula LC = [(RE-LE)/(R+L)]*100. The coefficients were subjected to a two-way analysis of variance with group and condition (repeated measures) as factors. This yielded a significant group-by-condition interaction ($p<0.001$). Subsequent decomposition using the LSD procedure revealed that in the NF condition the twenties group had a significantly larger ($p<0.05$) right-ear advantage than did the sixties group. In the FR condition the twenties group had a significantly ($p<0.001$) larger REA than both the forties and the sixties group. In the FL condition, the twenties group again had a significantly ($p<0.001$) larger ear effect, this time in favour of the left ear. No other group comparisons were significant.

In the FRE condition, performance at the right ear of the twenties group improved relative to the NF condition. That of the other two groups remained at much the same level as in the NF condition. In other words, the youngest group was able selectively to attend to the right ear, the effect summating with the "baseline" (NF) REA to produce a massive right-ear effect. The middle group (forties) was also able to improve their right-ear score but by not as much as the younger group, whereas the oldest group showed no improvement at either ear. This suggests that the ability to voluntarily pay attention to the right ear declines with age. Furthermore, in the FLE condition, only the youngest group showed a significant left-ear effect. Thus only the youngest group deployed attention to one or other ear in accordance with the experimental instructions.

Because mean scores obscure the performance of individuals, we have plotted in Figures 2-4 individual data points for each experimental condition. The diagonal line represents symmetry between left- and right-ear scores. Data points below the diagonal represent individuals with a right-ear score that is higher than their left-ear score. Conversely, data points above the diagonal represent individuals with a higher left- than right-ear score. It can be seen for the twenties group that in the NF condition (Figure 2) only one individual has a higher left- than right-ear score. In the FRE condition (Figure 3) all participants in the twenties group showed a considerably larger right ear score whereas in the FLE condition (Figure 4) only one individual showed a (very slightly) higher right-ear score. This shows very dramatically the effect of instructions to the twenties group to attend to one ear or the other. In comparison with the twenties group the shifts from below to above the diagonal are much less consistent for participants in the forties and sixties groups.

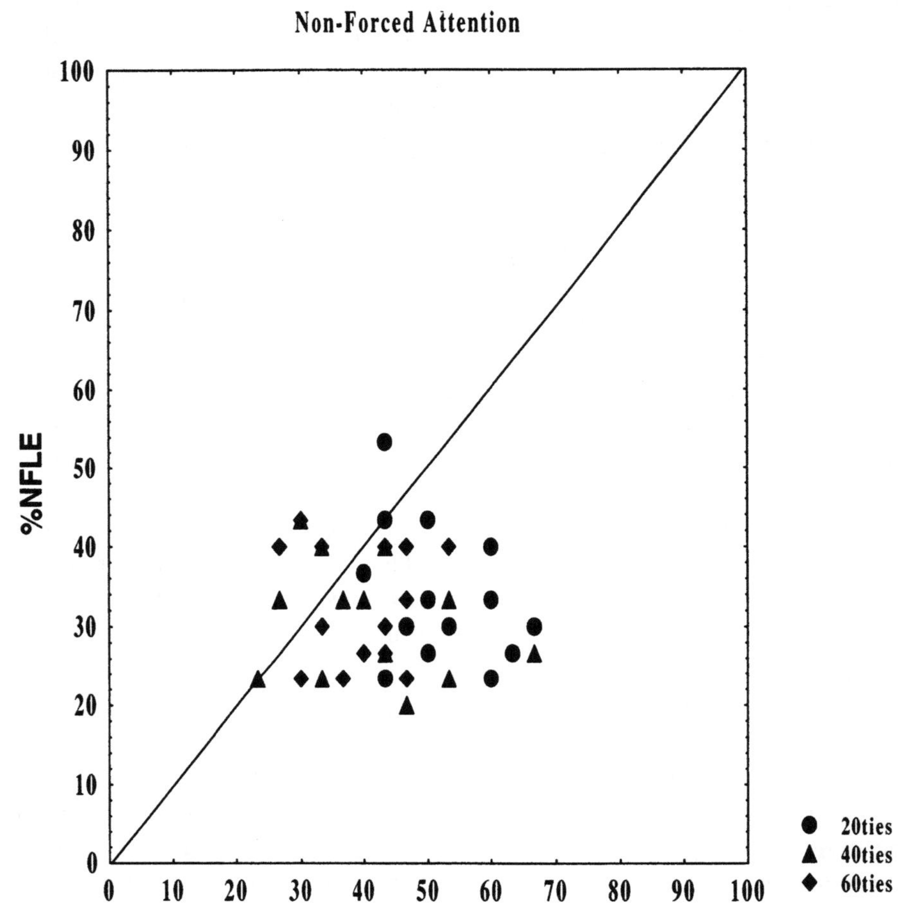

Figure 2. Right- and left-ear scores in the non-forced condition

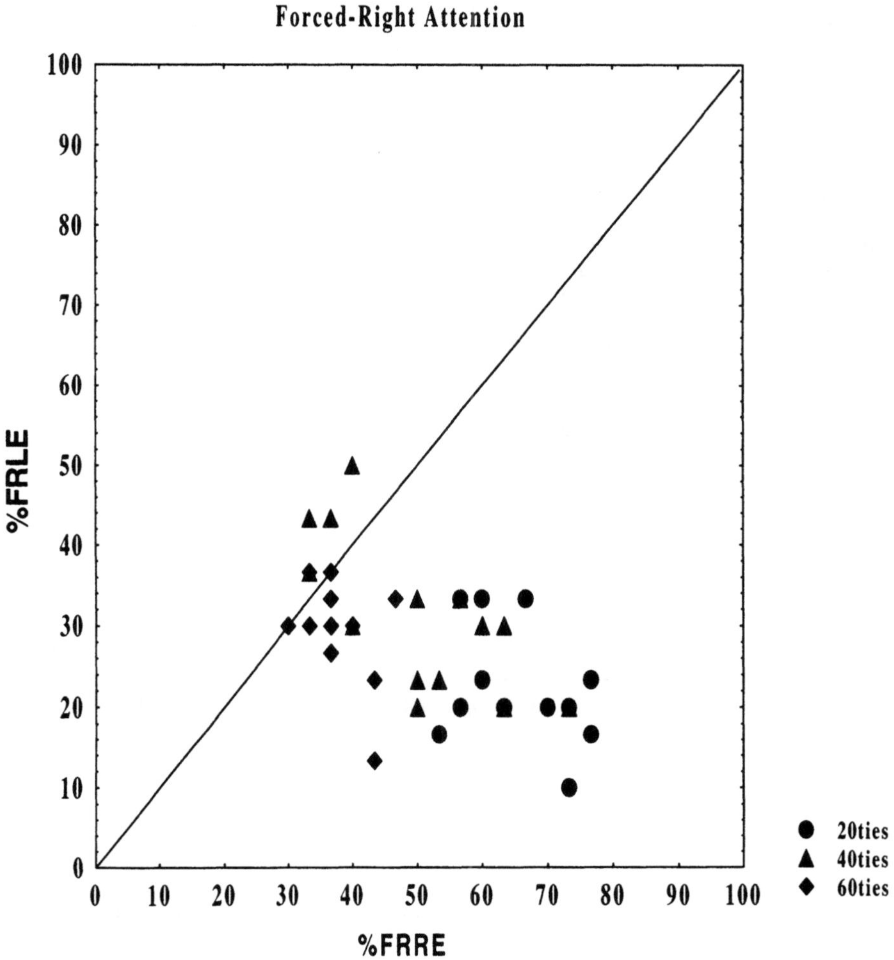

Figure 3. Right- and left-ear scores (intrusions) in the forced-right ear condition

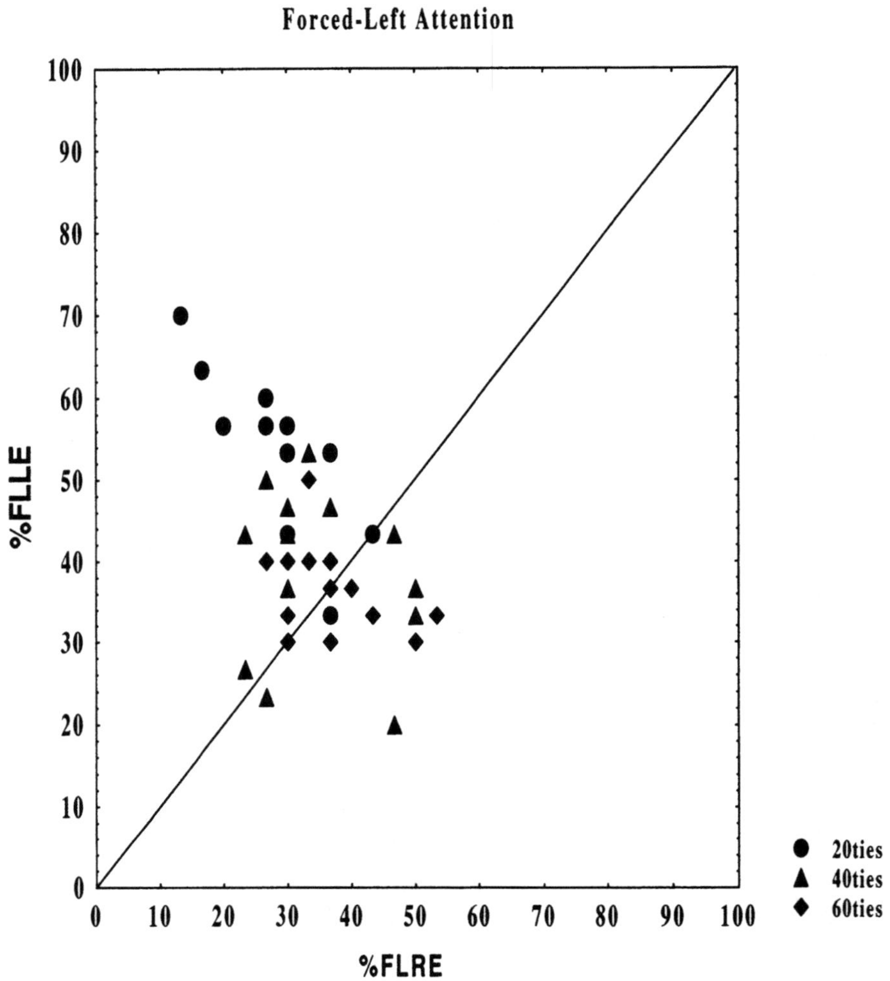

Figure 4. Right- and left-ear scores (intrusions) in the forced-left ear condition

The significant REA found for all three of our experimental groups in the NF condition of the dichotic-listening task presumably represents a superiority in auditory processing at the left side of the brain. Hugdahl et al. (1999) have shown, using PET scanning measures of blood flow, that monitoring CV syllables produces greater activation in the left superior temporal gyrus, compared to corresponding activation on the right side, and monitoring of musical sounds produces comparatively greater right-sided activation.

There has been much discussion in the neuroanatomical literature (for review see Beaton, 1997) concerning the functions of the region of the superior surface of the left temporal lobe known as the planum temporale (PT). This has often been regarded as a region for linguistic or phonological processing. Foundas, Leonard, Gilmore, Fennell, and Heilman (1994) have reported that 11 (right-handed) patients with a larger left planum as revealed by MR imaging were left-hemisphere dominant for speech as assessed by the Wada test whereas one (non-right handed) patient with a larger right planum was right-hemisphere dominant for speech. They concluded that their data "support the hypothesis that the planum temporale plays an important role in language dominance and that anatomic asymmetries as determined by MRI scans may predict language laterality" (although from their data it would be equally true to say that language laterality is predicted by handedness). Tzourio, Nkanga-Ngila and Mazoyer (1998) reported a significant correlation (one-tailed test) in 14 volunteers (including 5 left handers) between handedness and an index of asymmetry of the planum as revealed by MR imaging: the more asymmetrical the PT towards the left, the more right handed the subject. (The direction of asymmetry has to be worked out from the data provided, as the correlation appears to be negative but is reported as positive. Furthermore there is an inconsistency in the description of hand scores.) The authors also found that while the same participants listened to a story there was a significant positive correlation between degree of PET activation of the superior temporal gyrus on the left side and surface measurement of the left planum temporale.

Because the dichotic-listening technique is commonly used to assess language lateralization it might be expected that ear differences would correlate with neuroanatomical asymmetry. However, Jäncke and Steinmetz (1993) found no correlation between neuroanatomical asymmetry of the planum temporale and ear asymmetry in any of four verbal dichotic-listening tasks (including the same CV syllables as used in the present study). Nor was there a statistically significant association between side of ear advantage and direction of neuroanatomical asymmetry. Binder, Frost, Hammeke, Rao, and Cox (1996) using functional MRI found that the left planum temporale showed equal activation to words and tone sequences during passive listening and greater activation to tones than words during active listening. They suggested that the left planum is involved in auditory processing generally rather than linguistic processing specifically. This is not incompatible with a recent report of a region in the upper bank of the superior temporal sulcus, which is activated specifically by the human voice (Belin, Zatorre, Lafaille, Ahad, & Pike, 2000). However, it may be difficult to disentangle a specific contribution of the planum temporale (part of the superior surface of the temporal lobe) from that of the superior temporal gyrus (Tzourio et al., 1998). At the present time, therefore, it is unclear whether an REA on a dichotic-listening test reflects in any way structural aspects of hemispheric asymmetry. The most parsimonious explanation of the REA found for all groups in the NF condition is that it is related to a purely processing advantage at the left side.

There was a significant decline in overall accuracy of report from the youngest to the oldest group of participants. In comparison with the forties and sixties groups, the youngest group show a significantly greater right-ear effect due to a higher right-ear score. That is, right-ear scores declined with age but left-ear scores did not. In view of the above considerations it is likely that this derives from some left-hemisphere advantage over and above that due to the processing of the CV stimuli. The most probable explanation is that the younger participants have a bias towards attending to the right ear (see Mondor & Bryden, 1991,1992) and that this accentuates the pre-existing REA. It remains possible that there is a dramatic decline with age in the processing efficiency of the left hemisphere, at least for verbal material of the kind used here, but as no suggestion to this effect has appeared in the literature this seems an improbable explanation. The possibility of a selective hearing loss at the right ear can also be discounted because our control stimuli revealed no hearing problem with binaural (as opposed to dichotic) stimuli. Furthermore, the data of Chung, Mason, Gannon, and Willson (1983) provide no grounds for the idea that sensitivity of the right ear declines more rapidly with age than that of the left ear. We conclude that the age effect at the right ear is a reflection of a declining tendency to enhance right-ear recall through attentional mechanisms. This is consistent with the pattern of findings obtained for the FLE condition (see below).

At first sight our finding that the right but not the left ear declines with age contradicts the results of Johnson et al. (1979) who reported a left-ear decline with age. However, in their experiment participants "were specifically instructed to write down first what they heard in one ear, then to write down what they heard in the other ear, and told that which ear's worth of information was written down first was a matter of their own personal choice" (p. 110). Their findings suggest, therefore, that elderly persons adopt a right-ear-first order of report unless otherwise constrained. Our findings are consistent with theirs in so far as we have shown that even when requested to attend to the left ear, the sixties group were less able than younger people to do this.

For the two oldest groups, performance at each ear was very similar in both the NF and FRE conditions. Although these two groups show a slight shift towards the left ear in the FLE condition, the twenties group show a much more pronounced shift towards the left ear in this condition. It seems that the ability to flexibly shift attention towards one or other ear is much greater in the twenties group than in either of the two older groups. It is this, rather than any structural feature, that we believe characterizes age-related changes in dichotic ear asymmetry.

Summarizing our results overall, we have found, firstly, that with increasing age there is a decline in peg-moving times for both left and right hands with no difference between them in the rate of decline. This suggests that over the time period studied here there is a symmetrical overall decline in performance of both left and right hemispheres (that is, in overall cortical functioning, although the possible role of biomechanical factors cannot be ruled out). Second, as one grows older, interhemispheric integration, as measured by cross-hand finger localization, is compromised. With increasing age there appears to be a reduction in extent or fidelity of tactile information transfer across the corpus callosum. This would be consistent with suggestions of callosal thinning with advancing years, at least in those regions of the callosum that support tactile transfer. Thirdly, there is a reduction with age in dichotic ear asymmetry for CV syllables due to a decline in performance at the right ear. There is also a reduction in the ability to strategically direct attention to one or other ear. The right-ear effect is unlikely to be due to a reduction in interhemispheric transfer of auditory information because, with CV syllables as stimuli, this would show up at the left rather than the right ear. It is possible, however, that callosal mechanisms are implicated in what we have identified as a strategic effect. A number of authors have put forward the idea that the corpus callosum plays a critical role in the lateral control of attention (Kinsbourne, 1970; 1975; Holtzman, Sidtis, Volpe, Wilson, & Gazzaniga, 1981; Mayer, Koenig & Panchaud, 1988) and evidence that the callosum is related to the capacity to sustain attention has been provided by Rueckert, Sorensen, and Levy (1994). We suggest, therefore, that a reduction in callosal efficiency may lie at the heart of the effects that we have described. Further, it seems to us not inconceivable that a similar explanation might apply to other phenomena associated with aging, including an apparent reduction in choice of the left hand for a variety of everyday activities. If there is a bias, for whatever reason, towards attending to the right side of personal or extra-personal space, then it may become more and more difficult to resist this bias with increasing age.

13. REFERENCES

Aboitiz, F., Scheibel, A., & Zaidel, E. (1992). Morphometry of the sylvian fissure and the corpus callosum with emphasis on sex differences. *Brain, 115*, 1521-1541.

Absoud, E.M., & Harrop, S.N. (1984). Hand injuries at work. *Journal of Hand Surgery, 9*, 211-215.

Aggleton, J. P., Kentridge, R.W., & Good, J.M.M. (1994). Handedness and musical ability: a study of professional study of professional orchestral players, composers and choir members. *Psychology of Music, 22*, 148-156.

Alexander, M.P. & Warren, R.L. (1988). Localization of callosal auditory pathways: A CT case study. *Neurology, 38*, 802-804.

Allen, L., Richey, M., Chai, Y., & Gorski, R. (1991). Sex differences in the corpus callosum of the living human being. *The Journal of Neuroscience, 11*, 933-942.

Anderson, B., & Rutledge, V. (1996). Age and hemisphere effects on dendritic structures. *Brain, 119*, 1983-1990.

Anderson, M.G. (1989). Lateral preference and longevity. *Nature, 341*, 112.

Andersson, B., & Hugdahl, K . (1987). Effects of sex, age, and forced attention on dichotic-listening in children: A longitudinal study. *Developmental Neuropsychology, 3*, 191-206.

Annett, M. (1970). A classification of hand preference by association analysis. *British Journal of Psychology, 61*, 303-321.

Annett, M. (1973). Laterality of childhood hemiplegia and the growth of speech and intelligence. *Cortex, 9*, 4-33.

Annett, M. (1979). Family handedness in three generations predicted by the right-shift theory. *Annals of Human Genetics, 42*, 479-491.

Annett, M. (1993). The fallacy of the argument for reduced longevity in left handers. *Perceptual and Motor Skills, 76*, 295-298.

Annett, M. (1985). *Left, right, hand and brain: The right shift theory.* London: Erlbaum Associates.

Annett, M. (1994). Handedness as a continuous variable with dextral shift: Sex, generation and family handedness in subgroups of left- and right handers. *Behavior Genetics, 24*, 51-63.

Annett, M. (1995). The right shift theory of a genetic balanced polymorphism for cerebral dominance and cognitive processing. *Cahiers de Psychologie, 14*, 427-480.

Asbjørnsen, A. E., & Hugdahl, K. (1995). Attentional effects in dichotic-listening. *Brain and Language, 49,* 189-201.

Ashton, G.C. (1982). Handedness: An alternative hypothesis. *Behavior Genetics, 12,* 125-147.

Bakan , P. (1973). Left handedness and alcoholism. *Perceptual and Motor Skills, 36,* 514.

Bashore, T. R. (1981). Vocal and manual reaction time estimates of interhemispheric transmission time. *Psychological Bulletin, 89,* 352-368.

Beaton, A. A. (1979). Hemispheric emotional asymmetry in a dichotic-listening task. *Acta Psychologica, 43,* 103-109.

Beaton, A. A. (1985). *Left side, right side: A review of laterality research.* London: Batsford and New Haven:Yale University Press.

Beaton, A. (1997). The relation of planum temporale asymmetry and morphology of the corpus callosum to handedness, gender and dyslexia: A review of the evidence. *Brain and Language, 60,* 255-322.

Beaton, A.A., & Coleman, J. (1998). Finger tapping asymmetry in professional musicians. (*Unpublished data*).

Beaton, A. A., & Puddifer, A. (1999). Finger localization and phonemic awareness in dyslexic children. (*In preparation*).

Beaton, A. A., Williams, L., & Moseley, L .G. (1994). Handedness and hand injuries. *Journal of Hand Surgery (British and European Volume) 19B,* 158-161.

Belin, P., Zatorre, R., Lafaille, P, Ahad, P. & Pike, B. (2000). Voice selective areas in human auditory cortex. *Nature, 403*, 309-311.

Bentin, S., Sahar, A., & Moscovitch, M. (1984). Intermanual information transfer in patients with lesions in the trunk of the corpus callosum. *Neuropsychologia 22*, 601-611.

Berlucchi, G., Aglioti, S., & Tassinari, G. (1994). The role of the corpus callosum and bilaterally distributed motor pathways in the synchronization of bilateral upper-limb responses to lateralized light stimuli. In S.P. Swinnen, J. Massion, H. Heuer, P. Casaer (Eds.), *Interlimb coordination: Neural, dynamical, and cognitive constraints* (pp. 209-227). San Diego: Academic Press.

Berlucchi, G., Aglito, S., Marzi, C. A., & Tassinari, G .(1995). Corpus callosum and simple visuomotor integration. *Neuropsychologia, 33,* 923-936.

Beukelaar, L.J., Kroonenberg, P. M. (1986). Changes over time in the relationship between hand preference and writing hand among left handers. *Neuropsychologia, 24,* 301-303.

Bever, T.G , Chiarello, R.J. (1974). Cerebral dominance in musicians and non-musicians. *Science, 185*, 537-539.

Binder, J. R., Frost, J. A., Hammeke, T. A., Rao, S. M., & Cox, R. W. (1996). Function of the left planum temporale in auditory and linguistic processing. *Brain, 119,* 1239-1247.

Bishop, D.V.M. (1980). Handedness, clumsiness and cognitive ability. *Developmental Medicine and Child Neurology, 22,* 569-579.

Bishop, D.V.M. (1989). Does hand proficiency determine hand preference? *British Journal of Psychology, 80,* 191-199.

Bisiacchi, P., Marzi, C. A., Nicoletti ,R., Carena, G., Mucignat, C., & Tomaiuolo, F. (1994). Left-right asymmetry of callosal transfer in normal human subjects. *Behavioural Brain Research, 64,* 173-178.

Bleier, R., Houston, L., & Byne, W. (1986). Can the corpus callosum predict gender, age, handedness or cognitive differences? *Trends in Neurosciences, 9,* 391-394.

Borod, J. C., & Goodglass, H. (1980). Lateralization of linguistic and melodic processing with age. *Neuropsychologia, 18,* 79-83.

Brackenridge, C.J. (1981). Secular variation in handedness over ninety years. *Neuropsychologia, 19,* 459-462.

Bradshaw, J. & Rogers, L. (1993). *The evolution of lateral asymmetries, language, tool use, and intellect.* New York: Academic Press.

Brito, G.N.O., Brito, L.S.O., Paumgartten, J.R., & Lins, M.F.C. (1989). Lateral preferences in Brazilian adults. *Cortex, 25,* 403-415.

Brown, J. W., & Jaffe, J. (1975). Hypothesis on cerebral dominance. *Neuropsychologia, 13,* 107-110.

Bryden, M.P. (1978). Strategy effects in the assessment of hemispheric aymmetry. In Underwood, G. (Ed). *Strategies of information processing.* New York: Academic Press, 117-149.

Bryden, M. P. (1988). An overview of the dichotic-listening procedure and its relation to cerebral organization. In K. Hugdahl (Ed.), *Handbook of dichotic-listening, theory, methods and research.* Chichester: Wiley.

Bryden, M. P., McManus, I. C., & Bulman-Fleming, M. B. (1994a). Evaluating the empirical support for the Geschwind-Behan-Galaburda model of cerebral lateralization. *Brain and Cognition, 26,* 103-167.

Bryden, M. P., McManus, I. C., & Bulman-Fleming, M. B. (1994b). GBG, BMB, R & L, X & Y ....reply to commentaries. *Brain and Cognition ,26,* 312-326.

Bryden, M. P., Munhall, K., & Allard, F. (1983). Attentional biases and the right-ear effect in dichotic listening. *Brain and Language, 18,* 236-248.

Burke, H. L., & Yeo, R. A. (1994). Systematic variations in callosal morphology: The effects of age, gender, hand preference and anatomic asymmetry. *Neuropsychology, 8,* 563-571.

Byne, W., Bleier, R., & Houston, L. (1988). Variations in human corpus callosum do not predict gender: A study using magnetic resonance imaging. *Behavioral Neuroscience, 102*, 222-227.

Byrne, B. (1974). Handedness and musical ability. *British Journal of Psychology, 65,* 279-281.

Chiarello, C. (1980). A house divided? Cognitive functioning with callosal agenesis. *Brain and Language, 11*, 125-158.

Chiarello, C., & Maxfield, L. (1996). Varieties of interhemispheric inhibition, or how to keep a good hemisphere down. *Brain and Cognition, 30*, 81-108.

Christman, S. (1993). Handedness in musicians: Bimanual constraints on performance. *Brain and Cognition, 22,* 266-272.

Chung, D., Mason, K., Gannon, & Willson ,G. N. (1983). The ear effect as a function of age and hearing loss. *Acoustical Society of America, 73,* 1277-1282.

Clark, L. E., & Knowles, J. B. (1973). Age differences in dichotic-listening performance. *Journal of Gerontology, 28*, 173-178.

Clark, D.P., Scott, R.N., & Anderson, I.W.R. (1985). Hand problems in an accident and emergency department. *The Journal of Hand Surgery (British and European Volume)*, 10B, 297-299.

Cook, N . (1984). Homotopic callosal inhibition. *Brain and Language, 23*, 116-125.

Corballis, M.C. (1989). Laterality and human evolution. *Psychological Review, 96,* 492-505.

Corballis, M. C., Lee, K., McManus, I. C., & Crow, T. J. (1996). Location of the handedness gene on the X and Y chromosomes. *American Journal of Medical Genetics (Neuropsychiatric Genetics), 67,* 50-52.

Coren, S. (1989). Left handedness and accident-related injury risk. *American Journal of Public Health, 79,* 1040-1041.

Coren, S., & Halpern, D. (1991). Left handedness: A marker for decreased survival fitness. *Psychological Bulletin, 109*, 90-106.

Coren, S. & Porac, C. (1977). Fifty centuries of right handedness: the historical record. *Science, 198,* 631-632.

Costa, L.D., Vaughan, H.G., Levita, E. & Farber, N. (1963). Purdue pegboard as a predictor of the presence and laterality of cerebral lesions. *Journal of Consulting Psychology, 27,* 133-137.

Cowell, P. E., Allen, L. S., Zalatimo, N. S., & Denenberg, V. H. (1992). A developmental study of sex and age interactions in the human corpus callosum. *Developmental Brain Research, 66*, 187-192.

Cowell, P. & Hugdahl, K. (2000). Individual differences in neurobehavioral measures of laterality and interhemispheric function as measured by dichotic listening. *Developmental Neuropsychology*. (In press)

Daniel, W. F., & Yeo, R .A. (1991). Handedness and accident proneness. *American Journal of Public Health, 81,* 1346.

Daniel, W. F., & Yeo, R. A. (1992). Handedness, accidents and mental health. *American Journal of Public Health, 82*, 1691-1692.

Daniel, W. F., & Yeo, R. A. (1994). Accident proneness and handedness. *Biological Psychiatry, 35,*496.

Dargent-Paré, C., De Agostini, M., Mesbah, M., & Dellatolas, G. (1992). Foot and eye preference in adults. *Cortex, 28*, 343-351.

Davis, A. , Annett, M. (1994). Handedness asymmetries as a function of twinning, age and sex. *Cortex, 30*, 105-111.

Davidson, I., & Noble, W. (1993). Discussion and criticism. *Current Anthropology, 34,* 165-171.

De Agostini, M., Khamis, H.A., Ahui, A.M., & Dellatolas, G. (1997). Environmental influences in hand preference. *Brain and Cognition, 35*, 151-167.

Dellatolas, G., Moreau, T., Jallon, P., & Lellouch, J. (1993). Upper limb injuries and handedness plasticity. *British Journal of Psychology, 84*, 201-205.

Dellatolas, G., Tubert-Bitter, P., Castresana, A., Mesbah, M., Giallonardo, T., Lazaratou, H., & Lellouch, J. (1991). Age and cohort effects in adult handedness. *Neuropsychologia, 29,* 255-261.

Dennis, M. (1976). Impaired sensory and motor differentiation with corpus callosum agenesis: A lack of callosal inhibition during ontogeny? *Neuropsychologia, 14*, 455-46.

Dimond, S. J., Scammell, R. E., Brouwers, E. Y. M., & Weeks, R. (1977). Functions of the centre section (trunk) of the corpus callosum in man. *Brain, 100,* 543-562.

Doraiswamy, P. M., Figiel, G. S., Husain, M. M., McDonald, W. M., Shah, S. A., Boyko, O. B., Ellinwood, E. H., & Krishnan, E. R. R. (1991). Aging of the human corpus callosum: Magnetic resonance imaging in normal volunteers. *Journal of Neuropsychiatry and Clinical Neurosciences, 3*, 392-397.

Elbert, T., Pantev, C., Wienbruch, C., Rockstroh, B., & Taub, E. (1995). Increased cortical representation of the fingers of the left hand in string players. *Science, 270,* 305-307.

Ellis, R., & Oscar-Berman, M. (1989). Alcoholism, aging and functional cerebral asymmetries. *Psychological Bulletin, 106*, 128-147.

Fagard, J., Morioka, M., & Wolff, P. H. (1985). Early stages in the acquisition of a bimanual motor skill. *Neuropsychologia, 23*, 535-543.

Ferbert, A., Prior, A., Rothwell, J. C., Day, B. L., Colebatch, J. G., Marsden, C. D. (1992). Interhemispheric inhibition of the human motor cortex. *Journal of Physiology, 453,* 525-546.

Ferriss, G., Dorsen, M. M. (1975). Agenesis of the corpus callosum.1. Neuropsychological Studies, *Cortex, 11*, 95-122.

Fleminger, J. J., Dalton, R., & Standage, K. F. (1977). Age as a factor in the handedness of adults. *Neuropsychologia, 15*, 571-473.

Florence, S.L., Taub, H.B., & Kaas, J.H. (1998). Large-scale sprouting of cortical connections after peripheral injury in adult Macaque monkeys. *Science, 282*, 1117-1121.

Foundas, A. L., Leonard, C. M., Gilmore, R., Fennell, Heilman, K. M. (1994). Planum temporale asymmetry and language dominance. *Neuropsychologia, 32*, 1225-1231.

Franz, E. A., Eliassen, J. C., Ivry, R. B., & Gazzaniga, M. S. (1996). Dissociation of spatial and temporal coupling in the bimanual movements of callosotomy patients. *Psychological Science, 7*, 306-310.

Fudin, R., Renniger, L., Lembessis, E., & Hirshon, J. (1993). Sinistrality and reduced longevity: Reichler's 1979 data on baseball players do not indicate a relationship. *Perceptual and Motor Skills, 76*, 171-182.

Gabbard, C., & Iteya, M. (1996). Foot laterality in children, adolescents and adults. *Laterality, 1*, 199-205.

Galin, D., Diamond, R., & Herron, J. (1977). Development of crossed and uncrossed tactile localization on the fingers. *Brain and Language, 4*, 588-590.

Gates, A., & Bradshaw, J.L. (1977). The role of the cerebral hemispheres in music. *Brain and Language, 4*, 403-431.

Gazzaniga, M.S., Bogen, J.E. & Sperry, R.W. (1967). Dyspraxia following division of the cerebral commissures. *Archives of Neurology, 16*, 607-612.

Geffen, G., Walsh, A., Simpson, D., & Jeeves, M. (1980). Comparison of the effects of transcortical and transcallosal removal of intra-ventricular tumours. *Brain, 103*, 773-788.

Geffen, G., Nilsson, J., & Quinn, K. (1985). The effect of lesions of the corpus callosum on finger localization. *Neuropsychologia, 23*, 497-514.

Geschwind, N., & Behan, P. (1982). Left handedness: Association with immune disease, migraine and developmental learning disorder. *Proceedings of the National Academy of Sciences of the U.S.A., 79*, 5097-5100.

Geschwind, N., & Galaburda, A. (1987). *Cerebral lateralization*. Cambridge, MA: MIT Press.

Gilbert, A.N., & Wysocki, C.J. (1992). Hand preference and age in the United States. *Neuropsychologia, 30*, 601-608.

Goldstein, G., & Shelly, C. (1981). Does the right hemisphere age more rapidly than the left? *Journal of Clinical Neuropsychology, 3*, 65-78.

Gordon, H. (1978). Left hemisphere dominance for rhythmic elements in dichotically-presented melodies. *Cortex, 14*, 58-70.

Graham, C. J., Dick, R., Rickert, V., & Glenn, R. (1993). Left handedness as a risk factor for unintentional injury in children. *Pediatrics, 92*, 823-826.

Guiard, Y. (1987). Asymmetric division of labour in human skilled mimanual action: The kinematic chain as a model. *Journal of Motor Behavior, 19*, 486-517.

Halpern, D. F., Coren, S. (1988). Do right handers live longer? *Nature, 333*, 213.

Halpern, D. F., & Coren ,S. (1990). Laterality and longevity: Is left handedness associated with a younger age at death? In S. Coren (Ed), *Left handedness: Behavioral implications and anomalies* (pp.509--545). Amsterdam: Elsevier.

Halpern, D. F., Coren, S. (1991). Handedness and life span. *The New England Journal of Medicine, 324*, 998.

Halpern, D. F., & Coren, S. (1993). Left handedness and life span: A reply to Harris. *Psychological Bulletin, 114*, 235-241.

Harburg, E. (1981). Handedness and drinking-smoking types. *Perceptual and Motor Skills, 52*, 279-282.

Hardyck, C., & Petronovich, L.F. (1977). Left handedness. *Psychological Bulletin, 84*, 385-404.

Harris, L. (1980). Left handedness: Early theories, facts and fancies. In J. Herron (Ed.), *The neuropsychology of left handedness* (pp.3-78). San Diego: Academic Press.

Harris, L. (1990). Cultural influences on handedness: Historical and contemporary theory and evidence. In S. Coren (Ed), *Left handedness: Behavioral implications and anomalies*. Amsterdam: Elsevier. 195-258.

Harris, L. (1993). Do left handers die sooner than right handers? Commentary on Coren and Halpern's (1991) "Left handedness: A marker for decreased survival fitness". *Psychological Bulletin, 114*, 203-234.

Hassler, M., & Gupta, D. (1993). Functional brain organization, handedness and immune vulnerability in musicians and non-musicians. *Neuropsychologia, 31*, 655-660.

Hayakawa, K., Konishi, Y., Matsuda, T., Kuriyama, M., & Konishi, K. (1989). Development and aging of brain midline structures: Assessment with MR imaging. *Neuroradiology, 172*, 171-177.

Hemenway, D., Azrael, D., Rimm, E., Feskanich, D., & Willett, W. (1994). Risk factors for wrist fracture: Effect of age, cigarettes, alcohol, body height, relative weight, and handedness on the risk for distal forearm fractures in men. *American Journal of Epidemiology,140* ,361-367.

Hering, R., Catarci, T., & Steiner, T. (1995). Handedness in musicians. *Functional Neurology, 10*, 23-26.

Hewes, G.W. (1973). Primate communication and the gestural origin of language. *Current Anthropology, 14*, 5-24.

Hicks, R. A., Johnson, C., Cuevas, T., Deharo, D., & Bautista, J. (1994). Do right-handers live longer? An updated assessment of baseball player data. *Perceptual and Motor Skills, 78,* 1243-1247.

Hicks, R. A., Pass, K., Freeman, H., Bautista, J., & Johnson, C. (1993). Handedness and accidents with injury. *Perceptual and Motor Skills, 77,* 1119-1122.

Hill, C., Riaz, M., Mozzam, A., & Brennen, M. D. (1998). A regional audit of hand and wrist injuries. *Journal of Hand Surgery (British and European Volume), 23B,* 196-200.

Hollis, L. J., & Watson, D. P. (1993). The relationship between handedness, mechanism of injury and which hand injured. *Journal of Hand Surgery (British and European Volume), 18B*, 394.

Holloway, R. L., & de Lacoste, M. C . (1986). Sexual dimorphism in the human corpus callosum: An extension and replication study. *Human Neurobiology, 5*, 87-91.

Hugdahl, K. (1995). Dichotic-listening: Probing temporal lobe functional integrity. In R.J. Davidson & K. Hugdahl (Eds.), *Brain asymmetry* (pp.123-156). Cambridge, MA: MIT Press.

Hugdahl, K., Brønnick, K., Kyllingsbaek, S., Law, I., Gade, A., & Paulson, O.B. (1999). Brain activation during dichotic presentations of consonant-vowel and musical instruments stimuli: A 150-PET study. *Neuropsychologia, 37*, 431-440.

Hugdahl, K., & Andersson, L. (1986). The "forced-attention paradigm" in dichotic-listening to CV-syllables: A comparison between adults and children. *Cortex, 22,* 417-432.

Hugdahl, K., Carlsson, G., Uvebrant, P., & Lundervold, A. (1997). Dichotic-listening performance and intracarotid injections of amobarbital in children and adolescents. *Archives of Neurology 54*, 1494-1500.

Hugdahl, K., Satz, P., Mitrushina, M., & Miller, E. (1993). Left handedness and old age: Do left handers die earlier? *Neuropsychologia, 31,* 325-333.

Hugdahl, K., Zaucha, K., Satz, P., Mitrushina, M., & Miller, E. (1996). Left handedness and age: Comparing writing/drawing and other manual activities. *Laterality ,1,* 177-183.

Inglis, J. (1965). Dichotic-listening and cerebral dominance. *Acta Otolaryngologica, 60,* 231-238.

Ireland, W. (1880). Notes on left handedness. *Brain, 3,* 207-214.

Irwin, P. (1984). Greater brain response of left handers to drugs. *Neuropsychologia, 23*, 61-67.

Jakobson, L. S., Servos, P., Goodale, M. A., & Lassonde, M. (1994). Control of proximal and distal components of prehension in callosal agenesis. *Brain, 117,* 1107-1113.

Jäncke, L., Schlaug, G., & Steinmetz, H. (1997). Hand skill asymmetry in professional musicians. *Brain and Cognition, 34*, 424-432.

Jäncke, L., Staiger, J., Schlaug, G., Huang, Y., & Steinmetz, H. (1997). The relationship between corpus callosum size and forebrain volume. *Cerebral Cortex, 7,* 48-56.

Jäncke, L., & Steinmetz, H. (1993). Auditory lateralization and planum temporale asymmetry. *NeuroReport, 5*, 169-172.

Jeeves, M. A . (1979). Some limits to interhemispheric integration in cases of callosal agenesis and partial commissurotomy. In I. Steele Russell, M.W. Van Hof, & G. Berlucchi (Eds.), *Structure and function of the cerebral commissures* (pp.449-474). London: MacMillan.

Jeeves, M. A.(1990). Agenesis of the corpus callosum. *Handbook of neuropsychology,* Vol 4. 99-115.

Jeeves, M. A. (1994). Overview. In Lassonde M, Jeeves M A (Eds). Callosal agenesis: A natural split brain? *Advances in Behavioural Biology, 42,* 285-299.

Jeeves, M. A., & Moes, P. (1996). Interhemispheric transfer time differences related to aging and gender. *Neuropsychologia, 34,* 627-636.

Jeeves, M. A., & Silver, P.H. (1988). The formation of finger grip during prehension in an acallosal patient. *Neuropsychologia, 26,* 153-159.

Jeeves, M. A., Silver, P. H., & Jacobson, I .(1988). Bimanual co-ordination in callosal agenesis and partial commissurotomy. *Neuropsychologia, 26*, 833-850.

Jeeves, M. A., Silver, P. H., & Milne, A. B. (1988). Role of the corpus callosum in the development of a bimanual motor-skill. *Developmental Neuropsychology, 4,* 305-323.

Jenny, A. B. (1979). Commissural projections of the cortical hand motor area in monkeys. *Journal of Comparative Neurology, 188*, 137-146.

Johnson, R. C., Cole, R. E., Bowers, J. K., Foiles, S. V., Nokaido, A., Patrick, J., Woliver, P., & Woliver, R. E. (1979). Hemispheric efficiency in middle and later adulthood. *Cortex, 15*, 109-119.

Jones, E. G., & Powell, T. P. S. (1969). Connexions of the somatic sensory cortex of the rhesus monkey II- contralateral cortical connexions. *Brain, 92*, 717-730.

Jones, E.G., & Pons, T.P. (1998). Thalamic and brainstem contribution to large scale plasticity of primate somatosensory cortex. *Science, 282*, 1120-1125.

Jones, F. W. (1944). *The principles of anatomy as seen in the hand.* London: Baillère, Tindall & Cox.

Kelso, J.A.S., Southard, D.L., & Goodman, D. (1979). On the co-ordination of two-handed movements. *Journal of Experimental Psychology: Human Perception and Performance, 5,* 221-238.

Kelso, J. A. S., Putnam, C. A., & Goodman, D. (1983). On the space-time structure of human interlimb co-ordination. *Quarterly Journal of Experimental Psychology, 35A,* 347-375.

Kimura, D. (1961). Cerebral dominance and the perception of verbal stimuli. *Canadian Journal of Psychology, 15,* 156-165.

Kocel, K. (1980). Age-related changes in cognitive abilities and hemispheric specialization. In J. Herron (Ed.), *Neuropsychology of left handedness* (pp. 293-303). New York: Academic Press.

Kramer, M. A., Albrecht, S., & Miller, R. A. (1985). Handedness and the laterality of breast cancer in women. *Nursing Research, 34,* 333-337.

Kuhlemeier, K. V. (1991). Longevity and left handedness. *American Journal of Public Health, 81,* 513.

Lassonde, M. (1994). Disconnection syndrome in callosal agenesis. In Lassonde, M., & Jeeves, M. A. (Eds.). *Callosal agenesis: A natural split brain? Advances in Behavioural Biology,* Vol 42, pp 275-284.

Lassonde, M., Sauerwein, H., Geoffrey, G., & Décarie, M. (1986). Effects of early life and late transection of the corpus callosum in children: A study of tactile and tactuomotor transfer and integration. *Brain, 109,* 953-967.

Lembessis, E., & Fudin, R. (1994). Sinistrality and reduced longevity: Reply to Coren and Halpern's reply. *Perceptual and Motor Skills, 78,* 579-582.

London, W. P. (1986). Month of hospitalization of left handed substance abusers. *Neuropsychologia, 24,* 455-456.

London, W.P., Kibbee, P., & Holt, L. (1985). Handedness and alcoholism. *Journal of Nervous and Mental Disease, 173,* 570-572.

Lutsep, H., Wesinger, C., & Gazzaniga, M. (1995). Cerebral and callosal organisation in a right hemisphere dominant "split brain" patient. *Journal of Neurology, Neurosurgery and Psychiatry 59,* 50-54.

MacDonald, I., Beaton, A., & Folkard, S. (1999). The effect of extensive practice on tapping asymmetry. *(In preparation).*

McGee, M.G. & Cozad, T. (1980). Population genetic analysis of human hand preference: Evidence for generation differences, familial resemblance and maternal effects. *Behavior Genetics, 10,* 263-275.

McManus, I.C. (1985). Handedness, language dominance and aphasia: a genetic model. *Psychological Medicine:* Monograph Supplement 8.

McManus, I.C., & Bryden, M.P. (1992). The genetics of handedness, cerebral dominance and lateralization. In I. Rapin & S. Segalowitz (Eds.),

Handbook of neuropsychology: 10, Developmental neuropsychology (pp.115-144). Amsterdam: Elsevier.

McKeever, W. F., Sullivan, K., Ferguson, S., & Rayport, M .(1981). Typical cerebral hemisphere disconnection deficits following corpus callosum section despite sparing of the anterior commissure. *Neuropsychologia, 19*, 745-755.

McManus, I .C., Kemp, R. I., & Grant, J. (1986). Differences between fingers and hands in tapping ability: Dissociation between speed and regularity. *Cortex, 22*, 461-473.

Mayer, E., Koenig, O., & Panchaud, A. (1988). Tactual extinction without anomia: Evidence of attentional factors in a patient with a partial callosal disconnection. *Neuropsychologia, 26*, 851-868.

Meals, R. A. (1979). The laterality of fractures and dislocations with respect to handedness. *Clinical Orthopaedics and Related Research, 143*, 158-161.

Meerwaldt, J. D. (1983). Disturbances of spatial perception in a patient with agenesis of the corpus callosum. *Neuropsychologia, 21*, 161-165 .

Merckelbach, H., Muris, P., & Kop, W. J. (1994). Handedness, symptom reporting and accident susceptibility. *Journal of Clinical Psychology, 50*, 389-392.

Merzenich, M. (1998). Long-term change of mind. *Science, 282*, 1062-1063.

Messerli, P., Pegna, A., & Sordet, N. (1995). Hemispheric dominance for melody recognition in musicians and non-musicians. *Neuropsychologia, 33*, 395-405.

Meudell, P., & Greenhalgh, M . (1987). Age related differences in left and right hand skill and in visuo-spatial performance: Their possible relationships to the hypothesis that the right hemisphere ages more rapidly than the left. *Cortex, 23*, 431-445.

Meyer, B. U., Roricht, S., Grafin, E., Kruggel, F., & Weindl, A . (1995). Inhibitory and excitatory interhemispheric transfers between motor cortical areas in normal humans and patients with abnormalities of the corpus callosum. *Brain, 118*, 429-440.

Milner, A. D., Jeeves, M. A., Silver, P. H., Lines, C. R., & Wilson, J. (1985). Reaction times to lateralized visual stimuli in callosal agenesis: Stimulus and response factors. *Neuropsychologia, 23*, 323-331.

Milner, A. D., & Jeeves, M. A. (1979). A review of behavioural studies of agenesis of the corpus callosum. In I. Steele Russell, M.W. Van Hof, & G. Berlucchi, (Eds), *Structure and function of the cerebral commissures* (pp. 429-448). London: MacMillan.

Milner, B, Taylor, L., & Sperry, R,W. (1968). Lateralized suppression of dichotically presented digits after commissural section in man. *Science, 164*, 184-185.

Mitrushina, M., Fogel, T., D'Elia, L., Uchiyama, C., & Satz, P. (1995). Performance on motor tasks as an indication of increased behavioural asymmetry with advancing age. *Neuropsychologia, 33*, 359-364.

Moes, P., Jeeves, M. A., & Cook, K. (1995). Bimanual coordination with aging: Implications for interhemispheric transfer. *Developmental Neuropsychology, 11*, 23-40.

Monaghan, P., Metcalfe, N., & Ruxton, G. (1998). Does practice shape the brain? *Nature, 394*, 434.

Mondor, T. A., & Bryden, M. P. (1991). The influence of attention on the dichotic REA. *Neuropsychologia, 29,* 1179-1190.

Mondor, T. A., & Bryden, M. P. (1992). On the relation between auditory spatial attention and auditory perceptual asymmetries. *Perception and Psychophysics, 52,* 393-402.

Nebes, R. D. (1990). Hemispheric specialization in the aged brain. In C. Trevarthen (Ed), *Brain circuits and functions of the mind:* Essays in honour of Roger W. Sperry (pp. 364-370).

Nebes, R. D., Madden, D. J., & Berg ,W. D. (1983). The effect of age on hemispheric asymmetry in visual and auditory identification. *Experimental Ageing Research, 9,* 87-91.

Noble, W., & Davidson, I. (1996). *Human evolution, language and mind: A psychological and archaeological inquiry.* Cambridge: Cambridge University Press.

Nowicka, A., Grabowska, A., & Fersten, E. (1996). Interhemispheric transmission of information and functional asymmetry of the human brain. *Neuropsychologia, 34,* 147-151.

Oldfield, R.C. (1969). Handedness in musicians. *British Journal of Psychology, 60,* 91-99 .

Obler, L. K., Woodward, S., & Albert, M. L. (1984). Changes in cerebral lateralization in aging? *Neuropsychologia, 22*, 235-240.

Packer, G.J., & Shaheen, M.A. (1993). *The Journal of Hand Surgery (British and European Volume), 18B,* 511-514.

Pantev, C., Oostenveld, R., Engelien, A., Ross, B., Roberts, L., Hoke, M. (1998). Increased auditory cortical representation in musicians. *Nature, 292*, 811-815.

Pelletier, J., Habib, M., Lyon-Caen, O., Salamon, G., Poncet, M., & Khalil, R. (1993). Functional and magnetic resonance imaging correlates of callosal involvement in multiple sclerosis. *Archives of Neurology, 50*, 1077-1082.

Penhune, V.B., Zatorre, R.J. & Feindel, W.H. (1999). The role of auditory cortex in retention of rhythmic patterns as studied in patients with temporal lobe removals including Heschl's gyrus. *Neuropsychologia, 37,* 315-331.

Perelle, I. B., Ehrman, L., & Manowitz, J. W. (1981). Human handedness: The influence of learning. *Perceptual and Motor Skills, 53,* 967-977.

Peters, M. (1976). Prolonged practice of a simple motor task by preferred and nonpreferred hands. *Perceptual and Motor Skills, 43,* 447-450.

Peters, M. (1980). Why the preferred hand taps more quickly that the non-preferred hand : Three experiments on handedness. *Canadian Journal of Psychology, 34,* 62-71.

Peters, M .(1981). Attentional asymmetries during concurrent bimanual performance. *Quarterly Journal of Experimental Psychology, 33A,* 95-103.

Peters, M . (1983). RT to tactile stimuli presented ipsi-and contralaterally to the responding hand. *Quarterly Journal of Experimental Psychology, 35A,* 397-410.

Peters, M . (1985). Constraints in the performance of bimanual tasks and their expression in unskilled and skilled subjects. *Quarterly Journal of Experimental Psychology, 37A,* 171-196.

Peters, M. (1995). Handedness and its relation to other indices of cerebral lateralization. In R.J. Davidson, & K. Hugdahl (Eds.), *Brain asymmetry* (pp.183-214). Cambridge, MA: MIT Press.

Peters, M., & Perry, R. (1991). No link between left handedness and maternal age and no elevated accident rate in left handers. *Neuropsychologia, 29,* 1257-1259.

Plenger, P. M., Breier, J. I., Wheless, Ridley, T. D., Papanicolaou, A. C., Brookshire, B., Thomas, A., Curtis, V., & Willmore, L. J. (1996). Lateralization of memory for music: Evidence from the intracarotid sodium amobarbital procedure. *Neuropsychologia, 34,* 1015-1018.

Poffenberger, A. T. (1912). Reaction time to retinal stimulation with special reference to the time lost in conduction through nerve centres. *Archives of Psychology, 23,* 1-73.

Porac, C. (1993a) Are age trends in adult hand preference best explained by developmental shifts or generational differences? *Canadian Journal of Experimental Psychology, 47,* 697-713.

Porac, C. (1993b). Hand preference and the incidence of accidental unilateral hand injury. *Neuropsychologia, 31,* 355-362.

Porac, C. (1996). Hand and foot preference in young and older adults: A comment on Gabbard and Iteya. *Laterality, 1,* 207-213.

Porac, C., Coren, S., Duncan, P. (1980). Life-span age trends in laterality. *Journal of Gerontology, 35,* 715-721.

Porac, C., Coren, S., & Searleman, A. (1986). Environmental factors in hand preference formation: Evidence from attempts to switch the preferred hand. *Behavior Genetics, 16*, 250-261.

Porac, C., Rees, L., & Buller, T. (1990). Switching hands: A place for left hand use in a right hand world. In S. Coren (Ed.), *Left handedness: Behavioral implications and anomalies*(pp. 259-289). Amsterdam : Elsevier.

Preilowski, B. (1972). Possible contribution of the anterior forebrain commissures to bilateral motor co-ordination. *Neuropsychologia, 10*, 267-277.

Preilowski, B. (1975). Bilateral motor interaction: Perceptual-motor performance of partial and complete split-brain patients. In K. J. Zulch, D. Creutzfeld, & G. C. Galbraith (Eds.), *Cerebral localization* (pp.115-132). New York: Springer Verlag.

Provins, K. A. (.1956) . "Handedness" and skill. *Quarterly Journal of Experimental Psychology, 8*, 79-95.

Provins, K. A. (1958). The effect of training and handedness on the performance of two simple motor tasks. *Quarterly Journal of Experimental Psychology, 10*, 28-39.

Provins, K. A. (1967). Motor skills, handedness and behaviour. *Australian Journal of Psychology, 19*, 137-150.

Provins, K. A. (1997). Handedness and speech: A critical reappraisal of the role of genetic and environmental factors in the cerebral lateralization of function. *Psychological Review, 104*, 554-571 .

Rao, S. M., Bernardi, L., Leo, G. J., Ellington, L., Ryan, S. B., & Burg, L. S. (1989). Cerebral disconnection in multiple sclerosis. *Archives of Neurology, 46*, 918-920.

Rauch, R. A., Jinkins, J. R. (1994). Analysis of cross-sectional area measurements of the corpus callosum adjusted for brain size in male and female subjects from childhood to adulthood. *Behavioural Brain Research, 64*, 65-78.

Reinvang, I., Bakke, S., Hugdahl, K., Karlsen, N., & Sundet, K. (1994). Dichotic-listening performance in relation to callosal area on the MRI scan. *Neuropsychology, 8*, 445-450.

Reynolds, D. M., & Jeeves, M. A . (1977). Further studies of tactile perception and motor coordination in agenesis of the corpus callosum. *Cortex, 13*, 257-272.

Riege, W., Metter, E., & Williams, M. (1980). Age and hemispheric asymmetry in nonverbal tactual memory. *Neuropsychologia, 18*, 707-710

Risse, G., Le Doux, J. Springer,S.P., Wilson, D.H., & Gazzaniga, M.S. (1978). The anterior commissure in man: Functional variation in a multisensory system. *Neuropsychologia, 16*, 23-32.

Rogerson, P. A. (1993). On the relationship between handedness and longevity. *Social Biology, 40*, 283-287.

Rubens, A. B., Froehling, B., Slater, G., & Anderson, D. (1985). Left ear suppression on verbal dichotic tests in patients with multiple sclerosis. *Annals of Neurology, 18,* 459-43.

Rueckert, L., Sorensen, L., & Levy, J. (1994). Callosal efficiency is related to sustained attention. *Neuropsychologia, 32,* 159-173.

Salamy, A. (1978). Commissural transmission: Maturational changes in humans. *Science, 200*, 1409-1410.

Satomi, K., Kinoshita, Y., & Hirakawa, S. (1991). Disturbances of cross-localization of fingertips in a callosal patient. *Cortex, 27,* 327-331.

Schlaug, G., Jäncke, L., Huang, Y., Staiger, J., & Steinmetz, H. (1995). Increased corpus callosum size in musicians. *Neuropsychologia, 33*, 1047-1055.

Schnitzler, A., Kessler, K., & Benecke, R. (1996). Transcallosally mediated inhibition in interneurons within human primary motor cortex. *Experimental Brain Research, 112*, 381-391 .

Schutz, A., & Preißl, H. (1996). Basic connectivity of the cerebral cortex and some considerations on the corpus callosum. *Neuroscience and Biobehavioral Reviews, 20,* 567-570

Segalowitz, S. J., & Brown, D. (1991). Mild head injury as a source of developmental disabilities. *Journal of Learning Disabilities, 24*, 551-559.

Sexton, M. A., & Geffen, G. (1979). Development of three strategies of attention in dichotic monitoring. *Developmental Psychology, 15,* 299-310.

Sidtis, J. J. (1988). Dichotic listening after commissurotomy. In Hugdahl, K. (Ed). *Handbook of dichotic listening: Theory, methods and research.* Chichester: Wiley. 161-184.

Silver, P., & Jeeves, M. (1994). Motor coordination in callosal agenesis. In M. Lassonde & , M. A. Jeeves (Eds.), *Callosal agenesis: A natural split brain? Advances in Behavioural Biology,* Vol 42, pp.207-219.

Stelmach, G. E., Amrhein, P., & Goggin, N. (1988). Age differences in bimanual coordination. *Journal of Gerontology, 43*, 18-23.

Strauss, E. (1988). Dichotic listening and sodium amytal: Functional and morphological aspects of hemispheric asymmetry. In K. Hugdahl (Ed), *Handbook of dichotic listening, theory, methods and research* (pp.117-138). Chichester: Wiley.

Sugishita, M., Otomo, K., Yamazaki, K., Shimizu, H., Yoshioka, M., & Shinohara, A. (1995). Dichotic listening in patients with partial section of the corpus callosum. *Brain, 118*, 417-427.

Swanson, J., Ledlow, A., & Kinsbourne, M. (1978). Lateral asymmetries revealed by simple reaction time. In M. Kinsbourne (Ed.), *Asymmetrical function of the brain* (pp.274-291). Cambridge: Cambridge University Press.

Trankell, A. (1955). Aspects of genetics in psychology. *American Journal of Human Genetics, 7,* 264-276.

Toth, N. (1985). Archeological evidence for preferential right handedness in the lower and middle Pleistocene and its possible implications. *Journal of Human Evolution, 14,* 607-614 .

Trope, I., Fishman, B., Gur, R., Sussman, N., & Gur, R. (1987). Contralateral and ipsilateral control of fingers collowing callosotomy. *Neuropsychologia, 25,* 287-291.

Verfaellie, M., & Heilman, K. M .(1990) Hemispheric asymmetries in attentional control: Implications for hand preference in sensorimotor tasks. *Brain and Cognition, 14,* 70-80.

Tzourio, N., Nkanga-Ngila,B., & Mazoyer, B. (1998). Left planum temporale surface correlates with functional dominance during story listening. *NeuroReport, 9,* 829-833.

Volpe, B., Sidtis, J., Holtzman, J., Wilson, D., & Gazzaniga, M. (1982). Cortical mechanisms involved in praxis: Observations following partial and complete section of the corpus callosum in man. *Neurology, 32,* 645-50.

Weis, S., Kimbacher, M., Wenger, E., & Neuhold, A. (1993). Morphometric analysis of the corpus callosum using MR: Correlation of measurements with aging in healthy individuals. *Neuroradiology, 14,* 637-645.

Weller, M., & Latimer-Sayer, D. T. (1985). Increasing right hand dominance with age on a motor skill task. *Psychological Medicine, 15,* 867-872.

Wilkes, R. (1956). A social and occupational study of injured hands. *British Journal of Industrial Medicine, 13,* 119-130.

Witelson, S. F. (1989). Hand and sex differences in the isthmus and genu of the human corpus callosum: A postmortem morphological study. *Brain, 112,* 799-835.

Witelson, S. F. (1991). Sex differences in neuroanatomical changes with ageing. *The New England Journal of Medicine, 325,* 211-212.

Wood, E. K. (1988). Less sinister statistics from baseball records. *Nature, 335,* 212.

Woodruff, P. W. R., McManus, I. C., & David, A. S. (1995). Meta-analysis of corpus callosum size in schizophrenia. *Journal of Neurology, Neurosurgery and Psychiatry,* 58, 457-461.

Yakovlev, P. I., & Lecours, A. R. (1967). The myelogenetic cycles of regional maturation of the brain. In A. Monkowski (Ed), *Regional development of the brain in early life* (pp.3-70). Oxford: Blackwell.

Yeo, R.A., Gangstad, S.W., & Daniel, W.F. (1993). Hand preference and developmental instability. *Psychobiology, 21,* 161-168.

Zaidel, D., & Sperry, R. W. (1977). Some long term motor effects of cerebral commissurotomy in man. *Neuropsychologia, 15,* 193-204.

Zatorre, R. J. (1989). Perceptual asymmetry on the dichotic fused words test and cerebral speech lateralization determined by the carotid sodium amytal test. *Neuropsychologia, 27,* 1207-1219.

II

HANDEDNESS: MEASUREMENT AND OBSERVATIONS

Chapter 6

The Quantification and Definition of Handedness: Implications for Handedness Research

Steven C. Schachter
Harvard Medical School, USA

"No index of group difference, no matter how reliable, is a useful laterality measure unless it can be shown that the task is related to underlying brain asymmetry." (Schwartz & Kirsner, 1984).

A major problem in contemporary neuroscience is to explain the basis and origins of lateralized cognitive abilities such as language function. One approach has been to evaluate the anatomy of cerebral dominance with gross dissection, microscopic examination, and radiological imaging (Campain & Minckler, 1976; Chi, Dooling, & Gilles, 1977; Falzi, Perrone, & Vignolo, 1982; Geschwind & Levitsky, 1968; Rubens, Mahowald, & Hutton, 1976; Wada, Clarke, & Hamm, 1975; Weinberger, Luchins, Morihisa, & Wyatt, 1982; Witelson & Pallie, 1973). More recently, functional imaging studies and other non-invasive techniques have directly assessed lateralized cortical functions such as language and emotion (Canli, Desmond, Zhao, Glover, & Gabrieli, 1998; Cerf, Lebihan, Van de Moortele, Mac Leod, & Faurion, 1998; Frost, Binder, Springer, Hammeke, Bellgowan, Rao, & Cox, 1999; Pujol, Deus, Losilla, & Capdevila, 1999; Schneider, Grodd, Weiss, Klose, Mayer, Nagele, & Gur, 1997; Watanabe, Maki, Kawaguchi, Takashiro, Yamashita, Koizumi, & Mayanagi, 1998). However, defining cerebral asymmetries anatomically or functionally does not explain how they arose.

M.K. Mandal, M.B. Bulman-Fleming and G. Tiwari (eds.), Side Bias: A Neuropsychological Perspective, 155-174. © 2000 Kluwer Academic Publishers.

A complementary approach has been to correlate cerebral asymmetries with easily measured behaviours such as left handedness (Bear, Schiff, Saver, Greenberg, & Freeman, 1986; Chui & Damasio, 1980; Deuel & Moran, 1980; Hochberg & Le May, 1975; Kertesz, Black, Polk & Howell, 1986; Koff, Naeser, Pieniadz, Foundas, & Levine, 1986; LeMay, 1977; LeMay & Culebras, 1972; LeMay & Kido, 1978). These behaviours are then studied in order to identify associations with other biological attributes that, in turn, may shed light on the origins of cerebral dominance (Schachter, 1993a).

Although handedness studies have played a role in this endeavour, their full value has yet to be realized. First, different patterns of handedness (other than strong left or strong right handedness) have not been correlated with anatomic or functional cerebral asymmetries. Second, there has been a lack of consistency and standardization in the methods used to measure, quantify, and define handedness. This chapter will examine these methodological issues in greater detail.

1. METHODS OF HANDEDNESS MEASUREMENT AND QUANTIFICATION

There are two general methods used to assess handedness: 1) recording hand *preference* for everyday tasks, usually by interview or self-report on a handedness questionnaire, and 2) measuring hand *performance* for a specific manual task, such as moving pegs. The cross-correlations of preference and performance measures have been studied and reviewed (Bishop, 1989; Chapman & Chapman, 1987; Johnstone, Galin & Herron, 1979; Provins & Magliaro, 1993; Raczkowski, Kalat, & Nebes, 1974; Rigal, 1992). The vast majority of handedness studies in the adult literature use preference measures and therefore performance measures will not be further discussed.

1.1 Handedness questionnaires

Handedness preference may be ascertained with questionnaires that assess hand usage for commonly performed tasks. These questionnaires are completed by the subject or by the investigator by observing the subject perform the task. Many handedness questionnaires have been developed; the two most often used are the Edinburgh Handedness Inventory [EHI] (Oldfield, 1971) and the Annett questionnaire (Annett, 1970).

When Oldfield initially developed the EHI, he submitted a 20-item questionnaire that was based on a thesis of M. Humphrey of Oxford University (Humphrey, 1951) to 1,100 undergraduate psychology students in England and Scotland. After analyzing the responses, Oldfield selected 10 items, which became known as the EHI. The EHI contains the handedness items in a column on the left side of the page. There are two additional columns (Left and Right) that are left blank for the subject to fill in. The participant places a + in the Left or Right column for each item according to which hand is *usually* preferred. When one hand is *always* used for an item, or "where the preference is so strong that you [the participant] would never try to use the other hand unless absolutely forced to," (Oldfield, 1971) then the participant puts ++ in the appropriate column. If there is no preference, a + is put in both columns.

The scoring method recommended by Oldfield was based on an index of handedness that had previously been used in studies of hand performance: (Right minus Left)/(Right plus Left). Oldfield called the result of the scoring method the "laterality quotient" (LQ; range -100 to +100). In Oldfield's words , " . . . to calculate the L.Q., all that has to be done is to add all the +'s for each hand, subtract the sum for the left from that for the right, divide by the sum of both and multiply by 100" (Oldfield, 1971). A similar scoring method has also been used with Annett's questionnaire (Lindesay, 1987).

An alternate method for scoring the EHI was first described by White and Ashton (White & Ashton, 1976) and has been used since in several other studies (Bryden, 1977; Messinger, Messinger, & Graham, 1988; Schachter, Ransil, & Geschwind, 1987). In one method, each of the five possible responses to an item was given an associated column and score: "always left" (score: -10), "usually left" (-5), "no preference" (0), "usually right" (+5) and "always right" (+10) (Schachter, Ransil, & Geschwind 1987). Scores for the 10 items are summed and the total is called the "laterality score" (LS). The range varies from -100 to +100 by integral units of 5. The result is quickly and easily tabulated, and reliable on retest (Ransil & Schachter, 1994).

Both the LQ and LS scoring methods agree in the separation of participants above and below a score of 0. That is, they are equally sensitive to a participant's overall *direction* of handedness. However, they are not equally sensitive to *degree* of handedness. Schachter et al. emphasized this distinction with the following examples (Schachter et al., 1987).

Example 1

Laterality Quotient

	Left hand	**Right hand**
Writing		++
Drawing		++
Throwing		++
Scissors		++
Toothbrush		++
Knife		++
Spoon		++
Broom (upper hand)		++
Striking match		++
Opening box		++

Laterality Score*

	Always LH (-10)	**Usually LH (-5)**	**No preference (0)**	**Usually RH (+5)**	**Always RH (+10)**
Writing					vv
Drawing					vv
Throwing					vv
Scissors					vv
Toothbrush					vv
Knife					vv
Spoon					vv
Broom (upper hand)					vv
Striking match					vv
Opening box					vv

*LH = left hand, RH = right hand

In Example 1, the participant always uses the right hand for all ten items. To calculate the LQ, one counts 20 plusses; therefore, LQ is [(20 - 0)/20] X 100 = +100. Similarly, for the LS method, one observes that each of the answers is scored +10; therefore, the LS is also +100. In this example, the participant is behaviourally strongly right handed, and both the LQ and the LS are in agreement.

Consider Example 2.

Example 2

Laterality Quotient

	Left hand	**Right hand**
Writing		+
Drawing		+
Throwing		+
Scissors		+
Toothbrush		+
Knife		+
Spoon		+
Broom (upper hand)		+
Striking match		+
Opening box		+

Laterality Score*

	Always LH (-10)	**Usually LH (-5)**	**No preference (0)**	**Usually RH (+5)**	**Always RH (+10)**
Writing				v	
Drawing				v	
Throwing				v	
Scissors				v	
Toothbrush				v	
Knife				v	
Spoon				v	
Broom (upper hand)				v	

Striking match		v
Opening box		v

*LH = left hand, RH = right hand

In Example 2, the participant usually, but not always, uses the right hand for all items. This participant is not as strongly right handed as the one in Example 1 as indicated by the responses to the ten handedness tasks. However, the LQ of the participant in Example 2 is +100 ([(10 - 0)/10] X 100), which is the same as the LQ of the participant in Example 1, whereas the LS is +50 (10 X 5). Therefore, the LS more accurately reflects the difference in *degree* of right handedness between the participants in Examples 1 and 2 compared to LQ.

These two examples highlight a weakness of the LQ scoring method. As long as a participant usually or always uses their right hand for all 10 items, the LQ score is invariably +100, an observation first pointed out over 25 years ago (McMeekan & Lishman, 1975). Put differently, the LQ scoring method is not sensitive to the *degree* of handedness preference for participants who usually or always use the same hand for all 10 items. Essentially, participants whose LQ = +100 may have an LS between +50 to +100. Participants with an LS of +50 to +70 appear to differ biologically from participants whose LS = +100 (see below); therefore, the LQ method may be insensitive to such differences.

Finally, consider Example 3.

Example 3

	Left hand	**Right hand**
Writing		++
Drawing		++
Throwing		++
Scissors		++
Toothbrush		++
Knife		++
Spoon		++
Broom (upper hand)	+	
Striking match		++
Opening box		++

Laterality Score*

	Always LH (-10)	Usually LH (-5)	No preference (0)	Usually RH (+5)	Always RH (+10)
Writing					vv
Drawing					vv
Throwing					vv
Scissors					vv
Toothbrush					vv
Knife					vv
Spoon					vv
Broom (upper hand)		v			
Striking match					vv
Opening box					vv

*LH = left hand, RH = right hand

The participant in Example 3 always uses the right hand for nine items and usually the left hand for the tenth item. Yet this participant has an LQ of +89 ([(18 - 1)/19] X 100). This result is lower than the LQ for the participant in Example 2 (+100), who usually uses the right hand for every item, which is counter-intuitive. However, the results of the LS are more rational: the participant in Example 3 has an LS of +85 ([(9 X 10) + (-5)]), which is greater than the LS of the participant in Example 2 (+50); therefore, the LS in this example again more accurately reflects the *degree* of preference than the LQ.

Based on the advantages of the LS scoring method that became apparent during a project in the early 1980s (Schachter, 1993a; Schachter et al., 1987), Norman Geschwind recommended this scoring method to others, who subsequently incorporated it into their work (Messinger et al., 1988; Meyers & Janowitz, 1985; Tan, 1988). In fact, Tan called the LS the "Geschwind score" (Tan, 1988; Tan, 1993). A similar scoring modification for Annett's questionnaire has been described (Briggs & Nebes, 1975).

1.2 Self-described handedness

Many investigators ask their study participants to simply state whether they consider themselves to be right handed, left handed, or ambilateral. Because the frequency of self-described ambilaterality is lower than 5% in most studies (Schachter et al., 1987), ambilateral participants are either dropped from the analysis, or are combined with self-described left handers or right handers.

Self-described handedness (SDH) is usually less sensitive than quantified assessments for determining a participant's degree of right handedness, because self-described right handers have quantified handedness scores, such as LSs, over a large part of the handedness range. In one study, among participants with LS from 0 to +70, 85% were self-described right handers (Schachter et al., 1987). In addition, SDH does not correlate well with actual performance on manual dexterity tasks (Benton, Meyers, & Polder, 1962; Satz, Achenbach, & Fennell, 1967). Another measure of self-described handedness, the "hand used for writing," is often used in studies because of its unambiguous meaning (Halpern, Haviland, & Killian, 1998; Peters & Perry, 1991). However, this form of SDH may also less sensitive than quantified measures (Schachter et al., 1987), and may differ among participants as a function of age (De Agostini, Khamis, Ahui, & Dellatolas, 1997). For example, Plato et al. noted that 11.8% of adult white men under 40 taken from an aging study wrote with the left hand compared with 3.5% of men over 60 (Plato, Fox, & Garruto, 1984). Similarly, in another study, participants under 30 years of age used the left hand for writing more than twice as often as participants over 30 (Schachter et al., 1987).

2. THE ANALYSIS OF QUANTIFIED HANDEDNESS MEASURES: STATISTICAL CONSIDERATIONS

The distribution of handedness scores from questionnaires (e.g., the EHI) is strikingly asymmetric about the midpoint of the handedness range and usually bimodal, resulting in a J-shaped curve (Briggs & Nebes, 1975; Dellatolas, Annesi, Jallon, Chavance, & Lellouch, 1990; Johnstone et al., 1979; Oldfield, 1971; Schachter et al., 1987). Because of their asymmetry and bimodality, handedness distributions cannot usually be analyzed with parametric measures of central tendency such as means. Yet many handedness studies utilize means to characterize population handedness without indicating whether the population handedness was normally distributed (Dellatolas et al., 1990; Nass, Baker, Speiser, Virdis, Balsamo,

Cacciari, Loche, Dumic, & New, 1987; Pennington, Smith, Kimberling, Green, & Haith, 1987; Schur, 1986; Searleman & Fugagli, 1987; Weinstein & Pieper, 1988; White & Ashton, 1976).

Another methodological issue pertains to the size of the experimental and control groups in handedness studies. Many studies are not powered with enough participants to show significant differences between the experimental and control groups. For example, suppose an investigator hypothesizes that the prevalence of right and left handedness (however measured and defined) are 10% and 30%, respectively, in a control group and an experimental group (e.g., patients with rheumatoid arthritis). In that hypothetical study, a minimum of 71 participants in each group would be necessary to demonstrate statistical significance at the 0.05 level with a power of 0.8 (20% chance of missing the effect) (Fleiss, 1981). Further, if the frequency of left handedness is less than 3-fold higher in the experimental group than the control group, or the frequency of right handedness in the control participants is greater than 10%, then even greater numbers of participants are necessary to demonstrate a statistical difference between the two groups. However, relatively few studies analyze enough participants. For example, one study found the frequency of left handedness among 14 participants in the experimental group and 14 participants in the control group to be statistically the same and concluded that left handedness was not elevated in the experimental group (Bender, Puck, Salbenblatt, & Robinson, 1983). With such a low number of participants, left handedness would have to be 6 times more frequent in the experimental group (with a power of 0.8) to show statistical significance at the 0.05 level (assuming 10% frequency of left handedness in the control group), which is extremely rare.

Given these considerations, studies should include sufficient numbers of participants based on the anticipated difference in frequencies of left and right handedness. Further, handedness distributions should be tested for normality, and non-parametric statistics should be used when handedness distributions are not normal.

3. THE DEFINITION OF HANDEDNESS BASED ON QUANTIFIED MEASURES

As shown above, the LS method defines specific ranges of the handedness spectrum to a higher degree of specificity than the LQ, and SDH is less sensitive than the LS method. Therefore, it is clear that the definitions of "right handedness, " "mixed handedness," and "left handedness" assume different meanings depending on the methods used to measure and quantify handedness. This problem has plagued handedness research for several decades.

When handedness is measured with a questionnaire, definitions are based on arbitrary divisions of the handedness range into discrete segments. Geschwind and Behan used LQ = +100 and -100 as the criteria for right and left handedness, respectively (Geschwind & Behan, 1982; Geschwind & Behan, 1984). Other authors define left handedness as LQ less than 0, after Oldfield (Behan & Geschwind, 1985; Benbow, 1986; Cosi, Citterio, & Pasquino, 1988; Dellatolas et al., 1990; Geschwind & Behan, 1982; Koff et al., 1986; Messinger et al., 1988; Meyers & Janowitz, 1985; Oldfield, 1971; Schur, 1986; Smith, 1987; Weinstein & Pieper, 1988). This definition has the advantage of classifying the same participants as left handed irrespective of whether LS or LQ is used, because both scoring methods agree in the separation of participants above and below a score of 0 (Schachter, 1993b). In addition, this approach readily distinguishes laterality of hand preference and agrees well with SDH. As a result, this definition simplifies cross-study comparison (Dellatolas et al., 1990).

However, definitions of handedness such as those described above may exclude participants from analysis because only a limited portion of the entire handedness range is evaluated. Such definitions typically emphasize *direction*, but not *degree*, of hand preference. For instance, if an investigator defines right and left handedness as LQ = +100 and -100, respectively, than those participants with LQs between these extremes will not be included. Potentially useful information may be lost if a biological trait is associated with handedness scores that are purposefully excluded from the analysis (see below).

The term "non-right handedness" (NRH) has been used in the literature to designate participants who are not strongly right handed. Use of the term NRH emphasizes *degree* of hand preference rather than *direction* of handedness. Many authors define NRH as an LS or LQ less than or equal to +60 to +70 (Bear et al., 1986; Betancur, Velez, Cabanieu, Le Moal, & Neveu, 1990; Dellatolas et al., 1990; Schachter et al., 1987; Urion, 1988), after Schachter et al. (Schachter et al., 1987), who selected +70 as a cutoff point based on the cross-correlation of LS and SDH in 1,117 professionals. Among self-described right handers (N = 958), the distribution of LS was 86% over +70, 13% 0 to +70 and 1% less than 0. Further, among those with

LS over +70 (N = 826), 99% were self-described right handers. Therefore, participants with LS over +70 were defined as right handed, and participants with LS less than or equal to +70 were designated as NRH. The subgroup of NRH participants with LS 0 to +70 has often been termed weak right handers, or mixed-handers.

There are other methods of defining handedness based on the EHI that present additional problems. For instance, one group of investigators (Pennington et al., 1987) measured LQ on the EHI and defined handedness in a way that would classify a participant with LS of +80 as mixed handed while a participant with LS of +45 would be defined as right handed. One author used the EHI, but chose to ignore the scores and defined a left hander as anyone who expressed preference for any item with the left hand (McGee & Cozad, 1980); another used a different scoring method than the one recommended by Oldfield (Spivak, Segal, Mester, & Weizman, 1998). Another group focused on a subset of questionnaire items – writing, hammering, toothbrushing, throwing, striking a match, and using a racquet – and classified participants as mixed handed if they had no hand preference for any of these 4 items (Henderson, Stephens, & Gale, 1996).

4. IMPLICATIONS FOR HANDEDNESS RESEARCH

The variety of handedness questionnaires, scoring methods, and definitions for right, mixed, and left handedness used in the literature have impeded the contributions that handedness research could potentially make to the understanding of the biological basis of cerebral dominance. Perhaps the single most important portion of the handedness spectrum that has been obscured because of these methodological issues is the range between the two extremes, and in particular, the weak-right-handed range (e.g., LS 0 to +70). Just as self-described ambilateral participants are usually not separately analyzed because of low numbers, participants in the weak-right-handed range are often not separately analyzed in handedness studies. The two most common reasons are that the scoring method and/or handedness definitions do not identify participants in this range and there is a relatively low percentage of participants in this range (14% of all participants in one study (Schachter et al., 1987).

The following studies illustrate the importance of separately analyzing participants whose laterality scores are in the weak-right-handed range.

Collectively, these studies emphasize that *degree* of hand preference may be more important than *direction* of hand preference with respect to certain biological correlations of handedness.

In a study of handedness and hair colour using the EHI and LS, there was no significant correlation of blond hair with LS when a dichotomous analysis was used (LS < 0 vs LS > 0) (see Table 1) (Schachter et al., 1987). However, the percentage of blond participants among those with LS 0 to +70 was almost two and one-half times greater than in non-blonds (28% vs 12%). As a result, blond hair colour correlated with LS when a trichotomous analysis was used (LS < 0, vs LS 0 to + 70, vs LS > +70) because the weak-right-handed range strongly contributed to the statistical association.

Table 2. Association between hair colour and LS (Schachter et al., 1987)

	LS <0	LS 0 to +70	LS >+70	LS <0	LS >0
Blond hair	16%	28%	56%	16%	84%
Nonblond hair	12%	12%	76%	12%	88%

Further, as shown in Table 2, the percentage of participants with a history of a learning disability (LD) who had an LS between 0 and +70 was almost 2.5 times greater than the corresponding percentage of participants without a history of LD (32% vs. 13%). Again, there was a significant correlation of LD with LS when a trichotomous analysis was used (LS < 0, vs LS 0 to + 70, vs LS > +70), but not with a dichotomous analysis, (LS < 0 vs LS > 0), because the weak-right-handed range strongly contributed to the statistical association.

Table 3. Association between learning disability (LD) and LS (Schachter et al., 1987)

	LS <0	LS 0 to +70	LS >+70	LS <0	LS >0
History of LD	22%	32%	46%	22%	78%
No history of LD	12%	13%	75%	12%	88%

In another study, the LS distributions among women exposed to diethylstilbestrol (DES) *in utero* (DES daughters) were compared with those of controls (Schachter, 1994). When the LS ranges 0 to +70 and +70 to +100 were combined and the numbers of participants above and below LS = 0 were compared, there was no significant difference between the groups (Table 3). However, as in the previous two examples, there was a significant correlation of DES exposure and LS when a trichotomous analysis was used (LS < 0, vs LS 0 to + 70, vs LS > +70), because the weak-right-handed range strongly contributed to the statistical association.

Table 4. Association between intrauterine DES exposure and LS (Schachter, 1994)

	LS <0	**LS 0 to +70**	**LS >+70**	**LS <0**	**LS >0**
DES daughters	18%	56%	26%	18%	82%
Controls	14%	32%	54%	14%	86%

A growing number of other studies support the importance of separating out weakly right-handed participants. Lahita evaluated the association of handedness and systemic lupus erythematosus (SLE) using the LQ scoring method (Lahita, 1988). When he divided the handedness range into three segments: -100 to 0, 0 to +50, and >+50, he found that the percentage of patients with SLE who had an LQ between 0 and +50 was double the frequency in normal controls. Bakan et al. showed that ambilateral participants were twice as likely as right handers to report pregnancy and birth complications (Bakan, Dibb, & Reed, 1973). Lindesay compared handedness scores of 94 homosexual and 100 heterosexual men. The distribution of handedness among the homosexual men was significantly shifted from strong right handedness to weak right handedness (Lindesay, 1987). These results have been replicated in one study (Holtzen, 1994), but not another (Satz, Miller, Selnes, Van Gorp, D'Elia, & Visscher, 1991). Dellatolas et al. found stuttering was more than doubled in those with handedness scores equivalent to LS of +70 or less compared with LS over +70 (Dellatolas et al., 1990). Schulter and Papousek showed that weak-right-handed participants had a different pattern of brain lateralization on bilateral electrodermal activity (EDA) than strong right handers, and emphasized the importance of precisely controlling degree of handedness in studies of bilateral EDA (Schulter & Papousek, 1998). Hicks et al. noted that university students who inconsistently used their hands for items on a handedness scale were over 6 times more likely to experience injurious accidents than were students whose hand use was consistent (i.e., very strong left or right handers) (Hicks, Inman, Ching, Bautista, Deharo, & Hicks,

1998). Henderson et al. found that dental undergraduates and orthodontists have an increased prevalence of mixed handedness compared to the general population (Henderson et al., 1996), consistent with another study (Ransil & Schachter, 1998; Schachter & Ransil, 1996). Finally, ambidexterity has been associated with dissociation (Kunzendorf & Marsden, 1991), combat-related post-traumatic stress disorder (Spivak et al., 1998), and male sex offenders (Joseph, Schwartz, & Schachter, 1997).

Handedness studies of patients with schizophrenia have further demonstrated the value of careful analysis of the entire spectrum of the handedness range. In a review of the literature, Satz and Green concluded that there is an atypical leftward shift in the handedness distribution of patients with schizophrenia, resulting in an elevated frequency of mixed handedness (Satz & Green, 1999). To the authors, these results suggested an abnormality among schizophrenic patients affecting the left hemisphere, which is consistent with anatomical work (Kwon, McCarley, Hirayasu, Anderson, Fischer, Kikinis, Jolesz, & Shenton, 1999; Shapleske, Rossell, Woodruff, & David, 1999).

These studies underscore the importance of evaluating each segment of the handedness range, especially the weak-right-handed range. Several of these analyses lost their statistical power when participants from the weak-right-handed range were combined with strong right handers; that is, when a dichotomous definition of handedness was used with 0 as the midpoint. One interpretation of these observations is that weakly right-handed participants are biologically distinguishable from strongly right-handed participants, though further research is needed to pursue this speculation, particularly studies of brain anatomy that compare weakly right-handed participants to strongly right-handed participants.

5. SUMMARY

Handedness studies clearly have a significant role to play in the study of cerebral dominance, and careful attention to study design will further increase their value to researchers. Specifically, handedness studies should include an appropriate number of experimental and control participants to demonstrate statistical significance and must control for age, as well as other factors not discussed including gender, educational background, and familial handedness (McGee & Cozad, 1980). A sensitive, validated measuring instrument (e.g., the EHI, rather than SDH) should be administered and scored quantitatively with a method such as the LS, which produces a score that is sensitive to both the degree and direction of handedness. The distribution of the entire range of scores should be tested for normality, and segments of the handedness range should be selected and assessed for differences between experimental and control groups using the appropriate statistics.

6. REFERENCES

Annett, M. (1970). A classification of hand preference by association analysis. *British Journal of Psychology, 61*, 303-321.

Bakan, P., Dibb, G., & Reed, P. (1973). Handedness and birth stress. *Neuropsychologia, 11*, 363-366.

Bear, D., Schiff, D., Saver, J., Greenberg, M., & Freeman, R. (1986). Quantitative analysis of cerebral asymmetries. Fronto-occipital correlation, sexual dimorphism and association with handedness. *Archives of Neurology, 43*, 598-603.

Behan, P., & Geschwind, N. (1985). Dyslexia, congenital anomalies, and immune disorders: the role of the fetal environment. *Annals of the New York Academy of Sciences, 457*, 13-18.

Benbow, C. P. (1986). Physiological correlates of extreme intellectual precocity. *Neuropsychologia, 24*, 719-725.

Bender, B. G., Puck, M. H., Salbenblatt, J. A., & Robinson, A. (1983). Hemispheric organization in 47, XXY boys. *Lancet, 1*, 132.

Benton, A. L., Meyers, R., & Polder, G. J. (1962). Some aspects of handedness. *Psychiatria et Neurologia, 144*, 321-337.

Betancur, C., Velez, A., Cabanieu, G., Le Moal, M., & Neveu, P. J. (1990). Association between left handedness and allergy: a reappraisal. *Neuropsychologia, 28*, 223-227.

Bishop, D. V. M. (1989). Does hand proficiency determine hand preference? *British Journal of Psychology, 80*, 191-199.

Briggs, G. G., & Nebes, R. D. (1975). Patterns of hand preference in a student population. *Cortex, 11*, 230-238.

Bryden, M. P. (1977). Measuring handedness with questionnaires. *Neuropsychologia, 15*, 617-624.

Campain, R., & Minckler, J. (1976). A note on the gross configurations of the human auditory cortex. *Brain and Language, 3*, 318-323.

Canli, T., Desmond, J. E., Zhao, Z., Glover, G., & Gabrieli, J. D. (1998). Hemispheric asymmetry for emotional stimuli detected with fMRI. *Neuroreport, 9*, 3233-3239.

Cerf, B., Lebihan, D., Van de Moortele, P. F., Mac Leod, P., & Faurion, A. (1998). Functional lateralization of human gustatory cortex related to handedness disclosed by fMRI study. *Annals of the New York Academy of Sciences, 855*, 575-578.

Chapman, L. J., & Chapman, J. P. (1987). The measurement of handedness. *Brain and Cognition, 6*, 175-183.

Chi, J. G., Dooling, E. C., & Gilles, F. H. (1977). Gyral development of the human brain. *Annals of Neurology, 1*, 86-93.

Chui, H. C., & Damasio, A. R. (1980). Human cerebral asymmetries evaluated by computed tomography. *Journal of Neurology, Neurosurgery, and Psychiatry, 43*, 873-878.

Cosi, V., Citterio, A., & Pasquino, C. (1988). A study of hand preference in myasthenia gravis. *Cortex, 24*, 573-577.

De Agostini, M., Khamis, A. H., Ahui, A. M., & Dellatolas, G. (1997). Environmental influences in hand preference: An African point of view. *Brain and Cognition, 35*, 151-167.

Dellatolas, G., Annesi, I., Jallon, P., Chavance, M., & Lellouch, J. (1990). An epidemiological reconsideration of the Geschwind-Galaburda theory of cerebral lateralization. *Archives of Neurology, 47*, 778-782.

Deuel, R. K., & Moran, C. C. (1980). Cerebral dominance and cerebral asymmetries on computed tomogram in childhood. *Neurology, 30*, 934-938.

Falzi, G., Perrone, P., & Vignolo, L. A. (1982). Right-left asymmetry in anterior speech region. *Archives of Neurology, 39*, 239-240.

Fleiss, J. S. (1981). *Statistical methods for rates and proportions*. New York : Wiley.

Frost, J. A., Binder, J. R., Springer, J. A., Hammeke, T. A., Bellgowan, P. S., Rao, S. M., & Cox, R. W. (1999). Language processing is strongly left lateralized in both sexes. Evidence from functional MRI. *Brain, 122*, 199-208.

Geschwind, N., & Behan, P. (1982). Left handedness: Association with immune disease, migraine, and developmental learning disorder. *Proceedings of the National Academy of Sciences, U S A, 79*, 5097-6100.

Geschwind, N., & Behan, P. O. (1984). Laterality, hormones, and immunity. In N. Geschwind & A. M. Galaburda (Eds.), *Cerebral dominance: The biological foundations* (pp. 211-224). Cambridge, MA : Harvard University Press.

Geschwind, N., & Levitsky, W. (1968). Human brain: Left-right asymmetries in temporal speech region. *Science, 161*, 186-187.

Halpern, D. F., Haviland, M. G., & Killian, C. D. (1998). Handedness and sex differences in intelligence: Evidence from the medical college admission test. *Brain and Cognition, 38,* 87-101.

Henderson, N. J., Stephens, C. D., & Gale, D. (1996). Left handedness in dental undergraduates and orthodontic specialists. *British Dental Journal, 181*, 285-288.

Hicks, R. A., Inman, G., Ching, P., Bautista, J., Deharo, D., & Hicks, G. J. (1998). Consistency of hand use and accidents with injury. *Perceptual and Motor Skills, 87,* 851-854.

Hochberg, F. H., & Le May, M. (1975). Arteriographic correlates of handedness. *Neurology, 25*, 218-222.

Holtzen, D. W. (1994). Handedness and sexual orientation. *Journal of Clinical and Experimental Neuropsychology, 16,* 702-712.

Humphrey, M. D. (1951). Consistency of hand usage: A preliminary enquiry. *British Journal of Educational Psychology, 21*, 214-225.

Johnstone, J., Galin, D., & Herron, J. (1979). Choice of handedness measures in studies of hemispheric specialization. *International Journal of Neuroscience, 9,* 71-80.

Joseph, A. B., Schwartz, B. K., & Schachter, S. C. (1997). Increased nonrighthandedness among incarcerated male sex offenders. *Neurology, 48 (Suppl A),* 50.

Kertesz, A., Black, S. E., Polk, M., & Howell, J. (1986). Cerebral asymmetries on magnetic resonance imaging. *Cortex, 22*, 117-127.

Koff, E., Naeser, M. A., Pieniadz, J. M., Foundas, A. L., & Levine, H. L. (1986). Computed tomographic scan hemispheric asymmetries in right- and left handed male and female subjects. *Archives of Neurology, 43*, 487-491.

Kunzendorf, R. G., & Marsden, D. (1991). Dissociation in ambidextrous students. *Perceptual and Motor Skills, 73*, 778.

Kwon, J. S., McCarley, R. W., Hirayasu, Y., Anderson, J. E., Fischer, I. A., Kikinis, R., Jolesz, F. A., & Shenton, M. E. (1999). Left planum temporale volume reduction in schizophrenia. *Archives of General Psychiatry, 56,* 142-148.

Lahita, R. G. (1988). Systemic lupus erythematosus: Learning disability in the male offspring of female patients and relationship to laterality. *Psychoneuroendocrinology, 13*, 385-396.

LeMay, M. (1977). Asymmetries of the skull and handedness. Phrenology revisited. *Journal of Neurologica Sciences, 32*, 243-253.

LeMay, M., & Culebras, A. (1972). Human brain-morphologic differences in the hemispheres demonstrable by carotid arteriography. *New England Journal of Medicine, 287*, 168-170.

LeMay, M., & Kido, D. K. (1978). Asymmetries of the cerebral hemispheres on computed tomograms. *Journal o Computer Assisted Tomography, 2,* 471-476.

Lindesay, J. (1987). Laterality shift in homosexual men. *Neuropsychologia, 25,* 965-969.

McGee, M. G., & Cozad, T. (1980). Population genetic analysis of human hand preference: Evidence for generation differences, familial resemblance, and maternal effects. *Behavior Genetics, 10,* 263-275.

McMeekan, E. R. L., & Lishman, W. A. (1975). Retest reliabilities and interrelationship of the Annett hand preference questionnaire and the Edinburgh handedness inventory. *British Journal of Psychology, 66,* 53-59.

Messinger, H. B., Messinger, M. I., & Graham, J. R. (1988). Migraine and left handedness: Is there a connection? *Cephalalgia, 8,* 237-244.

Meyers, S., & Janowitz, H. D. (1985). Left handedness and inflammatory bowel disease. *Journal of Clinical Gastroenterology, 7,* 33-35.

Nass, R., Baker, S., Speiser, P., Virdis, R., Balsamo, A., Cacciari, E., Loche, A., Dumic, M., & New, M. (1987). Hormones and handedness: Left hand bias in female congenital adrenal hyperplasia patients. *Neurology, 37,* 711-715.

Oldfield, R. C. (1971). The assessment and analysis of handedness: The Edinburgh inventory. *Neuropsychologia, 9,* 97-113.

Pennington, B. F., Smith, S. D., Kimberling, W. J., Green, P. A., & Haith, M. M. (1987). Left handedness and immune disorders in familial dyslexics. *Archives of Neurology, 44,* 634-639.

Peters, M., & Perry, R. (1991). No link between left handedness and maternal age and no elevated accident rate in left handers. *Neuropsychologia, 29,* 1257-1259.

Plato, C. C., Fox, K. M., & Garruto, R. M. (1984). Measures of lateral functional dominance: Hand dominance. *Human Biology, 56,* 259-275.

Provins, K. A., & Magliaro, J. (1993). The measurement of handedness by preference and performance tests. *Brain and Cognition, 22,* 171-181.

Pujol, J., Deus, J., Losilla, J. M., & Capdevila, A. (1999). Cerebral lateralization of language in normal left handed people studied by functional MRI. *Neurology, 52,* 1038-1043.

Raczkowski, D., Kalat, J. W., & Nebes, R. (1974). Reliability and validity of some handedness questionnaire items. *Neuropsychologia, 12,* 43-47.

Ransil, B. J., & Schachter, S. C. (1994). Test-retest reliability of the Edinburgh Handedness Inventory and global handedness preference measurements, and their correlation. *Perceptual and Motor Skills, 79,* 1355-1372.

Ransil, B. J., & Schachter, S. C. (1998). Inventory-derived task handedness preferences of nine professions and their associations with self-report global handedness preferences. *Perceptual and Motor Skills, 86*, 303-320.

Rigal, R. A. (1992). Which handedness: Preference or performance? *Perceptual and motor skills, 75*, 851-866.

Rubens, A. B., Mahowald, M. W., & Hutton, J. T. (1976). Asymmetry of the lateral (sylvian) fissures in man. *Neurology, 26,* 620-624.

Satz, P., Achenbach, K., & Fennell, E. (1967). Correlations between assessed manual laterality and predicted speech laterality in a normal population. *Neuropsychologia, 5*, 295-310.

Satz, P., & Green, M. F. (1999). Atypical handedness in schizophrenia: Some methodological and theoretical issues. *Schizophrenia Bulletin, 25*, 63-78.

Satz, P., Miller, E. N., Selnes, O., Van Gorp, W., D'Elia, L. F., & Visscher, B. (1991). Hand preference in homosexual men. *Cortex, 27*, 295-306.

Schachter, S. C. (1993a). Clinical studies of anomalous structural and functional cerebral dominance: An overview of problems and progress. In A. M. Galaburda (Ed.), *Dyslexia and development: Neurobiological aspects of extra-ordinary brains* (pp.358-365). Cambridge, MA: Harvard University Press.

Schachter, S. C. (1993b). Studies of handedness and anomalous dominance: Problems and progress. In A. M. Galaburda (Ed.), *Dyslexia and development: Neurobiological aspects of extra-ordinary brains* (pp. 269-296). Cambridge, MA: Harvard University Press.

Schachter, S. C. (1994). Handedness in women with intrauterine exposure to diethylstilbestrol. *Neuropsychologia, 32*, 619-23.

Schachter, S. C., & Ransil, B. J. (1996). Handedness distributions in nine professional groups. *Perceptual and Motor Skills, 82*, 51-63.

Schachter, S. C., Ransil, B. J., & Geschwind, N. (1987). Associations of handedness with hair colour and learning disabilities. *Neuropsychologia, 25*, 269-76.

Schneider, F., Grodd, W., Weiss, U., Klose, U., Mayer, K. R., Nagele, T., & Gur, R. C. (1997). Functional MRI reveals left amygdala activation during emotion. *Psychiatry Research, 76*, 75-82.

Schulter, G., & Papousek, I. (1998). Bilateral electrodermal activity: Relationships to state and trait characteristics of hemisphere asymmetry. *International Journal of Psychophysiology, 31,* 1-12.

Schur, P. H. (1986). Handedness in systemic lupus erythematosus. *Arthritis and Rheumatism, 29*, 419-20.

Schwartz, S., & Kirsner, K. (1984). Can group differences in hemispheric asymmetry be inferred from behavioural laterality indices? *Brain and Cognition, 3,* 57-70.

Searleman, A., & Fugagli, A. K. (1987). Suspected autoimmune disorders and left handedness: Evidence from individuals with diabetes. Crohn's disease and ulcerative colitis. *Neuropsychologia, 25,* 367-374.

Shapleske, J., Rossell, S. L., Woodruff, P. W., & David, A. S. (1999). The planum temporale: A systematic, quantitative review of its structural, functional and clinical significance. *Brain Research Review, 29,* 26-49.

Smith, J. (1987). Left handedness: Its association with allergic disease. *Neuropsychologia, 25,* 665-674.

Spivak, B., Segal, M., Mester, R., & Weizman, A. (1998). Lateral preference in post-traumatic stress disorder. *Psychological Medicine, 28,* 229-232.

Tan, Ü. (1988). The relationship between nonverbal intelligence, familial sinistrality and Geschwind scores in right handed female subjects. *International Journal of Neuroscience, 43,* 177-182.

Tan, Ü. (1993). Normal distribution of hand preference and its bimodality. *International Journal of Neuroscience, 68,* 61-65.

Teszner, D., Tzavaras, A., Gruner, J., & Hécaen, H. (1972). L'asymétrie droite-gauche du planum temporale; A propos de l'étude anatomique de 100 cerveaux. *Revista de Neurologia, 146,* 444-449.

Urion, D. K. (1988). Nondextrality and autoimmune disorders among relatives of language-disabled boys. *Annals of Neurology, 24,* 267-269.

Wada, J. A., Clarke, R., & Hamm, A. (1975). Cerebral hemispheric asymmetry in humans. Cortical speech zones in 100 adults and 100 infant brains. *Archives of Neurology, 32,* 239-246.

Watanabe, E., Maki, A., Kawaguchi, F., Takashiro, K., Yamashita, Y., Koizumi, H., & Mayanagi, Y. (1998). Non-invasive assessment of language dominance with near-infrared spectroscopic mapping. *Neuroscience Letters, 256,* 49-52.

Weinberger, D. R., Luchins, D. J., Morihisa, J., & Wyatt, R. J. (1982). Asymmetrical volumes of the right and left frontal and occipital regions of the human brain. *Annals of Neurology, 11,* 97-100.

Weinstein, R. E., & Pieper, D. R. (1988). Altered cerebral dominance in an atopic population. *Brain, Behavior and Immunity, 2,* 235-241.

White, K., & Ashton, R. (1976). Handedness assessment inventory. *Neuropsychologia, 14,* 261-264.

Witelson, S. F., & Pallie, S. (1973). Left hemisphere specialization for language in the newborn. *Brain, 96,* 641-646.

Chapter 7

Factor Structures of Hand Preference Questionnaires: Are "Skilled" and "Unskilled" Factors Artifacts?

Yukihide Ida[1], Manas K. Mandal[2] and M.P. Bryden[3†]
[1]Osaka Gakuin Junior College, Japan :[2] Indian Institute of Technology, India : [3]University of Waterloo, Canada

The selection of activities to be included in a handedness questionnaire is of critical importance in the classification of handedness (see also Chapter 6, this volume). This is especially true when comparing data from different cultures, because the same activity may or may not be a good item for a hand preference questionnaire in different cultures. For example, "writing" is often considered one of the fundamental items to measure handedness and has been included in most questionnaires. However, the use of the left hand for writing is suppressed in Japan and thus, inclusion of writing in a questionnaire leads to a lower measured prevalence of left handers among Japanese. In two recent studies with Japanese participants, the prevalence of left handers was estimated to be 5% (Hatta & Kawakami, 1995) and 4.7% (Ida & Bryden, 1996), but the prevalence of people who write with the left hand was 1.2% and 1.4%, respectively. In societies under strong cultural pressure against the use of the left hand, some activities are culturally biased and thus, should be excluded from a questionnaire.

Factor analysis seems to be a useful tool to select activities to measure hand preference. For example, Bryden (1977) recommended five activities as the items of a handedness questionnaire, based on the results of factor analysis for the Crovitz-Zener Test (Crovitz & Zener, 1962) and Oldfield

M.K. Mandal, M.B. Bulman-Fleming and G. Tiwari (eds.). Side Bias: A Neuropsychological Perspective, 175-190.

Test (Oldfield, 1971). The selected activities (writing, throwing, drawing, using scissors, and using a toothbrush) were those that showed a much heavier loading on the first factor than on the second or the third factor. Hatta and Nakatsuka (1975) excluded "writing" from their factor analysis of handedness questionnaire data from a Japanese sample. "Writing" did not show a high loading on their two factors – an expected finding because when a given variable shows no variance, it will not contribute to the factor structure.

In general, the majority of early studies that used factor analysis interpreted the first factor to denote what is commonly described as handedness, and viewed handedness as a unitary variable. In contrast, recent studies have proposed a different view. Multifactorial analysis with a hand preference questionnaire indicates that handedness is not unidimensional. Beukelaar and Kroonenberg (1983) investigated hand preferences for a variety of activities with a questionnaire of 51 items and classified the activities based on a cluster analysis. They suggested that the obtained clusters of activities could be characterized by the muscle groups and joints involved in performing the tasks and that hand preference should be multidimensional. Healey, Liederman, and Geschwind (1986) obtained four factors with a factor analysis of 55 hand-preference items and suggested that manual preference could be governed by more than one neural system and that these systems might be lateralized independently. Based on factor analyses of a 60-item and a 33-item hand-preference questionnaire (different versions of Waterloo Hand-Preference Questionnaire), Steenhuis and Bryden (1989) concluded that hand preference is multidimensional with two major factors and a variable number of minor factors. They identified two groups of activities through these analyses: "skilled" unimanual ones such as "writing" and "throwing", and "unskilled" unimanual ones such as "picking up objects". Although the number and nature of assumed dimensions are different among different studies, these two factors, "skilled" and "unskilled", were repeatedly identified as primary factors in recent studies (e.g., Ida & Bryden, 1996; Singh & Bryden, 1994; Steenhuis, Bryden, Schwartz, & Lawson, 1990). Regardless of cultural differences, especially in the prevalence of left handedness, factor analysis revealed the existence of these two factors not only in North American samples (Steenhuis & Bryden, 1989; Steenhuis et al., 1990) but also in Asian samples (Ida & Bryden, 1996; Singh & Bryden, 1994). These facts strongly suggest the generality of these two factors.

Regardless of whether hand preference is uni- or multi-dimensional, factor analysis seems to be one of the most useful tools for the selection and classification of activities to construct a valid and reliable questionnaire for hand preference judging from the above mentioned studies. However, use of factor analysis itself for handedness data has an important statistical problem.

McManus (1996) suggested that "skilled" and "unskilled" factors may be artifacts resulting from a violation of the assumption, in factor analysis, that measures should be normally distributed. Item scores of hand preference tend to be bimodal with different proportions of right and left responses, and "skilled" items such as "writing" and "throwing" are usually highly lateralized and "unskilled" items such as "picking up objects" are less lateralized. He pointed out that "skilled" and "unskilled" factors might be spurious, so-called "difficulty factors". "Difficulty factors" are obtained when tests with largely different difficulties are subjected to conventional factor analysis, and are frequently seen when measures are binomial. Thus, factor structures and factors obtained from factor analysis may be artifacts due to different distributions of responses among items of a hand preference questionnaire.

One possible source of confounding in factor analyses is the fact that skilled activities tend to be highly lateralized and unskilled activities tend to be less lateralized and it is therefore unclear whether the obtained factor structures are artifacts due to distribution differences among items or reflect the true components of hand preference. In other words, the inclusion of highly lateralized but unskilled activities and less lateralized but skilled activities into hand preference questionnaire is needed to examine the two alternative assumptions. For example, if highly lateralized but unskilled activities show high loadings on the same factor as the other less lateralized and unskilled activities, it should support the existence of the two factors. But if highly lateralized but unskilled activities show high loadings on the same factor as the other highly lateralized and skilled activities, the obtained factor structure would be thought to be artifactual, and a result of distribution differences among items.

In a recent study on cultural differences in hand preference between India and Japan (Mandal, Ida, Harizuka, & Upadhaya, 1999), Indian participants showed more right handedness for unskilled than for skilled activities and the trend was reversed in Japanese participants . These data could be used to examine whether the "skilled" and "unskilled" factors really exist.

In the present chapter, reanalysis of data from two studies that were originally conducted to examine cultural differences in handedness using factor analysis will be carried out to clarify how factor structures of hand preference questionnaires depend on the degree of lateralization of their items and examine the role of "skilled" and "unskilled" factors in determining factor structure. The first set of data from Canadian and Japanese samples showed a typical factor structure that contained the "skilled" and "unskilled" factors (Ida & Bryden, 1996). The second set of data is from the study by Mandal et al. (1999). As mentioned above, that study showed that some unskilled activities were highly lateralized in India. It should be noted that the present analysis concerns not only the interpretation of factor structure but also the appropriateness of factor analysis as a tool for finding components and choosing items of hand-preference questionnaires.

1. ANALYSIS

Data analyzed here were from two studies that were originally conducted for cultural comparisons of hand preference. In both studies, hand preference was estimated with a questionnaire and factor analysis showed the existence of multiple factors. Because different questionnaires were utilized in these studies, the results of analysis will be presented separately.

The basic idea for the present analysis was that the effect of lateralization of activities measured with hand-preference questionnaires on factor structure could be better understood when the means of item scores were contrasted with factor loadings, especially by a 2-factor solution. Thus, tables for mean scores and factor loadings by 2-factor solution were prepared, and the items of questionnaires in the tables were arranged according to their mean score. Further, the items were classified into two groups, more lateralized and less lateralized items, and differences in the number of items that have higher loadings on the first factor than the second factor were tested between the groups statistically.

1.1 Data from Japanese and Canadian samples

The first study (Ida and Bryden, 1996) included data on 655 participants from Japan (333 men and 322 women) and 620 subjects form

Canada (245 men and 375 women). The participants were undergraduate students and were instructed to indicate their preference for each item on a 5-point scale (1 = left always, 2 =l left usually, 3 = equal, 4 = right usually, 5 = right always). Data on Japanese and Canadian participants were subjected separately to principal-components factor analysis with varimax rotation for a 2-factor solution (SPSS 7.5 for Windows). Although a hand-preference questionnaire used in this study included 66 items, 47 items were submitted to the factor analysis. The remaining 19 items were excluded from the analysis because of high omission rate (see Ida & Bryden, 1996).

1.1.1 Results for a Japanese sample

Two factors obtained by this analysis accounted for 50.5% of the total variance: the first factor, 41.7%; the second factor, 8.8%. As shown in Table 1, "skilled" items such as "peel", "knife", "hammer", and "dart" had high loadings on the first factor. In contrast, items with high loadings on the second factor such as "push buzzer", "insert coin", "turn water tap", "pick up small object", and "pull switch" were thought to be "unskilled" ones.

In Table 1, items are arranged in order of decreasing means. It is clear from this table that items with higher means tend to have higher loadings on the first factor than on the second factor. Thus, items were classified into two groups, those with higher means (more lateralized items) and those with lower means (less lateralized items), and the existence of connections between means and factors was examined. As a result, 19 of 24 more lateralized items and only 2 of 23 less lateralized items had higher loadings on the first than the second factor [χ^2 (1, $\underline{N}$ = 47) = 23.6, $\underline{p} < .001$].

Table 1. Means (M), standard deviations (SD) and a two-factor solution (Factor I & II) for a handedness questionnaire in a Japanese sample by Ida and Bryden (1996)

Item	M	SD	I	II	Item	M	SD	I	II
Write	4.9	0.49	.47	.06	Bowl	4.4	0.89	.67	.37
Draw	4.9	0.58	.57	.07	Insert coin	4.4	0.77	.27	.69
Peel	4.8	0.73	.88	.18	Turn water tap	4.3	0.81	.25	.69
Scissors	4.8	0.73	.74	.21	Pull pin	4.3	0.87	.46	.62
Cooking knife	4.8	0.80	.88	.16	Turn jar lid	4.3	1.06	.39	.52

Item	M	SD	I	II	Item	M	SD	I	II
Dial	4.8	0.51	.20	.48	Insert pin	4.3	0.86	.42	.66
Can opener	4.8	0.77	.76	.25	Push buzzer	4.2	0.82	.14	.71
Hammer	4.7	0.81	.84	.24	Comb	4.2	0.94	.59	.34
Screwdriver	4.7	0.74	.68	.35	Turn knob	4.2	0.84	.15	.67
Dart	4.7	0.87	.83	.23	Clean desk	4.1	0.84	.39	.68
Match	4.7	0.77	.70	31	Point	4.1	0.84	.21	.66
Tennis Racquet	4.7	0.87	.76	.24	Wave hand	4.0	0.81	.15	.63
Spoon	4.7	0.82	.68	.24	Take from shelf	4.0	0.88	.34	.56
Throw	4.6	0.90	.73	.26	Black-board	4.0	0.88	.35	.62
Deal cards	4.6	0.95	.52	.37	Turn page	4.0	0.89	.08	.57
Touch – tone Telephone	4.6	0.67	.09	.60	Pick up small object	3.9	0.80	.23	.69
Shuffle	4.6	1.10	.52	.22	Pull switch	3.9	0.81	.19	.69
Wash Dish	4.6	0.91	.64	.36	Crumple paper	3.9	0.88	.28	.60
Bottle opener	4.5	0.91	.65	.41	Push door	3.8	0.88	.15	.65
Turn key	4.5	0.77	.38	.64	Hold cup	3.8	1.05	.22	.50
Eraser	4.5	0.96	.69	.27	Hold glass	3.7	0.99	.20	.58
Insert key	4.5	0.79	.36	.65	Heavy bag	3.7	1.02	.09	.36
Pull key	4.5	0.79	.35	.64	Brush of lint	3.6	0.81	.14	.61
Toothbrush	4.4.	1.06	.73	.26					

It should be noted that "write" and "draw" showed relatively low loadings on the first, supposedly "skilled" factor. This could be attributed to too little variability in responses to the items as shown by extremely high means for these items.

1.1.2 Results for a Canadian sample

The two factors obtained in the analysis of the Canadian sample accounted for 57.7% of the total variance: the first factor, 50.8%; the second factor, 6.9%. Similarly to the Japanese sample, "skilled" items such as "write", "peel", "draw", "hammer", and "dart" showed high loadings on the first factor (Table 2). On the second factor, such items as "push buzzer", "turn knob", "clean desk", "pull switch", and "turn water tap" had high loadings. These items could be assumed to be "unskilled".

Table 2. Means, standard deviations and a two-factor solution for a handedness questionnaire in a Canadian sample by Ida and Bryden (1996)

Item	M	SD	I	II	Item	M	SD	I	II.
Scissors	4.7	0.87	.70	.29	Pull key	4.2	1.02	.40	.62
Bowl	4.6	0.99	.82	.26	Insert coin	4.2	0.96	.36	.64
Dart	4.6	1.02	.84	.31	Insert pin	4.1	1.04	.59	.57
Write	4.6	1.11	.86	.22	Pull pin	4.1	1.03	.46	.66
Draw	4.6	1.12	.85	.23	Clean desk	4.0	1.05	.42	.69
Peel	4.6	1.06	.86	.31	Blackboard	4.0	1.02	.41	.67
Hammer	4.6	1.06	.85	.33	Turn jar lid`	3.9	1.31	.33	.52
Throw	4.6	1.01	.79	.25	Push buzzer	3.9	0.98	.20	.71
Can opener	4.6	0.95	.51	.29	Wave hand	3.9	0.93	.34	.59
Cooking knife	4.5	1.10	.78	.30	Turn knob	3.9	0.96	.21	.71
Tennis racquet	4.5	1.04	.79	.27	Point	3.9	0.95	.33	.65
Match	4.5	1.07	.75	.33	Hold cup	3.8	1.08	.34	.59
Eraser	4.4	1.12	.81	.31	Pick up small object	3.8	0.96	.32	.67
Spoon	4.4	1.17	.80	.33	Shuffle	3.8	1.27	.51	.28
Screwdriver	4.4	0.98	.65	.50	Pull switch	3.8	0.95	.26	.69
Dial	4.4	1.00	.48	.59	Take from shelf	3.8	1.02	.44	.52
Bottle opener	4.4	1.08	.66	.49	Hold glass	3.7	1.00	.26	.59

Item	M	SD	I	II	Item	M	SD	I	II.
Toothbrush	4.4	1.11	.74	.38	Heavy bag	3.7	1.11	.29	.48
Insert key	4.3	0.97	.48	.60	Turn page	3.7	1.18	.16	.51
Turn key	4.3	0.99	.46	.60	Turn water tap	3.7	0.91	.16	.69
Deal cards	4.3	1.29	.55	.39	Crumple paper	3.6	0.88	.22	.56
Wash dish	4.2	1.14	.65	.37	Push door	3.5	0.87	.08	.65
Touch-tone telephone	4.2	1.04	.44	.61	Brush off lint	3.5	0.82	.16	.64
Comb	4.2	1.10	.62	.47					

As for the relation between means and factors, 20 of 24 more lateralized items and only 2 of 23 less lateralized items had higher loadings on the first than the second factor [χ^2 (1, N = 47) = 26.3, p < .001]. The finding that highly lateralized items tend to have higher loadings on the first, "skilled" factor than the second, "unskilled" factor in both Japanese and Canadian samples suggests a strong connection between factor structure and the degree of lateralization of questionnaire items.

1.2 Data from Japanese and Indian samples

The second study (Mandal et al., 1999) included data on 400 participants from India (223 men and 177 women) and 502 participants from Japan (245 men and 257 women). The participants were undergraduate students. A 32-item Waterloo Handedness Questionnaire (Steenhuis & Bryden, 1989) was used for this study. Participants were instructed to indicate their preference for each item on a 7-point scale (1 = left always, 2 = left mostly, 3 = left usually but right sometimes, 4 = equal, 5 = right usually but left sometimes, 6 = right mostly, 7 = right always). Data for Japanese and Indian participants were subjected separately to principal-components factor analysis with varimax rotation for a 2-factor solution (SPSS 7.5 for Windows).

1.2.1 Results for a Japanese sample

The obtained two factors accounted for 51.0% of the total variance: the first factor, 41.0%; the second factor, 10.0%. As shown in Table 3, the first factor could be assumed to reflect preference for skilled activities. Items such as "hammer", "scissors", "sewing", "throwing", and "knife" showed high loadings on the first factor. The second factor could be identified with unskilled activities such as "pick up a marble", "pick up a screw", "pick up a book", "pick up a paper clip", and "pick up a pin".

In Table 3, items were arranged in decreasing order of means. Items with higher means appeared to have higher loadings on the first factor. Thus, items were classified into two groups, those with higher means (more lateralized items) and those with lower means (less lateralized items), and the existence of connections between means and factors was examined. As a result, all of 16 more lateralized items and only 5 of 16 less lateralized items had higher loadings on the first than the second factor [χ^2 (1, N = 32) = 16.8, $p < .001$].

Table 3. Means, standard deviations, and a two-factor solution for handedness questionnaire in a Japanese sample by Mandal et al. (1999)

Item	M	SD	I	II	Item	M	SD	I	II
Write name	6.9	0.79	.53	.10	Insert a pin	5.9	1.30	.53	.45
Hold a hammer	6.5	1.37	.86	.20	Hold an eraser	5.9	1.55	.67	.37
Hold a scissor	6.5	1.35	.84	.18	Throw a spear	5.8	1.66	.64	.03
Hold a needle for sewing	6.5	1.37	.83	.23	Hold a comb	5.7	1.41	.55	.33
Hold a knife	6.5	1.34	.80	.22	Swing an axe	5.5	1.52	.46	.04
Throw a ball	6.5	1.37	.82	.19	Pick up a coin	5.4	1.22	.18	.73
Right or left handed	6.4	1.37	.78	.25	Pick up a pin	5.3	1.25	.21	.79
Hold a racket	6.4	1.39	.79	.22	Pick up a marble	5.3	1.19	.16	.81
Throw a dart	6.4	1.39	.77	.20	Pick up a screw	5.3	1.27	.15	.81
Tighten a screw	6.3	1.18	.58	.32	Pick up a paper clip	5.2	1.23	.17	.80
Rest a bat on shoulder	6.3	1.49	.45	.20	Pet an animal	5.2	1.26	.24	.42
Flip a coin	6.2	1.22	.62	.37	Pick up a book	5.2	1.20	.19	.81
Hold your toothbrush	6.2	1.49	.67	.28	Pick up a glass	5.1	1.46	.28	.47

Item	M	SD	I	II	Item	M	SD	I	II
Use a pair of tweezers	6.1	1.37	.65	.30	Pick up a paper	5.1	1.29	.12	.64
Shoot a marble	6.0	1.33	.61	.38	Lift a heavy suitcase	5.1	1.58	.22	.43
Wind a stopwatch	6.0	1.37	.37	.21	Lift a heavy object	5.0	1.50	.19	.49

A relatively low loading of "write name" on the first factor is most likely attributable to the lack of variability of this item. This situation is similar to that shown by "write" and "draw" in Table 1. These results suggest that, in spite of the use of a different questionnaire, "skilled" items are highly lateralized and have high loadings on the first factor in a Japanese sample.

1.2.2 Results for an Indian sample

In the Indian sample, two factors accounted for 54.0% of the total variance: the first factor, 44.3%; the second factor, 9.7%. As shown in Table 4, items that had high loadings on the first factor comprised both "skilled" and "unskilled" items; those items such as "pick up a paper", "hold a heavy object", "pick up a glass of water", and "pick up a pin" were thought to be "unskilled" and other items such as "shoot a marble", "use a pair of tweezers", "hold an eraser", and "throw a dart" were thought to be "skilled". Loadings on the second factor were generally lower than those on the first factor. Items that showed high loadings on the second factor were mainly "skilled" ones such as "tighten a screw by hand", "throw a ball", "hold a knife", "hold your toothbrush", and "flip a coin".

Table 4. Means, standard deviations, and a two-factor solution for handedness questionnaire in an Indian sample by Mandal et al. (1999)

Item	M	SD	I.80	II	Item	M	SD	I	II
Hold a racket	6.7	1.27	.72	.07	Hold a hammer	6.3	1.87	.44	.46
Throw a spear	6.7	1.31	.71	.06	Pick up a book	6.2	1.53	.70	.18
Hold an eraser	6.7	1.34	.80	.27	Flip a coin	6.0	1.41	.36	.70
Pick up a paper	6.6	1.42	.90	.19	Pick up a coin	5.9	1.46	.36	.57
Swing an axe	6.6	1.44	.82	.25	Pet an animal	5.8	1.64	.40	.46
Use a pair of tweezers	6.6	1.20	.82	.37	Pick up a marble	5.8	1.86	.42	-.05

Item	M	SD	I.80	II	Item	M	SD	I	II
Hold a needle for sewing	6.6	1.51	.78	.32	Hold your toothbrush	5.7	1.61	.30	.73
Throw a dart	6.5	1.53	.80	.15	Insert a pin	5.7	1.96	.38	.27
Lift a heavy object	6.5	1.53	.89	.27	Pick up a screw	5.5.	1.65	.47	.34
Right or left handed	6.5	1.52	.71	.44	Hold a comb	5.4	1.60	.30	.60
Pick up a pin	6.5	1.62	.84	.25	Tighten a screw by hand	5.4	1.34	.11	.74
Pick up a glass	6.5	1.57	.87	.28	Throw a ball	5.3	1.44	.16	.74
Shoot a marble	6.4	1.72	.85	.32	Hold a knife	5.3	1.58	.25	.74
Wind a stopwatch	6.4.	1.66	.72.	.09	Pick up a paper clip	4.9	2.79	.08	.19
Lift a heavy suitcase	6.3	1.87	.36	.44	Rest a bat on shoulder	4.6	2.47	-.23	.54
Write name	6.3	1.85	.44	.46	Hold scissors	4.5	2.38	-.22	.49

The factor structure shown by this analysis was different from that for the Japanese data. The examination of the connection between means and factors showed the same trend as in the Japanese data. Fourteen of 16 more lateralized items and only 4 of 16 less lateralized items had higher loadings on the first than the second factor [χ^2 (1, N = 32) = 12.7, p < .001]. In this Indian sample, the factor structure seemed to be difficult to explain on the basis of differentiating "skilled" and "unskilled" factors. In contrast, the factor structure seemed to be predicatable simply by the degree of lateralization of each item, as indicated by its mean value.

2. DISCUSSION

The present analyses suggest that the different degrees of lateralization among the items of a hand-preference questionnaire can affect the factor structure obtained with conventional factor analysis. The effects on factor structure of different degrees of lateralization among items were found in all samples from Canada, Japan, and India, and with different questionnaires using different items and response categories. These results strongly suggest that the observed effects have general significance. Furthermore, the fact that highly lateralized but unskilled activities, such as picking behaviours, had higher loadings on the same factor as highly lateralized and skilled activities in Indian participants , supports the view that the obtained factor structure is an artifact that is due to violation of the assumption of multivariate normality in factor analysis (McManus, 1996). The effects of different degrees of lateralization of items on factor structure might be explained not by "skilled" and "unskilled" factors but simply by different distributions of responses among questionnaire items. Such a problem could be applicable not only to "skilled" and "unskilled" factors but also to other factors obtained with conventional factor analysis for similar measures of hand preference. As McManus (1996) pointed out, these factors could be "difficulty factors", and other multivariate techniques such as cluster analysis and association analysis might also produce artifacts with laterality data.

This point, of course, does not necessarily lead to rejection of the notion of multidimensionality of hand preference. Messinger and Messinger (1996) normalized the items of a hand preference questionnaire by choosing a transformation for each item before conducting factor analysis, and obtained a factor structure that suggested the multidimensionality of handedness. Elaboration of statistical methods could be one of the effective ways to demonstrate the existence of multiple factors of handedness.

The results of the present analyses also concern the suitability of factor analysis as a tool for choosing good items for hand preference questionnaires, because classification of items by factor analysis might merely reflect distribution differences among the items. The items concerning writing and drawing showed relatively low loadings on the first factor in both Japanese samples (Table 1 and 3). Based on the view that handedness is unidimensional - and because the first factor is considered to represent hand preference - these items should be excluded from hand preference questionnaires for Japanese as was done by Hatta and Nakatsuka (1975). However, this situation may be produced by the extremely biased distributions of responses to these items, not by a component common to hand-preference items. Since the extremely low prevalence of people who use the left hand for these activities could be attributed to strong cultural and social pressures, the exclusion of these activities from hand preference questionnaires may be appropriate for some types of studies such as those that investigate cultural differences in the prevalence of left handers. The exclusion of these activities, however, may be justified simply by the extremely biased distributions themselves. Further, Peters and Murphy

(1993) compared the results of factor analyses conducted for left- and right handers separately and those conducted for pooled data. They showed that the factor structures of these separate and pooled samples are quite different. In particular, they showed that artifacts in factor structure arise when samples known a priori to be different in their response to key items are lumped together in a single analysis. The Peters and Murphy analysis of Canadian data is similar in its implications to the analysis of, for instance, "writing" in the Japanese sample. Most Japanese write with the right hand and thus there is no variability in this item. Similarly, and by definition, Canadian individuals who are labeled as left handers write with the left and those labeled as right handers write with the right hand. Analyzed separately, "write" does not contribute to variability in either group but when pooled, the source of variability is precisely the different hand choice for writing in the two handedness groups, which will therefore show up as significant factor with a high factor loading.

Although the "skilled" and "unskilled" factors may be artifacts, a skilled/unskilled dimension could be important for classifying the items of hand-preference questionnaires. For example, it has been suggested that hand preference relates to localization of function in the cerebral hemispheres (e.g., Bryden, 1982). The typical pattern among right handers, that is, predominance of the left hemisphere for verbal functions and that of the right hemisphere for nonverbal or visuospatial functions, is less clear among left handers (e.g., Bryden & Steenhuis, 1991). According to the conventional view, only the preference for skilled activities could be regarded as markers for handedness and thus, the connection mentioned above between handedness and cerebral laterality has been shown mainly for skilled activities. As Steenhuis and Bryden (1989) suggested, the neural mechanisms for the selection and execution of sequenced motor behaviours in speech and manual praxis may be more effective when they are lateralized in one hemisphere, and a fundamental feature of skilled activities may be the need to execute a relatively complex sequence of motor behaviours. In that case, "skilled" items of hand preference would have a stronger connection with cerebral laterality, especially with the lateralization of language, than would "unskilled" items. Thus, hand preference measured with "skilled" items might have different relations to cerebral laterality than hand preference measured with "unskilled" items. A recent study (Ida, 1998) suggested that "skilled" and "unskilled" items might have different correlations with attentional bias toward the left or the right half of faces. This type of attentional bias was assumed to reflect a specialized function of the right hemisphere (Gilbert & Bakan, 1973).

Handedness can be measured by preference and skill. Normally, the preferred hand is the more skilled hand. But which is more intrinsic? If skill is the primary factor that results in handedness, skilled activities should be main objects to measure handedness with questionnaires. But if preference is the primary factor, "unskilled" items should , at the very least, be paid equal attention. McManus, Murray, Doyle, and Baron-Cohen (1992) found that children with autism showed population dominance for hand preference without the preferred hand being more skillful, and concluded that preference was causally prior to skill, an argument similar to that advanced by Peters (1983). In that case, the right/left hand is thought to become more skillful among right/left handers because they practised more with their preferred right/left hand. It might be possible that an original trend of handedness is reflected more on unskilled activities, although hand preference for those activities may be changed more easily by cultural pressure. Apart from the problem of "skilled" and "unskilled" factors, including unskilled activities into a hand preference questionnaire would be recommended from this viewpoint.

The degree of lateralization of hand preference items and the skilled/unskilled dimension of activities described by the items could be to some extent independent, although it may be a general trend that skilled activities are more lateralized than unskilled activities. It is possible that differential application of cultural pressure to skilled and unskilled activities modifies the relation between the lateralization of items and the skilled/unskilled dimension. As shown in Tables 3 and 4, picking behaviours in India and writing in Japan are both highly lateralized. These activities are thought to have been the targets of strong cultural pressure (Mandal et al., 1999). The use of the left hand for some skilled activities in Japan and for some unskilled activities in India may be more subjected to cultural pressure, respectively. This cultural difference may have caused a difference in the relation between the lateralization of items and the skilled/unskilled dimension. The cultural pressure against the use of the left hand seems to strengthen the relation in Japan and weaken the relation in India.

In conclusion, factor structures emerging from conventional factor analysis cannot be used as unqualified support for the multifactorial model of hand preference, and the suitability of conventional factor analysis to select good items to measure hand preference might be questionable. Finally, the skilled/unskilled dimension might not be applicable for identifying independent factors of hand preference, but it will be useful for classifying activities to measure hand preference.

3. REFERENCES

Beukelaar, L. J., & Kroonenberg, P. M. (1983). Towards a conceptualization of hand preference. *British Journal of Psychology, 74,* 33-45.

Bryden, M. P. (1977). Measuring handedness with questionnaires. *Neuropsychologia, 15,* 617-624.

Bryden, M. P. (1982). *Laterality: Functional asymmetry in the intact brain.* New York : Academic Press.

Bryden, M. P., & Steenhuis, R. E. (1991). Issues in the assessment of hand preference. In F. L. Kitterle (Ed.), *Cerebral laterality: Theory and research* (pp. 35-51). Hillsdale, NJ : Erlbaum.

Crovitz, H. F., & Zener, K. A. (1962). A group-test for assessing hand and eye dominances. *American Journal of Psychology, 75,* 271-276.

Gilbert, C., & Bakan, P. (1973). Visual asymmetry in the perception of faces. *Neuropsychologia, 11,* 355-362.

Hatta, T., & Kawakami, A. (1995). Patterns of handedness in modern Japanese: A cohort effect shown by re-administration of the H. N. Handedness Inventory after 20 years. *Canadian Journal of Experimental Psychology, 49,* 505-512.

Hatta, T., & Nakatsuka, Z. (1975). H. N. Handedness Inventory. In S. Ohno (Ed.), *Papers celebrating the 63rd birthday of Prof. Ohnishi* (pp. 224-247). Osaka: Osaka City University Press.

Healey, J. M., Liederman, J., & Geschwind, N. (1986). Hand preference is not a unidimensional trait. *Cortex, 22,* 33-53.

Ida, Y. (1998). Dual associations between hand preference and visual asymmetry in perception of faces. *Perceptual and Motor Skills, 87,* 1035-1041.

Ida, Y., & Bryden, M. P. (1996). A comparison of hand preference in Japan and Canada. *Canadian Journal of Experimental Psychology, 50,* 234-239.

McManus, I. C. (1996). Handedness. In J. G. Beaumont, P. M. Kenealy, & M. J. C. Rogers (Eds.), *The Blackwell dictionary of neuropsychology* (pp. 367-376). Cambridge, MA : Blackwell Publishers.

McManus, I. C., Murray, B., Doyle, K., & Baron-Cohen, S. (1992). Handedness in childhood autism shows a dissociation of skill and preference. *Cortex, 28,* 373-381.

Mandal, M. K., Ida, Y., Harizuka, S., & Upadhaya, N. (1999). Cultural difference in hand preference: Evidence from India and Japan. *International Journal of Psychology, 34,* 59-66.

Messinger, H. B., & Messinger, M. I. (1996). Factoring handedness data: II. Geschwind's multidimensional hypothesis. *Cortex, 32,* 375-381.

Oldfield, R. C. (1971). The assessment and analysis of handedness: The Edinburgh Inventory. *Neuropsychologia, 9,* 97-113.

Peters, M. (1983). Differentiation and lateral specialization in motor development. In G. Young, C. Corter, S. J. Segalowitz and S. Trehub (Eds.). *Manual specialization and the developing brain: Longitudinal studies* (pp. 141-159). New York: NY Academic Press:

Peters, M., & Murphy, K. (1993). Factor analyses of pooled hand questionnaire data are of questionable value. *Cortex, 29,* 305-314.

Singh, M., & Bryden, M. P. (1994). Factor structure of handedness in India. *International Journal of Neuroscience, 74,* 33-43.

Steenhuis, R. E., & Bryden, M. P. (1989). Different dimensions of hand preference that relate to skilled and unskilled activities. *Cortex, 25,* 289-304.

Steenhuis, R. E., Bryden, M. P., Schwartz, M., & Lawson, S. (1990). Reliability of hand preference items and factors. *Journal of Clinical and Experimental Neuropsychology, 12,* 921-930.

Chapter 8

Contributions of Imaging Techniques to Our Understanding of Handedness[1]

Michael Peters
University of Guelph, Canada

In spite of several earlier studies (e.g., Roland, Larsen, Lassen, & Skinhoj, 1980), the bulk of research in the area that relates handedness to brain imaging is, by and large, only about 10 years old. Nevertheless, a substantial amount of research has been published. Much of it can be described as "feeling out the potential" of the new techniques and it is premature to review the entire literature in detail. Here, the limitation of space dictates a cautious approach, singling out papers that are of particular interest either because of the way in which a method was used or because of substantial and reliable information that is germane to our central interest. This means that much important and relevant work will be neglected, to be considered in more exhaustive reviews.

1. SOME PRELIMINARY REMARKS ABOUT METHODS

Considering that work on handedness is over 100 years old we cannot claim much progress. To be sure, methods for describing handedness have

[1]This work was supported by a Natural Sciences and Engineering Research Council of Canada Grant (A 7054).
M.K. Mandal, M.B. Bulman-Fleming and G. Tiwari (eds.), Side Bias: A Neuropsychological Perspective, 191-222.

improved in the last decade, and various means of assessment are available, together with a much better understanding of the expressions of handedness in different cultures (Brito, Brito, Paumgartten, & Lin, 1989; Dellatolas, Tubert, Castresana, Mesbah, Giallonardo, Lazaratou, & Lellouch 1991; Gilbert & Wysocki, 1992; Hatta, 1995; Hoosain, 1990; Mandal, Pandey, Singh, & Asthana, 1992; Perelle & Ehrman, 1994). There are also a number of genetic models available (Annett, 1995; Collins, 1977; Corballis, Lee, McManus, & Crow, 1996; Laland, Kumm, Van Horn, & Feldman, 1995; McManus, 1991), which allow us to think about the genesis of handedness in an organized way. Nevertheless, the question of mechanism is still quite unresolved. In an earlier attempt, Peters (1983) suggested that small structural asymmetries interact with ongoing experience to lead to large functional and structural differences. This may or may not have some validity but there still is the question of the nature of structural differences that relate to handedness. What sort of information is available to us to answer this question ? The last few years have brought some methodological advances that could prove important for researchers of handedness.

The first, and historically oldest, line of evidence relates to direct anatomical studies. In many ways this remains the most powerful tool in examining cerebral asymmetries but it is only very recently that concerted efforts have been made to explore the full potential of neuroanatomical methods in the context of asymmetries. I am not entirely sure why this is; the advent of new methods in science often leads to a premature abandonment of more established methods - not because the older methods have been fully exploited but because scientists are attracted by the possibilities afforded by newer methods (Bulman-Fleming, Grimshaw, & Berenbaum, 2000). In the case of neuroanatomy, new and exciting ways of staining for functional transmitters and the opportunities afforded by methods such as immuno-histology have seduced individuals away from the more traditional methods. However, it is clear that traditional anatomical methods of examining the spacing and classification of neurons, and the exhaustive documentation of the complexity of synaptic connections, offer large attractions for work in structural asymmetries of the brain.

As far as correlations of the anatomical substrate and handedness are concerned, there have been several earlier attempts to identify anatomical asymmetries. I am thinking here of the earlier studies that consider the level at which the pyramidal tracts cross over, and where asymmetries are observed (Kertesz & Geschwind, 1971; Yakovlev & Rakic, 1966) or asymmetries in the size of the descending corticospinal tracts (e.g., Nathan, Smith, & Deacon, 1990). One intriguing observation, though based on only five cases, is that by Irving, Rebeiz, & Tomilson (1974), who report far more motor neurons on the right side of the spinal cord at the level of S3 than on the left side. If supported, this would mean that anatomical asymmetries at the spinal level might be more pronounced for the machinery guiding leg and foot movement than for that guiding hand and arm movement. However, it has to be admitted that even here, the effort that has been invested is too little to allow any firm conclusions. This is a pity, because the general problem of how to delineate areas on the left and right axis of the central nervous system that are to be compared is far more easily solved at the brainstem and spinal level than at the cerebral level.

At any rate, the first, and in my mind still the most important line of attack on the problem of cerebral asymmetries related to handedness lies in traditional anatomical methods, which can, of course, be combined with the recent imaging possibilities. An exemplary illustration of this approach is provided by the work of Amunts and colleagues (Amunts, Schlaug, Schleicher, Steinmetz, Dabringhaus, Roland, & Zilles, 1996), to be discussed below.

The second line of evidence represents the first of the imaging techniques. It is the electroencephalogram, wherein mass changes in local excitatory and inhibitory synaptic potentials are recorded. By imaging we mean that we can - in whatever form - represent activity in the brain with a topographic map. A commonly used method concerns sensory evoked potentials (EPs) and event-related potentials (ERPs). By definition, EPs are recorded when sensory neurons are activated by some input. The term event-related potential is applied to changes that take place relative to some general context, and ERP's are used in the study of potentials that, for instance, are recorded during different states of attention. Both depend on local activation of pyramidal cells and the recording of the electric currents that are generated by this activation. Much depends on the fact that pyramidal cells show a vertical alignment in the cortex, so that there is a directionality (dipole) to the activation of synapses on apical dendrites (which are closer to the surface) relative to the cell body. Naturally, because the outer layers of the cortex are not all aligned in relation to the skull, as would be the case in lissencephalic brains, there is a problem of inferring the source of potential changes that are picked up on the surface of the head, a problem for which solutions or approximations of solutions have been found.

The magnetoencephalogram (MEG) is conceptually related to EEG technology, with the exception that in the EEG technology, electrical potential changes are recorded whereas in the MEG technology use is made of the fact that neural activity creates small magnetic fields. The equivalent of the ERPs with EEG technology are the event-related fields (ERFs) in MEG technology. MEG has not supplanted the older EEG techniques, which have many attractions for researchers. Indeed computer technology has allowed some fairly powerful imaging in the context of ERP recordings, which offer the advantage of much greater simplicity and lower costs in contrast to MEG technology. MEG also has limitations of the magnetic field orientations that it can pick up. However, MEG offers great flexibility in concurrent recording of events, and relatively better spatial localization because the magnetic fields are not distorted by their passage through bone and scalp. Both EEG and its variants, and MEG, have the great advantage that they can detect neural events as soon as they happen, without any appreciable time lag. fMRI has been more appropriate for events in the range of seconds but recent modifications (cf. Friston, Fletcher, Josephs, Holmes, Rugg & Turner, 1998) have allowed an appreciably faster time resolution. A disadvantage for both EEG and MEG lies in the relatively poorer spatial resolution so that although the "when" of activity can be mapped with extreme precision, the "where" of activity cannot be mapped as precisely as, for example, in fMRI. We will come back to the aspect of spatial precision of imaging in the context of what constitutes the proper level of resolution for different kinds of activity. It goes without saying that the principal attraction of EEG and MEG methods lies in providing an image of brain function. However, and especially in the case of the mapping of sensory potentials, the method has also been used for the purpose of pinpointing cortical mapping of the external sensory surface.

Although the previous methods are able to detect actual changes in neural activity, PET and most fMRI methods respond to changes that are secondary to neural activity. If there is an increase in neural activation, there is an increase in local blood circulation and the location of this increase is made visible in PET and fMRI. PET appears to have a number of disadvantages relative to fMRI. First, it is invasive, with injection of the radioactive tracers and, second, it also is more costly to conduct PET experiments than fMRI studies. fMRI offers the advantage of considerably better spatial resolution. However, when considering functional activations, greater spatial resolution is not necessarily an advantage. To give an extreme example, the recording of activity of a single cell may be very informative in simple functional contexts (sensory mapping) but when distributed networks of cells collaborate in a function, information from a single component in the network may not be very informative. The level of resolution of an imaging application should reflect not what the technique can deliver but what is an appropriate level of resolution. Examples of "the right level of resolution" can be provided for many fields, and the determination of optimal resolution of imaging may be a fruitful field of research. It should be pointed out that the issue of "too fine a resolution" constitutes not a technical limitation but a conceptual/computational problem – the volume of active space can be defined by algorithm after data collection, and can be varied at will. There are some differences between the methods that are, perhaps, of great importance for the study of handedness and hand motor control. For instance, in Joliot et al.'s comparison of PET and MEG (Joliot, Crivello, Badier, Diallo, Tzourio, & Mazoyer, 1998), the supplementary motor area (SMA) shows up consistently in PET, but not in MEG. This is not just an idiosyncrasy of this particular study because PET, as Joliot et al. (1998) note, seems generally sensitive to SMA area activation (cf. also Kawashima, Yamada, Kinomura, Yamaguchi, Matsui, Yoshioka, & Fukuda, 1993; Seitz, Canavan, Yaguez, Herzog, Tellmann, Knorr, Huang, & Homberg, 1997) and may not only be more sensitive in this regard than MEG, but also than fMRI. Similarly, PET and fMRI may be relatively more sensitive in picking up ipsilateral activation than MEG. Before we conclude that fMRI and PET (and perhaps especially PET) are the methods of choice when activations in higher-order motor areas and ipsilateral activations in primary motor areas are to be investigated (Kawashima et al., 1993), much more paradigmatic work has to be done, along the lines presented by Joliot et al. (1998), in which the same tasks are used for the same participants . Such work is not glamorous and is very expensive, but also very necessary to further imaging research in functional cerebral asymmetry.

2. ANATOMICAL EVIDENCE OF ASYMMETRIES IN THE CORTICAL HAND AREAS

A paper by White et al. (White, Lucas, Richards, & Purves, 1994), based on 22 autopsy cases, reported an anatomical asymmetry of the dorsolateral area in the region of the central sulcus favouring the left hemisphere. In their study, only 63.6 % of the sample showed such an asymmetry, but the paper provided a hint for the anatomical base of handedness. However, further examination of the issue, with an expanded sample of 67 autopsy cases led White et al. (White, Andrews, Hulette, Richards, Groelle, Paydarfar, & Purves, 1997) to the conclusion that there was not discernible asymmetry between those areas in the left and right primary motor cortex that might be expected to represent the hand areas. Their conclusion was confirmed by cytoarchitectonic investigations. Of necessity, a reasonable determination of handedness was not possible in either one of the studies but this is not a reasonable explanation for the lack of an observed asymmetry because the sample of 67 cases would be expected to have contained a substantial majority of right-handed individuals. Of further interest is the observation by White et al. (1997) that the outflow from the cortical motor areas also did not show an asymmetry. For instance, the pyramidal tracts were said to be symmetrical. Whatever difficulty there might be to agree on the cytoarchitectonic delineations of the motor cortex, there should be fewer definitional problems for the cross-sectional area of the pyramidal tracts at the medullary level. Finally, White et al. also failed to detect lateral asymmetries at the level of level of lumbar and sacral enlargements, and, more importantly, there was no asymmetry in number or size of neurons in the ventral horn.

White et al.'s finding of a lack of asymmetry stands in contrast with several other anatomical papers in which some asymmetry was found. One of the earliest indications came from an examination of the level at which the pyramidal tracts cross over; in 87 % of cases, there was a priority for pyramidal tract fibres coming from the left hemisphere (Yakovlev & Rakic, 1966). However, Kertesz & Geschwind (1971) found that the crossover pattern did not correlate with handedness because the pattern favouring the left was also seen in their sample of left handers. In the latter study, an attempt was made to ascertain handedness, which was not the case in the former study. It is not entirely clear how crossover patterns relate to handedness. One possibility concerns simply an asymmetry in the rate of development in the left as opposed to the right hemisphere with earlier differentiation leading to the crossover pattern of left above right, with some intercalation. However, this is entirely speculative and it is possible that the frequencies favouring the left/right crossover patterns and their close relation to the prevalence of right- and left handedness are not linked to handedness at all. If frequency and kind of hand use has an impact on the size and numbers of neurons and their descending axons, the findings by Nathan et al. (1990) are relevant. Nathan et al. (1990) reported asymmetries in the sizes of the lateral and anterior corticospinal tracts. Their study showed a larger right lateral corticospinal tract, consistent with the idea that there is a size difference between the originating areas of the right and left cortices. Nathan et al. also described a larger anterior corticospinal tract on the right side. Because this tract is largely uncrossed, and using the same logic, one would have to assume that the originating area is larger in the RIGHT cerebral cortex. Additional evidence that could be compatible with structural asymmetries at the spinal level derives from Tan's work on the H-reflex (1985a,b). Tan established, in several studies, a relation between the recovery curve of the H-reflex and handedness. This relation can well be the result of descending influences on the H-reflex, but it is not incompatible with asymmetries at the local spinal level. The association of the recovery curves with H-reflex recovery is most likely the outcome of complex interactions between local functional circuits and the effects of descending influences. That no simple local factors are at work is emphasized by the observation that there are no differences in the velocity of the median and ulnar nerves in the left and right hands of left- and right handers (Tan, 1985c).

More direct evidence derives from studies conducted by Amunts et al. (1996), whose work disagrees with the second study by White (White et al., 1997) and colleagues. Recall that in the earlier study, White and colleagues had found evidence for, on average, a larger hand area in the left hemisphere, but that further work with larger samples (White et al., 1997) had not supported the earlier observations. Amunts and colleagues studied the brains of 31 male right handers and 14 male left handers. Taking advantage of the fact that MRI methods allow behavioural observations in the intact subject, they also ascertained handedness with questionnaire and performance measures. When such imaging studies are done, the selection of target areas presents a problem with structural MRI: "Which areas should be looked at and how can one be sure that one deals with the cortical hand area in the primary motor cortex"? A number of approaches have been taken in order to find landmarks that delineate the hand motor area of the primary motor cortex. Joliot et al. (1998) used all three major imaging methods, MRI, PET and MEG on the same set of participants , and they were thus able to compare the loci of activation generated by simple flexion and extension of the index finger, with pauses between the flexion and extension phases. They found relatively good agreement of coordinates for the primary motor areas, with some notable differences for higher order motor areas, to be discussed later. In terms of specific landmarks, the work of Yousry et al. (Yousry, Schmid, Alkadhi, Schmidt, Peraud, Buettner, & Winkler, 1997; Yousry, Schmid, Jassoy, Schmidt, Eisner, Reulen, Resier, & Lisser, 1995) may prove very helpful. Yousry et al.'s work used direct confirmation of the putative hand cortex area, as determined by MRI methods, by directly stimulating the cortex in neurosurgery patients. Yousry et al. suggest that a knob-like region on the precentral gyrus identifies the hand motor region of the primary motor cortex, a finding that has been confirmed by use of TMS (transcranial magnetic stimulation) to identify the hand motor area with subsequent imaging by MRI (Ro, Cheifet, Ingle, Shoup, & Rafal, 1999). Amunts et al. made use of coordinates that had been established with positron emission tomography and MRI. What makes their study especially valuable is that a blind procedure was adhered to: those who took the measurements did not know whose brain image they analyzed, and whether it was a right or left hemisphere that was being measured. This latter - and very important - methodological refinement was accomplished by providing mirror images and unreversed images so that a brain that was presented in a certain orientation did not allow the individual doing the measurements to say "this is the left half and this is the right half".

The principal measure taken by Amunts et al. was a depth measure of the central sulcus in an area defined by Talairach coordinates as the relevant area. The difference in the left and right central sulcus depths was expressed as a coefficient (100 x [(L - R)/(L + R)/2] and the resultant value was considered the asymmetry coefficient. This is much like the calculation of handedness coefficients. The formula expresses the expectation of greater central sulcus depth on the left side, leading to positive values for the quotient if the left values exceed the right values. Significant differences emerged in sections at the dorsal level, where the hand area is located. Amunts et al. made two principal observations. First, for the group of right handers, the dorsal coordinates yielded positive values (greater sulcus depth in the left hemisphere) and, second, for the left handers, the corresponding coordinates yielded negative values (greater sulcus depth in the right hemisphere). Also in keeping with general expectations was the generally less pronounced asymmetry for the left handers; the left and right dorsal central sulcus length differences were not nearly as pronounced as for the right handers. A more recent study by Amunts, Jäncke, Mohlberg, Steinmetz & Zilles (2000) provided substantial corroboration of the previous findings, but also qualified the asymmetries relative to sex; asymmetries in intrasulcal lengths were confined to males.

We thus have a clear contradiction between the analyses by White and coworkers, using direct measures on brains obtained by autopsy, and the analyses by Amunts and colleagues, who used the less direct MRI measures for the 1996 study. However, Amunts et al. also used direct cytoarchitectonic measures on actual autopsy-derived brains and the results for this subsample – comprising fewer cases – support the MRI conclusions. Foundas et al. (Foundas, Hong, Leonard, & Heilman, 1998) also used a sizable number of right-handed and left-handed participants, and they included sex of participants as a variable. Like Amunts et al., Foundas et al. found that the hand motor region of the primary motor cortex is larger on the left side for right handers. However, they did not find the converse for left handers; there were no statistically significant differences for their left-handed sample of six men and nine women. A variety of possibilities can be considered to account for this difference. First, Amunts et al. eliminated the additional variability introduced by using both sexes, and, second, if it is assumed that asymmetries for left handers are altogether smaller than for right handers, sampling artifacts can be a factor. An aspect that is somewhat problematic in the Foundas et al. data is the small surface area of the primary motor cortices. Because of the great variation in methods, it is never quite possible to make telling comparisons between studies, but the overall surface area described by Foundas et al. is surprisingly small (cf., for example, Sanes, Donoghue, Thangeraj, Edelman, & Warach, 1995).

We would be remiss not to mention another instance of a structural brain asymmetry that appears to relate to handedness. Snyder et al. (Snyder, Bilder, Wu, Bogerts, & Lieberman, 1995) have shown that the cerebellum shows some volume asymmetries that are a function of handedness. In agreement with some much behavioural data (but note the issue of subclassification), Snyder et al. found volume asymmetries in left handers to be less clearly expressed than in right handers. In right handers, there was a torsion-type asymmetry with the right anterior cerebellum having a higher volume than the left, and with a converse pattern for the posterior cerebellum. This is reminiscent of the overall cerebral cortex asymmetries that have been described for right- and left handers (Kertesz, Polk, Black, & Howell, 1990). Recent fMRI work has also provided evidence on functional asymmetries in the cerebellum, with a distinction between responses to bimanual as opposed to unimanual movements (Jäncke, Specht, Mirzazade & Peters, 1999).

Less directly related to handedness asymmetries than the work on specific motor areas, but also of general importance, are efforts to identify the relation between handedness and specific structures such as the corpus callosum. Work here seems to have taken several turns with disagreement on whether handedness does play a role in corpus callosum size and configuration (Habib, Gayraud, Oliva, Regis, Salamon, & Khalil, 1991; Steinmetz, Jäncke, Kleinschmidt, Schlaug, Volkmann, & Huang, 1992; Witelson, 1985).

3. EVIDENCE OF PLASTICITY: HAND MOTOR AREA ASYMMETRY IS SENSITIVE TO LONG-TERM DEVELOPMENTAL CHANGES AND TO SHORT-TERM MOTOR EXPERIENCE

In a more current set of studies, Amunts and coworkers (Amunts, Schmidt, Schleicher, & Zilles, 1997; Amunts, Istomin, Schleicher & Zilles, K.) confirmed preliminary findings made in the 1996 study, of a larger hand motor area on the left side. In addition they suggest that the reason for the asymmetry was not necessarily a difference in numbers of neurons, but rather a lesser packing density on the left side, with more of the volume occupied with synapses and axons than was the case for the right hemisphere. It is possible that other left-hemisphere regions share this reduced packing density (cf. Buxhoeveden & Casanova, in press; Seldon, 1981). A developmental perspective was added in terms of the degree of asymmetry found at different layers as a function of age. Here, the observation was that the differentiation of asymmetry reflected the ontogenetic course of hand-preference development. Thus, the layers below the granular layer, destined to be very much reduced in the motor cortex of normal individuals, show similar degrees of asymmetry for children and adults. However, the supergranular layers showed a greater degree of asymmetry for the adults. The subgranular layers 5 and 6, full of pyramidal cells, are first to differentiate in cortical development whereas the outer layers differentiate later.

It seems that the anatomical asymmetry of the hand areas in the cortex can be seen as including two different aspects. The first concerns the inherent asymmetry that is the result of a complex of gene/cell environment interactions, which determine lateral asymmetries in general. Such mechanisms are seen in as simple an organism as C. elegans (Wood & Kershaw, 1991). By "cell environment" we mean that the fate of cell lineages is not solely determined by the genes, but by an interaction between the genes and biochemical messages received from neighbouring cells. This type of asymmetry concerns the basic “set-up” of the system. The other aspect of asymmetry is reflected by the differentiation of the motor substrate as a function of motor experience, and this includes the relatively greater development of neural connections in those areas that are more frequently used. To the extent that it is the latter aspect of asymmetry that interacts with behaviourally expressed hand asymmetry, it is here that we are most likely to find a direct correlate between expressed hand preference and hand experience and the neural substrate.

A number of recent studies that concern motor experience are of interest here. Karni et al. (Karni, Meyer, Jezzard, Adams, Turner, & Ungerleider, 1995) monitored the area of the primary motor cortex that was activated by finger-tapping activities and found that the area activated as well as the intensity of activation was a function of learning. Thus, rather than showing an invariant area of activation that responded to certain finger activities, even the primary motor cortex showed change in response to experience. This kind of pattern was further confirmed by evidence obtained in a somewhat different way: Pascual-Leone et al. (Pascual-Leone, Wassermann, Sadato, & Hallett, 1995) delineated the primary motor cortex of blind Braille readers by use of transcranial magnetic stimulation. It turned out that the size of the mapped areas was a function of recent motor experience. Thus, the extent of the maps was larger directly after the Braille readers had activated their hands than it was after a day of rest. The Pascual-Leone study involved older participants (minimum age 44), and therefore shows that plasticity as a result of current motor experience can be observed even later in life.

A more direct line of evidence comes from a study by Amunts et al. (Amunts, Schlaug, Jäncke, Steinmetz, Schleicher, Dabringhaus, & Zilles, 1997), in which the length of the wall of the precentral gyrus adjacent to the central sulcus was measured in individuals with keyboard experience and control participants . Similar to the earlier Amunts et al. (1996) study, the anatomical measure was taken as a shorthand indicator of the extent of the hand area of the primary motor cortex. The results showed, in accordance with the earlier study, an asymmetry favouring the left hemisphere for all of the right-handed participants , controls and keyboard players alike. However, and pertinent to our discussion, the degree of asymmetry in the keyboard players was less than in the controls, due to a somewhat greater length in keyboard players of the *right hemisphere* sulcus as compared to that of the control participants. This suggests that the greater left-hand experience by right-handed keyboard players was reflected in a relative enlargement of the hand region in primary motor cortex of the right hemisphere. Although it is true that the right hands of keyboard players tend to be dominant in the majority of keyboard compositions (Peters, 1986), the left hands of experienced players do get considerably more exacting practice than do those of control participants . There is a remote possibility that the motor experience of the hands is not the sole causal factor in the observed asymmetries and that a certain amount of bias in selecting keyboard players plays a role. However, this possibility does not likely apply because Amunts et al. were also able to show that there was a negative correlation between the anatomical measure and the age of commencement of playing, so that the earlier the individual commenced playing, the larger the length of the posterior wall of the precentral gyrus. A more powerful argument yet against the "sample bias" argument stems from the findings of Elbert et al. (Elbert, Pantev, Wienbruch, Rockstroh, & Taub, 1995). These researchers looked at the primary sensory cortex of string players. Unlike keyboard players, who exercise the right hand more than the left, the fingering movements in string instruments are a matter for the left hand. The right hand is concerned with bow control and, although this is very demanding in terms of dynamic movement precision, the digits of the right hand are comparatively inactive in string players. Elbert et al. could show that the left hand area that represented the digits in the right hemisphere was larger in surface extent than the comparable area for the right hand in the left hemisphere. My assumption here is that the increased representation of the primary sensory cortex makes functional sense only if it is related also to a corresponding change in the primary motor cortex of the right hemisphere.

These studies show a considerable amount of plasticity in the hand motor areas, which is logically consistent with a larger extent of the left-hemisphere motor area for the right hand in right handers, but allows for plasticity in the right-hemisphere motor hand area for the left hand in right handers as well. This work also provides an important caution not to neglect motor experience in imaging studies.

Additional evidence for plasticity derives from observations made after amputation in humans. It has long been understood, on the basis of experimental work with nonhuman primates, that a considerable amount of plasticity exists with regard to reorganization of the somato-motor cortex after loss of limbs or parts of limbs. Use of fMRI techniques has shown that such plasticity is also observed in humans. Weiss et al. (Weiss, Miltner, Dillmann, Meissner, Huonker, & Nowak, 1998) could show that after amputation of a finger, there was a reorganization in the primary motor and sensory cortices, which was interpreted in terms of an invasion of processes from neurons serving the parts of the hand that were neighbouring the lost finger. When the entire hand area is lost, plasticity still follows the established patterns, in which areas that neighbour the somato-sensory mappings invade the "vacated" region.

The demonstration of plasticity at the level of the primary motor cortex raises the question of whether there are any pre-existing structural asymmetries that predate any appreciable influence of motor activity. Part of the answer of this question was provided by the comparison of cortical asymmetries in children and adults (Amunts et al., 1997a). Another part stems from a study that compares motor performance in hemiplegic children. There have been, of course, numerous studies of linguistic competence after early hemispherectomy, but little is available on related paradigms that are concerned with motor performance. It is for this reason that the study by Hiscock et al. (Hiscock, Hiscock, Benjamins, & Hillman, 1989) merits detailed description.

4. EVIDENCE FOR THE LACK OF EQUIPOTENTIALITY OF THE MOTOR SYSTEMS THAT GUIDE HAND MOVEMENT

One of the few studies that does address the point was done by Hiscock et al. (1989). Hemiplegic children in this study were tested on a variety of motor tests. Although a pegboard task yielded no differences other than the expected worse performance on the hemiplegic side, there was a significant superiority in the right hand of left hemiplegics, as compared to the left hand of right hemiplegics. Assuming equipotentiality of the two hemispheres for motor performance, one would not necessarily expect that it should matter which hand is intact. The evidence suggests that there is a predisposition for the left hemisphere to be superior in guiding performances of this kind. Exclusive experience with the left hand, driven by the intact right hemisphere, will not match the performance of the right hand, driven by the intact left hemisphere. We had noted that little other evidence is available on this point, but there is supporting work by Llorente et al. (Llorente, Satz, Brumm, & Philpott, 1998) on a case of "pathological left handedness". Normally, single-case studies would not be suitable for reviews of this kind but the exceptionally strong documentation of performance and case history in the Llorente case merits its inclusion. In this particular case, the left hand had become the preferred hand after an injury that affected the motor cortex for the right hand. On a finger-tapping task, the patient only reached a performance in the 13th percentile of that for normal control participants, suggesting that the ability of the right motor cortex to guide performance of the tapping movement was limited.

5. EVIDENCE ON FUNCTIONAL ASYMMETRIES IN THE CEREBRAL CORTEX THAT ARE RELATED TO HAND MOVEMENTS

In discussing the evidence, there is a dilemma in how to approach the issue of general limitations in methods. Should it be done in the specific context of describing a research study, or should the general issues be covered by themselves? Here, I have decided to summarize the more general problems encountered in doing functional studies with imaging technology.

5.1 Limited expectations

So far, the discussion has focused on structural asymmetries related to handedness. In functional studies, we need to distinguish between "trivial" and "substantial" asymmetries. A trivial asymmetry would consist of greater activation of the contralateral than the ipsilateral primary motor cortex during the performance of a simple unimanual task. A substantial asymmetry would consist of relatively greater activation in the left hemisphere when the right hand is active than in the right hemisphere when the left hand is active. The substantial asymmetries are important in the context of handedness, but it is also important not to expect that such asymmetries will be easy to demonstrate. In contrast to the systems that control speech (Tzourio, Crivello, Mellet, Nkanaga-Ngila, & Mazoyer, 1997), the motor-control system underlying the movements of the preferred and nonpreferred hand must, by definition, be far less asymmetrical in its capabilities. Notwithstanding the previous discussions of structural asymmetries, humans use both hands extensively and although there is a clearly observed population tendency to prefer one hand over the other, the system would be poorly designed if it did not also allow considerable quality of movement in the nonpreferred hand - not least because "preferred" and "nonpreferred" are somewhat artificial terms mostly defined for unimanual activities. In naturally occurring activities, both hands are concurrently active and the non-preferred hand supports and complements the movements of the preferred hand. This supportive action need not by any means be less complex or demanding, as is illustrated by the fingering movements of the left hand in the right-handed violin player. The available data do show strong motor competence of the nonpreferred hand. For example, if the preferred hand is lost, the nonpreferred hand is quite capable of learning tasks that used to be done with the preferred hand, and if a stroke disables the systems providing the final motor outflow control for one hand, the systems for the other hand are very much in the position to take over. So, although part of the occasional ambiguity in the findings that relate to functional asymmetries lies in the methodology, in which lateral differences tend to be based on difference measures that are open to some "noise", there is also the fact that – at least as far as the final outflow from the primary motor cortices is concerned – the left and right areas would be much more closely matched in their output capabilities than would, for instance, the left and right speech motor areas in adults.

6. WHAT THE SPECIFIC MOTOR TASKS USED CAN AND CANNOT TELL US

There are limitations in terms of the motor tasks that can be used in some of the imaging techniques. Whereas EEG and MEG technology allow considerable mobility in the participants , PET and MRI impose serious limitations. In the case of PET, there are some possibilities in terms of the choice of materials that are involved in the motor tasks, but mobility is limited. For instance, in one of the few studies that uses very complex motor activities, Seitz et al. (1997) had participants write on a tablet. Because participants are in the prone position and head movement has to be constrained, they could not directly observe the writing hand but were able to guide the hand by viewing a monitor. In the MRI setting, an additional constraint is posed by the fact that metal objects cannot be present in the apparatus. Conventional devices for recording motor performance often use metal parts, such as relays and switches. One way to get around the problem posed by the "no metal apparatus" rule is to use no recording devices at all. Thus, participants can be asked to tap repetitively with their fingers, or they can perform more complex tasks such as touching a sequence of fingers to the thumb.

Because of the great simplicity of producing tapping movements, which can be executed precisely, do not need visual monitoring, and lend themselves to well-controlled rate changes, the great majority of studies in which fMRI is used to look at cerebral activation do use tapping rather than any other task. The "metal" problem for fMRI can be circumvented by constructing manipulanda made entirely of plastics. Instead of using metal switches that close circuits, optic fibres can be used to record button presses, with the actual equipment used for registering performance located outside the MRI room.

We end this section with a comment on the need to record manual performance. In a number of the studies that will be discussed below, investigators did not actually record motor performance while performing imaging studies. The question is: does this invalidate the conclusions of such studies? If the tapping task required participants to tap at certain rates (either self-paced or paced by a flashing light or sound source) the answer is: no, the results are not invalidated. A very extensive literature shows that tapping rates up to 5 taps/sec are reproduced very accurately and with remarkably little variance. For those who wish to correlate exact movement phases with neural activation it should be of interest to note that although individuals can tap at 5 taps/sec or even faster, it is not possible for them to keep in accurate synchronization with a pacer at this rate. This is a somewhat esoteric but interesting point: when provided with a 200-msec-interval pacing beat (e.g., 5 taps/sec), participants can match the required interval with considerable accuracy after hearing only 3 or 4 pacing beats, but they cannot do so in any reasonable synchronicity with the pacing beat (Peters, 1989).

Thus, when tapping was asked to be done at a certain rate, rate-dependent changes in the imaging context can be taken at face value. For instance, Rao et al. (Rao, Bandettini, Binder, Bobholz, Hammeke, Stein, & Hyde, 1996) asked participants to tap at 1,2,3,4 and 5 Hz, as paced by a metronome. All of their participants showed a reliable rate effect in activation, such that activation was greatest at 5 Hz and least at 1 Hz. Although some investigators have failed to find such a rate effect, it seems quite certain now that the rate effect for tapping is real (cf. also Jäncke, Specht, Mirzazade, Loose, & Shah, (1998); Jäncke, Peters, Schlaug, Posse, Steinmetz, & Muller-Gartner, 1998). Of course, the precise mapping of movement and imaging-related changes (i.e., change in magnetic fields) can be of interest in itself when the method can track the neural changes related to tapping while they occur, as is the case in MEG work (cf. Volkmann, Schnitzler, Witte, & Freund, 1998). The issue of recording performance becomes more contentious if more complex tapping patterns are required. For instance, a common procedure requires participants to tap each finger against the thumb so that a sequence is produced. This task requires between- rather than within-digit sequencing. The question of whether any violation of the task, that is errors in the required sequence, is serious or not depends on the research objective. If the intent is merely to document areas of cerebral activation during such sequential movements, the occasional violation is not likely to make a difference. We proceed here from the assumption that it is the general intent to perform the motor performance as asked, as well as the neural substrate involved in immediate performance that are of importance. At this level, the imaging techniques are not likely to be sensitive enough to register errors.

If, however, complex patterns are required that have to be mastered by extensive practice, and if changes in cerebral activation during the learning process are to be documented, then recording of the actual motor performance is a requirement.

7. FUNCTIONAL ASYMMETRIES

It is not possible, within the limits of this chapter, to cover all relevant studies in detail. Having the choice of either mentioning all of the available work with brief comments, or to dwell more extensively on selected studies, I am choosing the latter option in order to emphasize the most salient aspect of this type of research.

Several studies suggest that there is a correlation between the degree of hand preference and the degree to which there are differences in left/right hemisphere activation. An fMRI study by Dassonville et al. (Dassonville, Zhu, Ugurbil, Kim, & Ashe, 1997) used a complex finger-tapping task, in which 7 right-handed and 6 left-handed participants had to tap with specific fingers according to instructions, which were presented by means of four stimulus circles, each corresponding to a finger. Depending on which of the stimulus annuli lit up, participants had to press on the buttons upon which each of the four fingers rested. Reaction times and errors in the choice of finger were recorded. Two conditions were used; one in which participants would know which finger would be required to move next and one in which participants could not predict which finger would be cued to move next. There were no differences in activation between these two conditions. Dassonville et al. (1997) also recorded hand preference with a standard questionnaire and used a laterality quotient derived from this questionnaire to characterize the degree of lateralization. The results showed a non-trivial asymmetry, in the sense that the degree of cortical activation contralateral to the dominant hand was greater than the degree of cortical activation contralateral to the non-dominant hand. What is of special interest is that this difference did not correspond to a performance difference. That is, the dominant hand did not perform this task more quickly than the non-dominant hand. In addition, there was also no significant difference in patterns of activation between the two handedness groups; if there was any hint of a difference at all it pointed in the direction of less activation for the non-dominant-hand/left-hemisphere combination for left handers. The investigators also found a correlation between the degree of lateral preference as established by preference questionnaire and the degree of ipsilateral activation. Thus, although the degree of preference did not show a correlation between dominant hand activity and contralateral cerebral activation, the ipsilateral cortex (ipsilateral to the moving hand) was relatively more active in those participants who were less strongly lateralized in their hand preference choices. In this particular study, the clearest and strongest relation between degree of hand preference and cerebral activation was found for activation of the primary motor cortex; there was no significant correlation between hand preference and activation in any one of five other areas where activity was recorded. However, when the activities in these areas were pooled (I assume a multiple regression was used - but it is not clear from the description), there was a significant correlation between degree of hand preference and activation, as strong as the one observed between hand preference and primary motor cortex ($r = .584$ vs. $r = .601$). The results of the Dassonville et al. study are in partial agreement with work by Kim et al. (Kim, Ashe, Georgopoulos, Merkle, Ellerman, Menon, Ogawa, & Ugurbil, 1993), but Kim et al. did identify differences between right- and left handers. Kim et al. found that, although the right motor cortex was activated mostly only by contralateral hand movements, the left motor cortex showed considerable activation to ipsilateral movements, and especially so for right handers. However, it must be noted that Kim et al. had only 5 right handers and one subject described as "ambidextrous"; that is, no left-handed participants as such.

A closely related set of findings stems from a MEG study by Volkmann et al. (1998). In this study there were 5 right-handed and 5 left-handed participants , who were carefully classified as to handedness (Jäncke, 1996). The motor task consisted of a variety of simple abduction and flexion movements of the digits, and wrist flexion. The researchers documented motor performance by recording the appropriate EMG signatures for the movements of the different digits and the wrist. The rate of movements was low; participants had to move once every 4 sec. Therefore, the ability of MEG technology to track movement rates was not used. The detailed documentation of EMG patterns revealed no difference between dominant and nondominant hand performance. In this, the Volkmann et al. study was similar to the Dassonville et al. study, because there were no obvious performance differences in the tasks that were used to study cerebral activation.

Most importantly, it was shown that the area of activation (though not the intensity) was larger for the primary motor cortex opposite the preferred hand. Volkmann et al. speculate that the greater extent of the area taken by the primary motor cortex that drives the preferred hand is due to greater spacing of the active neural elements, rather than differences in numbers of pyramidal cells in the two comparable cortical areas. This is in good agreement with the anatomical findings of Amunts et al. (1996) who suggest a greater overall motor area in the cortex contralateral to the preferred hand, but less-dense packing of neurons in that area. Although the MEG analysis of Volkmann et al. allows for other interpretations than packing density of pyramidal cells in order to account for the pattern of activation, it is tempting to interpret the Volkmann et al. findings within the context of the findings by Amunts and colleagues.

Like Dassonville et al., Volkmann et al. also reported a significant correlation between the degree of hand preference and the direction of cortical specialization: individuals with a performance profile that favoured the left hand tended to show a right cortical dominance whereas individuals with right hand dominance favoured the left cortical area. However, the existence of a significant linear correlation between hand performance asymmetry and cortical specialization does not allow, in the case of the Volkmann et al. study, the conclusion that there is a linear relation between degree of hand performance lateralization and cortical lateralization, as is the case for the Dassonville et al. data. In the Volkmann et al. study there are two clusters: one cluster of individuals with left-handed performance superiority who show a relatively larger right primary motor cortex and a cluster of individuals with right-handed performance superiority who show a relatively larger left primary motor cortex. There is no suggestion of any linear correlation between hand preference and cortical asymmetry within these clusters; the reported linear correlation simply arises because of the slope of the straight line that runs through the two clusters.

A point of interest in the Volkmann et al. study is the failure to demonstrate ipsilateral activations – even the pooling of data from both handedness groups failed to show a statistically significant ipsilateral activation, though the authors note a tendency for stronger ipsilateral activation during movements of the non-dominant hand. Jäncke et al. (1998a) similarly failed to find significant ipsilateral activation during unimanual movements at low rates. In contrast, Dassonville et al. did find ipsilateral activation and they reported that ipsilateral activation tended to be stronger in individuals who had less strongly polarized hand preference asymmetries. Kawashima et al. (1993) also found ipsilateral activation – possibly related to the sequential nature of the required movements. The Volkmann et al. study did not identify any significant differences between right- and left handers, and this appears to be contradicted by the Kim et al. (1993) study, which is reported to show a difference. However, in the Kim et al. study there was only one ambidextrous individual and generalizations are thus not possible. There was documentation for differences in location of the centres of activation when different digits moved, and Volkmann et al. observed that the mapping did not follow simple somatotopic principles. This supports the now widely held view (cf. Schieber & Hibbard, 1993) that the topography of the primary motor cortex is largely reflective of a mapping of kinetically meaningful muscle activation patterns rather than a simple somatotopic mapping of cortex to muscle.

Triggs et al. (Triggs, Calviano, & Levine, 1997) examined lateral asymmetries by means of transcranial magnetic stimulation (TMS). This technique, of course, allows much greater freedom in terms of movement because participants do not have to be confined in a scanning apparatus and there are no limitations as to the nature of manipulanda and material needed for motor tasks (see also Walsh & Rushworth, 1999, for a general discussion TMS applications in neuropsychology). Triggs et al. had participants perform a traditional peg-moving task (the Purdue pegboard test), a finger-tapping task, and a strength-of-grip task. They used a considerable number of participants who were selected carefully to exclude familial handedness (30 left handers and 30 right handers), and who were given a handedness questionnaire that was used to assign laterality quotients, such that participants could be assigned a location on a spectrum ranging from extreme left handedness to extreme right handedness. The most salient finding was that the thresholds for effective magnetic stimulation were lower for the motor cortex contralateral to the dominant hand for both left handers and right handers. Thus, thresholds were lower for the left motor cortex in right handers and lower for the right motor cortex in left handers. There were no overall significant differences in thresholds as a function of handedness - other than the expected Handedness x Side interaction. This suggests that the pattern in left handers was simply the reverse of that in right handers rather than showing a substantive difference between the two groups. Triggs et al. showed that there was a linear correlation between their performance measures and the TMS threshold: like Dassonville et al., they found that individuals with the strongest between-hand differences (in the case of Dassonville et al., these were preference differences) showed the most marked asymmetries in the criterion measure. Triggs et al. were also able to show that the choice of motor task made a difference. The strongest linear correlation was obtained for TMS thresholds and finger-tapping performance, with a somewhat less marked, but still statistically significant, correlation between TMS thresholds and peg-placing performance.

In contrast, there was only a weak relation between TMS threshold and grip strength. If there is a relation simply between the extent of the motor cortex and thresholds, then the Triggs et al. findings provide a good complement to the results by Amunts and colleagues, who find the dominant motor cortex for left- and right handers larger than the corresponding nondominant motor cortex. The large number of participants used by Triggs et al. allowed a distinction between individuals with consistent handedness (that is, who showed a consistent hand preference) and individuals who did not. Their results show that, although the TMS thresholds clearly distinguished between the dominant and non-dominant motor cortex for the "consistently handed", they failed to do so for the "inconsistently handed". This forces investigators to recognize the importance of handedness classification, and suggests that sampling error for smaller samples has to be taken seriously. For instance, in the case of the Volkmann et al. study, the small number of left handers could easily favour one or the other group of individuals. Another example is the actual reversion of directional findings for inconsistent right handers when a small number of such participants (Macdonell, Shapiro, Chiappa, Helmers, Cros, Day, & Shahani, 1991) was compared with a larger sample (Triggs, Calviano, Macdonell, Cros, & Chiappa, 1994).

In general, the issue of subclassification of handedness is more of a problem for left handers than right handers because for most classifications, there are only a relatively small number of inconsistent left handers. We note, however, that some investigators use a classification that assigns "inconsistent" status to most right handers. For instance, in the adaptation of Annett's handedness questionnaire, Witelson (1985) described all right handers who did not have a complete repertoire of right hand choices as mixed right handers. If Triggs et al. had used this rather extreme procedure, their results might have looked different. Peters (1992) has shown that, depending on the classification used to distinguish consistent from inconsistent handers, practically any kind of distribution can be obtained, from no inconsistent right handers (if only writing hand is used to classify) to a vast majority of right handers being classified as inconsistent in their preference choices (if large handedness questionnaires are used).

The actual cause of the lower TMS thresholds for preferred hand MEPs is not easy to isolate, because TMS constitutes a relatively coarse means of stimulating cerebral neurons. As Triggs et al. note, it could be that the topography of the primary motor cortices differs sufficiently to yield a different impact of TMS on the pyramidal motor neurons. It could also be that the networks feeding into the pyramidal neurons that provide the final outflow differ, or that differences in the overall areas of primary motor cortices is a factor (as suggested by anatomical findings and the MEG and fMRI studies). In summarizing the above findings, it can be said that there is evidence for non-trivial asymmetries with regard to hand preference, such that the cortex associated with the preferred hand is either larger in extent than the cortex that is associated with the non-dominant hand, shows greater activation in imaging studies or lower TMS thresholds in stimulation studies. With regard to handedness, the weight of the evidence shows only a "trivial" asymmetry, in the sense that there tends to be a reversal of patterns seen in right handers. However, this particular asymmetry remains somewhat of an open issue, because some of the expected differences may not arise at the level of the primary motor cortex. For both groups of left handers, the primary motor cortices are the final outflow to the preferred hands, and to the extent that motor competence of the preferred hand is similar for both groups, both in precision and speed of movements, one would not expect differences here. Differences between handedness groups might be found a) in terms of the cortex associated with the nondominant hand and b) in terms of higher order motor areas. The reason for a) is simple. The vast majority of work on motor performance of normal individuals shows that although the motor performance of the preferred hands of the handedness groups is matched, left handers as a group tend to perform better with the nonpreferred hand than right handers. Combining this aspect with the demonstrations that life motor experience can change the extent of primary motor cortices (i.e., Amunts et al., 1997b; Elbert et al., 1995), some differences between right- and left handers might be expected at this level. Indeed, if we assume - as we must - that all behaviour is reducible to neuronal activity, there must be some structural differences that separate right- and left handers at the level of the primary motor cortex. However, it is also true that these differences may be too subtle to be picked up with current methods.

The second level at which differences might be found, b), is in many ways more interesting than the first. We have outlined why differences at a) might be expected to be small. However, in terms of the supplementary motor area (SMA) and the premotor area (PMA) it is not clear exactly what to expect. In the case of right handers, there is some agreement that the roles of the cortical areas that feed into the primary motor cortex are not necessarily symmetrical on the left and right side. Some support for this derives from the paper by Jäncke et al. (1998a), who asked right handers to perform simple unimanual and simple bimanual sequences. Their participants showed significantly higher levels of activity in the SMA of the left hemisphere during concurrent bimanual sequencing movements as compared to the SMA of the right hemisphere during comparable bimanual activity. This suggests a non-trivial asymmetry in the sense suggested earlier, such that the left hemisphere and the right hemisphere do not show equivalent activation during comparable activities. A related finding of asymmetrical activation for bimanual activities was reported by Baraldi et al. (2000) who reported greater left motor-cortex activation during bimanual activity, and specifically a cluster of activity in the left superior parietal lobe. Finally, an asymmetry of activation was also reported by Stephan et al. (1999) who showed that repeated finger-to-thumb opposition movements in righthanders led to midline activity that had a leftward bias when the task was performed with the right hand, but there was activity to the left and right of the midline when the left hand performed the movements. It is quite likely that the observed asymmetries are a function of the specific movements used: Ehrsson et al. (2000) report different results for power-grip and precision-grip tasks. In their study, one-handed power grip was associated with contralateral activation whereas one-handed precision grip led to bilateral activation of the primary motor areas but also of the premotor and parietal areas.

Although one should perhaps avoid an overly sharp delineation between left and right in terms of apraxia, clinical evidence suggests a greater role in apraxia for such "support" areas on the left side. Some evidence in left handers (Kimura, 1993) suggests that there may be a relatively more important role for the right cortex "support" areas. However, other evidence suggests that left handers share some motor control mechanisms with right handers on the left side of the brain. The clearest evidence for this derives from behavioural work. In a series of studies in our laboratory (in preparation) that examined reaction time in the Simon Effect (this effect concerns compatibility: the right hand reacts more quickly to a visual stimulus presented to the right visual field than to one presented to the left visual field and vice versa), both left- and right handers were found to have faster reaction times with the right hand, to stimuli presented in the right visual field (and therefore to the left hemisphere) than with the left hand for the compatible left-hand conditions (in which the stimulus is presented to the right hemisphere). There is no simple explanation for this effect other than that left- and right handers share the same cerebral asymmetry that leads to faster reactions by the left hemisphere/right-hand combination than in the right hemisphere/left-hand combination.

8. CONCLUSIONS

The new imaging techniques, as well as the older, more traditional methods of visualizing structure and function in the brain promise substantial advances in our understanding of the neural underpinnings of handedness. However, a cautious evaluation suggests that the work has only just begun. With few exceptions, the existing research can only be considered exploratory in nature, with relatively small subject numbers, limited variation in the way in which handedness information is related to the imaging data, and a limited repertoire of motor tasks, with insufficient variation in the tasks. In addition, the analysis of activation patterns has been approached with a relatively limited range of statistical tools, and much change can be expected here as the sophistication of techniques advances. That the particular analysis method can reveal, or fail to reveal important patterns is quite evident. A recent example would be the study by Craik et al. (Craik, Moroz, Moscovitch, Stuss, Winocur, Tulving, & Kapur, 1999), in which a partial least-squares analysis (McIntosh, Bookstein, Haxby, & Grady, 1996) allowed the discovery of subtle patterns that did not emerge with statistical parametric mapping (SPM) procedures.

For now, it can be stated that functional and anatomical asymmetries exist between the left and right primary motor cortices of right handers. The work of Amunts and her colleagues also introduces an important developmental dimension into the discussion: the asymmetries emerge early in life, but there is an indication that the anatomical pattern of asymmetry is subject to change during development. The status for left handers is not as clear, and here future work will have to focus on subclassification - an area important for right handers as well, but crucial in the understanding of hand/brain relations in left handers. In terms of specific areas that need to be addressed, some thought has to be given to the meaning of different levels of activation. What does it mean when the primary motor cortex of the left hemisphere yields higher levels of activation than that of the right hemisphere, and how does this interact with different motor tasks? Activation in response to motor activities is not invariant because it is sensitive to practice (Toni, Krams, Turner, & Passingham, 1998) and thus the number of trials and the amount of experience with a task are factors that must play a role in activation patterns. From a conceptual point of view *lower* levels of activation in the left primary motor cortex as compared to the right primary motor cortex in right handers would have not come as a surprise (had they been observed) because it not unreasonable to think that the "more experienced" cortex needs fewer resources to perform a task than the "less experienced" cortex. Similarly, if a motor task elicits weak but widely distributed and scattered activation, this might lead to a failure to appreciate the fact that there is a response - as compared to local foci of high levels of activation that may be somewhat more trivial in terms of understanding motor control. Other factors that are likely of importance are the "effort" expended in moving and whether the movement is triggered by external sources or generated internally (Boecker, Kleinschmidt, Requardt, Hanicke, Merboldt, & Frahm, 1994). There is also considerable work to be done in finding out why some methods, such as PET, appear more prone to show activation in SMA than others, and under what circumstances SMA activity is and is not observed. A related question concerns the circumstances under which ipsilateral activity in the primary motor cortices can be observed, and the distinction between areas that show contralateral and unilateral activation. Finally, the techniques discussed will also allow a finer analysis of temporal sequences of events that culminate in the final outflow of motor commands from the primary motor cortex (Friston, Williams, Howard, Frackowiak, & Turner, 1998; Wildgruber, Erb, Klose, & Grodd, 1997). In terms of the understanding of differences between left- and right handers this last approach is of particular interest because it might reveal asymmetries in the events leading up to the final activation of primary motor cortex response.

9. REFERENCES

Amunts, K., Jäncke, L., Mohlberg, H., Steinmetz, H., & Zilles, K. (2000). Interhemispheric asymmetry of the human motor cortex related to handedness and gender. *Neuropsychologia, 38*, 304-312.

Amunts, K., Istomin, V., Schleicher, A., & Zilles, K. (1995). Postnatal development of the human primary motor cortex: A quantitative cytoarchitecture analysis. *Anatomy and Embryology, 192*, 557-571.

Amunts, K; Schlaug, G; Jäncke, L., Steinmetz, H., Schleicher, A., Dabringhaus, A., & Zilles, K ; (1997b). Motor cortex and hand motor skills: Structural compliance in the human brain. *Human Brain Mapping, 5*, 206-215.

Amunts, K., Schlaug, G., Schleicher, A., Steinmetz, H., Dabringhaus, A., Roland, P.E., & Zilles, K.C. (1996). Asymmetry in the human motor cortex and handedness. *Neuroimage, 4*, 216-22

Amunts, K., Schmidt, P. F., Schleicher, A., & Zilles, K. (1997a). Postnatal development of interhemispheric asymmetry in the cytoarchitecture of human area 4. *Anatomy and Embryology, 196*, 393-402.

Annett, M. (1995). The right shift theory of a genetic balanced polymorphism for cerebral dominance and cognitive processing. *European Journal of Cognitive Psychology, 14,* 427-480.

Baraldi, P., Porro, C.A., Serafini, M., Pagnoni, G., Murari, C., Corazza, R., & Nichelli, P. (2000). Bilateral representation of sequential finger movements in human cortical areas. *Neuroscience Letters, 269,* 95-98.

Boecker, H., Kleinschmidt, A., Requardt, M., Hanicke, W., Merboldt, K. D., & Frahm, J. (1994). Functional cooperativity of human cortical motor areas during self-paced simple finger movements. A high-resolution MRI study. *Brain, 117,* 1231-1239.

Brito, G.N., Brito, L.S., Paumgartten, F.J., & Lins, M.F. (1989). Lateral preferences in Brazilian adults: An analysis with the Edinburgh Inventory. *Cortex, 25,* 403-405.

Bulman-Fleming, M. B., Grimshaw, G. M., & Berenbaum, S. A. (2000). Achieving convergent evidence through divergent approaches. *Brain and Cognition,, 42,* 85-88.

Buxhoeveden, D., & Casanova, M. (in press). Comparative lateralization patterns in the language area of human, chimpanzee, and rhesus monkey brains. *Laterality.*

Collins, R. L., (1977). Origins of the sense of asymmetry: Mendelian and non-Mendelian models of inheritance. *New York Academy of Sciences, 299,* 283-305.

Corballis, M. C. (1997). The genetics and evolution of handedness. *Psychological Review, 104,* 714-727.

Corballis, M. C., Lee, K., McManus, I. C., & Crow, T. J. (1996). Location of the handedness gene on the X and Y chromosomes. *American Journal of Medical Genetics, 67,* 50-52.

Craik, F.I.M., Moroz, T.M., Moscovitch, M., Stuss, D.T., Winocur, G., Tulving, E., & Kapur S. (1999). In search of the self: A positron emission tomography study. *Psychological Science, 10,* 26-34.

Dassonville, P., Zhu, Z-H., Ugurbil, K., Kim, S-G., & Ashe, J. (1997). Functional activation in motor cortex reflects the direction and the degree of handedness. *Proceedings of the National Academy of Sciences (USA), 94,* 14015-14018.

Dellatolas, G., Tubert. P, Castresana, A., Mesbah, M., Giallonardo, T., Lazaratou, H., & Lellouch, J. (1991). Age and cohort effects in adult handedness. *Neuropsychologia, 29,* 255-261.

Ehrsson, H.H., Fagergren, E., Jonsson, T., Westling, G., Johansson, R.S., & Forssberg, H. (2000). Cortical activity in precision versus powergrip tasks: An fMRI study. *Journal of Neurophysiology, 83,* 528-536.

Elbert, T., Pantev., C., Wienbruch, C., Rockstroh, B., & Taub, E. (1995). Increased cortical representation of the fingers of the left hand in string players. *Science, 270*, 305-307.

Foundas, A. L., Hong, K., Leonard, C. M., & Heilman, K. M. (1998). Hand preference and magnetic asymmetries of the central sulcus. *Neuropsychiatry, Neuropyschology and Behavioral Neurology, 11*, 65-71.

Friston, K. J., Williams, S., Howard, R., Frackowiak, R. S., & Turner, R. (1996). Movement-related effects in fMRI time-series. *Magnetic Resonance Medicine, 35*, 346-355.

Friston, K. J., Fletcher, P., Josephs, O., Holmes, A., Rugg, M. D., & Turner, R. (1998). Event-related fMRI: characterizing differential responses. *Neuroimage, 7*, 30-40.

Frost, J.A., Binder, J.R., Springer, J.A., Hammeke, T.A., Bellgowan, P.S.F., Rao, S.M., & Cox, R.W. (1999). Language processing is strongly left lateralized in both sexes: Evidence from functional MRI. *Brain, 122*, 199-208.

Gilbert, A.N., & Wysocki, C.J. (1992). Hand preference and age in the United States. *Neuropsychologia, 30*, 601-608.

Habib, M., Gayraud, D., Oliva, A., Regis, J., Salamon, G., & Khalil, R. (1991). Effects of handedness and sex on the morphology of the corpus callosum: A study with brain magnetic resonance imaging. *Brain and Cognition, 16*, 41-61.

Hatta, T., & Kawakami, A. (1995). Patterns of handedness in modern Japanese: A cohort effect shown by re-administration of the H.N. Handedness Inventory after 20 years. *Canadian Journal of Experimental Psychology, 49*, 505-512.

Hiscock, C. K., Hiscock, M., Benjamins, D., & Hillman, S. (1989). Motor asymmetries in hemiplegic children: Iimplications for the normal and pathological development of handedness. *Developmental Neuropsychology, 5*, 169-186.

Hoosain, R. (1990). Left handedness and handedness switch amongst the Chinese. *Cortex, 26*, 451-454.

Irving, D., Rebeiz, J.J., & Tomilson, B.E. (1974). The numbers of limb motor neurons in the individual segments of the human lumbo-sacral cord. *Journal of Neurological Sciences, 21*, 203-212.

Jäncke, L. (1996). The hand performance test with a modified time limit instruction enables the examination of hand performance asymmetries in adults. *Perceptual and Motor Skills, 82*, 735-738.

Jäncke, L , Peters, M., Himmelbach, M., Nüsselt, T., Posse, S., & Steinmetz, H. (2000). fMRI study of bimanual coordination. *Neuropsychologia, 38*, 164-174.

Jäncke, L., Peters, M., Schlaug, G., Posse, S., Steinmetz, H., & Müller-Gartner, H. W. (1998). Differential magnetic resonance signal change in human sensorimotor cortex to finger movements of different rate of the dominant and subdominant hand, *Cognitive Brain Research, 6*, 279-284.

Jäncke, L., Specht, K., Mirzazade, S., & Peters, M. (1999). The effect of finger-movement speed of the dominant and the subdominant hand on cerebellar activation: A functional magnetic resonance imaging study. *Neuroimage, 9*, 497-507.

Jäncke, L., Specht, K, Mirzazade, S, Loose, R, & Shah, N. J. (1998). A parametric analysis of the 'rate effect' in the sensorimotor cortex: A functional magnetic imaging analysis, *Neuroscience Letters, 252*, 37-40.

Joliot, M., Crivello, F., Badier, J. M., Diallo, B., Tzourio, N., & Mazoyer, B. (1998). Anatomical congruence of metabolic and electromagnetic activation signals during a self-paced motor task: A combined PET-MEG study. *Neuroimage, 7*, 337-351.

Karni, A., Meyer, G., Jezzard, P., Adams, M. M., Turner, R., & Ungerleider, L. G. (1995). Functional MRI evidence for adult motor cortex plasticity during motor skill learning. *Nature, 377*, 155-158.

Kawashima, R., Yamada, K., Kinomura, S., Yamaguchi, T., Matsui, H., Yoshioka, S., & Fukuda, H. (1993). Regional cerebral blood flow changes of cortical motor areas and prefrontal areas in humans related to ipsilateral and contralateral hand movements. *Brain Research, 623*, 33-40.

Kertesz, A., & Geschwind, N. (1971) Patterns of pyramidal decussation and their relationship to handedness. *Archives of Neurology, 24*, 326-332.

Kertesz, A., Polk, M., Black, S. E., & Howell, J. (1990). Sex, handedness, and the morphometry of cerebral asymmetries on magnetic resonance imaging. *Brain Research, 530*, 40-48.

Kim, S.-G., Ashe, J., Georgopoulos, A. P., Merkle, H., Ellerman, J., Menon, R., Ogawa, S., & Ugurbil, K. (1993). Functional imaging of human motor cortex at high magnetic yield. *Journal of Neurophysiology, 69*, 297-302.

Kimura, D. (1993). *Neuromotor mechanisms in human communication.* Oxford: Oxford University Press.

Laland, K. L., Kumm, J., Van Horn, J. D., & Feldman, M. W. (1995). A gene-culture model of human handedness. *Behavior Genetics, 25*, 433-445.

Llorente, A.M., Satz, P., Brumm, V.L., & Philpott, L.M. (1998). Pathological left handedness : A case report examining the developmental course of the syndrome following head trauma. *Child Neuropsychology, 4*, 98-109.

MacDonell, R.A.L., Shapiro, B.E., Chiappa, K.H., Helmers, S.L., Cros, D., Day, B.J., & Shahani, B.T. (1991). Hemispheric threshold differences for motor evoked potentials produced by magnetic cortical stimulation. *Neurology, 41*, 1441-1444.

Mandal, M.K., Pandey, G., Singh, S.K, & Asthana, H.S. (1992). Degree of asymmetry in lateral preferences: Eye, foot, ear. *Journal of Psychology, 26*, 155-162.

McIntosh, A.R., Bookstein, F.L., Haxby, J.V., & Grady, C.L. (1996). Spatial pattern analysis of functional brain images using partial least squares. *Neuroimage, 3,* 143-157.

McManus, I. C. (1991). The inheritance of left handedness. In G. R. Bock & J. Marsh (Eds.), *Biological asymmetry and handedness, CIBA Symposium 162* (pp. 251-267). London: Wiley.

Nathan, P. W., Smith, M. C., & Deacon, P. (1990). The spinocortical tracts in man. *Brain, 113,* 303-324.

Pascual-Leone, L. A., Wassermann, E. M., Sadato, N., & Hallett, M. (1995). The role of reading activity on the modulation of motor cortical outputs of the reading hand in Braille readers. *Annals of Neurology, 38,* 910-915.

Perelle, I.B., & Ehrman, L. (1994). An international study of human handedness: The data. *Behavior Genetics, 24*, 217-227.

Peters, M. (1983). Differentiation and lateral specialization in motor development. In G. Young, C. Corter, S. J. Segalowitz & S. Trehub (Eds.), *Manual specialization and the developing brain: Longitudinal studies* (pp. 141-159). New York : Academic Press.

Peters (1992). How sensitive are handedness prevalence figures to differences in handedness classification procedures? *Brain and Cognition, 18,* 208-215.

Peters, M. (1986). Hand roles and handedness in music. *Psychomusicology, 6*, 29-34.

Peters, M. (1989). The relationship between variability of intertap intervals and interval duration. *Psychological Research, 51*, 38-42.

Porro, C. A., Francescato, M., P., Cettolo, V., Diamnon, M. E., Baraldi, P., Zuiani, C., Bazzochi, M, & Prampero, P. E. (1996). Primary motor and sensory cortex activation during motor performance and motor imagery: A functional magnetic resonance imaging study. *The Journal of Neuroscience, 16*, 7688-7698.

Rao, S. M., Bandettini, P. A., Binder, J. R., Bobholz, J. A., Hammeke, T. A., Stein, E. A., & Hyde, J. S. (1996). Relationship between finger movement rate and functional magnetic resonance signal change in human primary motor cortex. *Journal of Cerebral Blood Flow and Metabolism, 16,* 1250-1254.

Rao, S. M., Binder, J. R., Bandettini, P. A., Hammeke, T. A. Yetkin, F. Z., Stein, E.A., & Hyde, J. S., (1993). Functional magnetic resonance imaging of complex human movements. *Neurology, 43,* 2311-2318.

Ro, T., Cheifet, S., Ingle, H., Shoup, R., & Rafal, R. (1999). Localization of the human frontal eye fields and motor hand area with transcranial magnetic stimulation and magnetic resonance imaging. *Neuropsychologia, 37,* 225-231.

Roland, P.E., Larsen, B., Lassen, N.A., & Skinhoj, E. (1980). Supplementary motor area and other cortical areas in organization of voluntary movements in man. *Journal of Neurophysiology, 43,* 118-136.

Sanes, J. N., Donoghue, J. P., Thangeraj, V., Edelman, R. R., & Warach, S. (1995). Shared neural substrates controlling hand movements in human motor cortex. *Science, 268,* 1775-1777.

Schieber, M. H., & Hibbard, L. S. (1993). How somatopic is the motor cortex hand area? *Science, 261,* 489-492.

Seitz, R., Canavan, A. G. M., Yágüez, L., Herzog, H., Tellmann, L., Knorr, U., Huang, Y., & Hömberg, V. (1997). Representations of graphomotor trajectories in the human parietal cortex: Evidence for controlled processing and automatic performance. *European Journal of Neuroscience, 9,* 378-389.

Seldon, H. L. (1981). Structure of the human auditory cortex, I. Cytoarchitectonics and dendritic distributions. *Brain Research, 229,* 277-294.

Snyder, P. J., Bilder, R. M., Wu, H., Bogerts, B., & Lieberman, J. A. (1995). Cerebellar volume asymmetries are related to handedness: A quantitative MRI study. *Neuropsychologia, 33,* 407-419.

Steinmetz, H., Jäncke, L., Kleinschmidt, A., Schlaug, G., Volkmann, J., & Huang, Y. (1992). Sex but no hand difference in the isthmus of the corpus callosum. *Neurology, 42,* 749-752.

Stephan, K.M., Binkofski, F., Posse, S., Seitz, R.J., & Freund, H.J. (1999). Cerebral midline structures in bimanual coordination. *Experimental Brain Research, 128,* 243-249.

Tan, Ü. (1985a). Left-right differences in the Hoffmann reflex recovery curve associated with handedness in normal subjects. *International Journal of Psychophysiology, 3,* 75-78.

Tan, Ü. (1985b). Relationships between hand skill and the excitability of motorneurons innervating the postural soleus muscle in human subjects. *International Journal of Neuroscience, 26,* 289-300.

Tan, Ü. (1985c). Velocities of motor and sensory nerve conduction are the same for right and left arms in right- and left handed normal subjects. *Perceptual and Motor Skills, 2,* 625-626.

Toni, I., Krams, M., Turner, R., & Passingham, R. E. (1998). The time course of changes during motor sequence learning: A whole-brain fMRI study. *Neuroimage, 8*, 60-61.

Triggs, W. J., Calviano, R., & Levine, M. (1997). Transcranial magnetic stimulation reveals a hemispheric asymmetry correlate of intermanual differences in motor performance. *Neuropsychologia, 35*, 1355-1363.

Triggs, W. J., Calviano, R., Macdonell, R. A. L., Cros, D., & Chiappa, K. H. (1994). Physiological motor asymmetry in human handedness: Evidence from transcranial magnetic stimulation. *Brain Research, 636*, 270-276.

Tzourio, N., Crivello, Mellet, E., Nkanaga-Ngila, B., & Mazoyer, B. (1997). Functional anatomy of dominance for speech comprehension in left handers vs. right handers. *Neuroimage, 8*, 1-6.

Volkmann, J., Schnitzler, A., Witte, O. W., & Freund, H.-J. (1998). Handedness and asymmetry of hand representation in human motor cortex. *Journal of Neurophysiology, 79*, 2149-2154.

Walsh, V., & Rushworth, M. (1999). A primer of magnetic stimulation as a tool for neuropsychology. *Neuropsychologia, 37*, 125-135.

Weiss, T; Miltner, W.H.R., Dillmann, J., Meissner, W; Huonker, R., & Nowak, H. (1998). Reorganization of the somatosensory cortex after amputation of the index finger. *Neuroreport, 9*, 213-216.

White, L. E., Andrews, T. J., Hulette, C., Richards, A., Groelle, M., Paydarfar, J., & Purves, D. (1997). Structure of the human sensorimotor system. II: Lateral symmetry. *Cerebral Cortex, 7*, 31-47.

White, L. E., Lucas, G., Richards, A., & Purves, D. (1994). Cerebral asymmetry and handedness. *Nature, 368*, 197-198.

Wildgruber, D., Erb, M., Klose, U., & Grodd, W. (1997). Sequential activation of supplementary motor area and primary motor cortex during self-paced finger movement in human evaluated by functional MRI. *Neuroscience Letters, 227*, 161-164.

Witelson. S. F. (1985). The brain connection: The corpus callosum is larger in left handers. *Science, 229*, 665-668.

Wood, W. B., & Kershaw, D. (1991). Handedness asymmetry, handedness reversal and mechanisms of cell fate determination in nematode embryos. In G. R. Bock & J. Marsh (Eds.), *Biological asymmetry and handedness, CIBA Symposium 162* (pp. 143-159). London: Wiley.

Yakovlev, P. I., & Rakic, P. (1966). Patterns of decussation of bulbar pyramids and distribution of pyramidal tracts on two sides of the spinal cord. *Transactions of the American Neurological Association, 91*, 366-367.

Yousry T. A., Schmid U. D., Alkadhi H., Schmidt D., Peraud A., Buettner A., & Winkler P. (1997). Localization of the motor hand area to a knob on the precentral gyrus: A new landmark. *Brain, 120*, 141-57.

Yousry, T., Schmid, U. D., Jassoy, A. G., Schmidt, D., Eisner, W. E., Reulen, H-J., Resier, M. F., & Lissner, J. (1995). Topography of the cortical hand area: Prospective study with functional MR imaging and direct motor mapping at surgery. *Radiology, 195,* 23-29.

III

SIDE BIAS: FOOT, CRADLE, FACE AND ATTENTION

Chapter 9

Lateral Preference, Skilled Behaviour and Task Complexity: Hand and Foot

Pamela J. Bryden
Wilfrid Laurier University, Canada

The performance of most tasks with one hand, typically the right, is an intriguing human characteristic. Not only do people prefer to use one hand rather than the other, but also they usually perform tasks faster and more accurately with this hand. Interestingly, individuals also have a preferred foot, which generally outperforms the other foot on performance tasks (Peters, 1988). One of the notable aspects of lateral preference is that certain tasks tend to elicit stronger preferences than other tasks. For example, individuals have strong preferences for the hand they use to write, but show much weaker preferences, if any, for the hand they use to turn on a light switch. Likewise, the performance abilities of the preferred hand are superior for certain tasks as compared to the non-preferred hand. Similar findings have been reported for foot preferences. Individuals tend to show strong, consistent preferences for kicking, but are less consistent in their preference for which foot is used to step up onto a chair. As well, the magnitude of the differences in performance abilities between the feet appears to depend on the task being examined. Why certain tasks elicit stronger preferences and greater performance differences between the two sides is not well understood. One of the reasons often cited is the degree of skill: the more complex the task the stronger the preference and the greater

M.K. Mandal, M.B. Bulman-Fleming and G. Tiwari (eds.), Side Bias: A Neuropsychological Perspective, 225-248.

the preferred-hand advantage. But, does skill really underlie lateral preference and performance?

The aim of this chapter is to discuss the role of skill and task complexity in lateral preference and performance of both the hands and feet. First, the chapter will provide a brief review of the literature on lateral preference as well as lateral performance, focusing on the measurement and expression of handedness and footedness. Next, the chapter will discuss motor skill and how skill and complexity may underlie lateral preference, specifically handedness. Finally, the role of skill in manual performance will be addressed focusing on a series of studies that examined manual asymmetries and task complexity.

1. LATERAL PREFERENCE: HANDEDNESS AND FOOTEDNESS

1.1 The measurement and expression of hand preference

Handedness is perhaps the most studied human asymmetry (M. P. Bryden, 1982). Approximately 90% of the population shows a preference for the right hand (Annett, 1985), with the other ten percent preferring the left hand. The prevalence of right handedness appears to be somewhat dependent upon the culture or country examined (M. P. Bryden, A. Ardila, & O. Ardila, 1993; McManus & M. P. Bryden, 1993). As well, it appears that the prevalence of left handedness may be changing over time (see also Chapter 4, this volume). Turn-of-the-century estimates of left handedness are in the range of 4% to 7% (Rife, 1940), whereas data collected more recently suggest that the prevalence is closer to 12% (McManus, 1997). Handedness surveys have also found that the prevalence of left handedness appears to be greater in males than in females (M. P. Bryden, 1982).

The measurement of hand preference has long been a source of disagreement among researchers (see also Chapter 6, this volume). In the past, investigators often used crude behavioural measures, typically equating hand preference solely with the hand used to write. Luria (1965) even suggested that a person is a latent left hander if the left arm is uppermost when the arms are folded in front of the body. Now, researchers generally classify handedness using more quantitative measures such as questionnaires or surveys (M. P. Bryden, 1977; Oldfield, 1971). Hand preference questionnaires ask participants which hand they prefer to use to perform a number of tasks, such as writing, holding a hammer when nailing, holding a toothbrush, and turning on a light switch. Early forms of hand preference questionnaires asked participants only to identify the hand they prefer to use (i.e., response alternatives are either "right" or "left"). Other researchers (Annett, 1976; M. P. Bryden, 1977; Oldfield, 1971) have recognized that there is not only a direction to handedness (i.e., right or left handed) but there are also different degrees of handedness (i.e., strongly right handed versus ambidextrous). To assess the degree of handedness, experimenters ask participants to specify whether they "always" or "usually" use the right or left hand to perform the task in question.

One of the problems with preference measures, as in any self-report measure, is their inherent subjectivity. Not only is the meaning of the questions posed open to the reader's interpretation, but also, the individual must imagine or recall what they would do in a given circumstance in order to complete the questionnaire accurately. There is, nonetheless, a high degree of correspondence between questionnaire responses and observed hand preference in performance. M. Reib, G. Reib, and Freye (1998) recently reported an agreement of 95.4% between self-reported handedness and observed handedness. Hand preference questionnaires have also shown high reliability, as people appear to be consistent in their hand preference over time. Provins, Milner, and Kerr (1982) showed a significant correlation in responses to a 75-item preference questionnaire completed twice in six months. Dodrill and Thoreson (1993) examined the reliability of the Lateral Dominance Examination (Reitan & Davidson, 1974) over a five-year period. Items evaluating handedness were found to be highly consistent, with the most reliable item relating to the hand used for writing (100% consistency), suggesting that the Lateral Dominance Examination renders reliable results over a long period. Ransil and Schachter (1994) also demonstrated acceptable test-retest reliability using the Edinburgh Handedness Inventory (Oldfield, 1971). It should be noted that some items on hand preference questionnaires are less reliable than other items (Raczkowski, Kalat, & Nebes, 1974; Ransil & Schachter, 1994). Raczkowski et al. (1974) suggest that items concerning relatively infrequently performed activities are particularly unreliable. Equally, bimanual tasks (opening a jar, using a broom) are often less reliable in test-retest paradigms (Ransil & Schachter, 1994).

Hand preference and its measurement have clearly been the focus of a plethora of research, nonetheless there is controversy concerning the dimension(s) underlying hand preference. More specifically, it is not well understood why individuals show strong preference for some tasks and weaker preferences for others.

1.2 The measurement and expression of foot preference

Although the majority of research conducted on lateral preference has focused on handedness, research examining foot preference has recently come to be popular (see Chapter 10, this volume). The reason for this renewed interest in foot preference perhaps originates in the notion that foot preference may be a less biased measure of lateral preference than hand preference. Specifically, the preferred hand has had years of practice performing complex unimanual tasks such as writing. This is not true for the feet. The effects of experience and practice on lateral preference are especially important considering that cultural influences have been cited as partially responsible for the population shift towards right handedness (Annett, 1972).

First, it is important to consider what is meant by foot preference. Peters (1988) highlights role differentiation of the feet in his definition of foot preference. He suggests that "the foot used to manipulate an object or to lead out, as in jumping" is the preferred foot, whereas the foot that is used to "support the activities of the preferred foot by lending postural support and stabilizing support" is the non-preferred foot (p. 181). Thus, to give an example: in kicking, the preferred foot is the one used to kick the ball, rather than the one used to provide balance. MacNeilage (1993) also emphasizes the complementary roles of the two feet in his evolutionary theory. He argues, based on primate research, that the right side of the body has evolved for object manipulation mainly because the left side is needed for postural support. Previc (1991) reaches a similar conclusion about the importance of postural stabilization and support in the left limb from a neurodevelopmental perspective, arguing that cerebral lateralization can be traced to the asymmetric prenatal development of the ear and labyrinth. Briefly, Previc argues that lateralized shearing forces *in utero* create a left-otolith dominance, which gives rise to the general population trend toward a left-side preference for postural control and a right-side preference for voluntary movement. Researchers (e.g., Hart & Gabbard, 1998; Peters, 1988) have also defined footedness in terms of both bilateral and unilateral contexts. For example, standing on one foot while tapping is considered a bilateral task (Hart & Gabbard, 1996), whereas simply standing on one foot is considered a unilateral task (Hart & Gabbard, 1998).

Foot preference is generally measured in the same manner as hand preference, by using self-report questionnaires. Several foot preference questionnaires have been developed. Harris (1958) included two foot-preference items on the Harris Test of Lateral Dominance: kicking and picking up a pebble. Coren's (1993a) sixteen-item Lateral Preference Inventory included four items to determine foot preference: two manipulative and two postural questions. The Lateral Preference Inventory does not allow degree of footedness to be examined, because responses are recorded as "right", "left", or "either". More recent foot preference questionnaires (e.g., Elias & M. P. Bryden, 1998) have incorporated measures of strength or degree of foot preference, as well as increasing the number of items on the questionnaire. Some investigators have also developed behavioural measures of foot preference. J. Chapman, L. Chapman, and Allen (1987) developed a thirteen-item behavioural measure to determine foot preference. Participants were asked to perform various tasks, including kicking a ball, smoothing sand, and picking up marbles with their toes. Overall, the reliability of foot preference measures appears to be comparable with hand-preference reliability, although relatively little research has focused solely on the issue of foot preference reliability. Chapman et al. (1987), for example, report high test-retest reliability for their behavioural foot preference measure. Reliability of foot preference measures was examined in the study of the Lateral Dominance Examination (Reitan & Davidson, 1974) conducted by Dodrill and Thoreson (1993). It was found that 98% of participants showed exact agreement over five years on foot preference for the task of kicking a football, whereas only 81% showed exact agreement for the task of stomping on a bug. Again, tasks that are frequently performed have relatively higher test-retest reliability than novel, unfamiliar tasks.

Overall, the majority of people appear to be right-footed. Coren (1993a) reports that approximately 88% of females and 83% of males are right-footed as measured by his Lateral Preference Inventory. By taking into account the two roles of the feet (mobilization versus postural stability), it has been shown that most right-handed individuals are right-footed for mobilization action (kicking, smoothing sand) and left-footed for postural stability (Gabbard & Iteya, 1996), whereas left handers show essentially the opposite pattern, albeit less consistently. Augustyn and Peters (1986), using the two items from the Harris Test of Lateral Dominance (Harris, 1958), found that of right handers, 72% preferred the right foot exclusively, 1.5% preferred the left foot exclusively, and 26.5% had no preference. For left handers, 54.8% preferred the left foot exclusively, 18.7% preferred the right foot, and 26.5% used either foot. Augustyn and Peters (1986) concluded that right handers are more consistently right-footed than left handers are left-footed. Chapman et al. (1987) also reported this pattern of reduced consistency in left handers.

One interesting finding in the footedness literature is that right-footed individuals who use the left foot for stability in the bilateral context, "switch" feet when asked to balance on one leg. Hart and Gabbard (1996) suggest that the preferred limb may be used preferentially in the unilateral context for difficult tasks. When the mobilizing or manipulative aspect of footedness is examined in a unilateral context (i.e., sitting in a chair while foot tapping), individuals choose to use the same foot used for manipulative tasks in the bilateral context (Hart & Gabbard, 1998). In other words, right handers prefer the right foot to perform manipulative tasks in both unilateral and bilateral contexts, and choose the right foot for unipedal stabilizing. Why individuals "switch" their preference for stabilizing is unknown, although it has been hypothesized that task complexity may play a role (Hart & Gabbard, 1998).

Defining a preference for one foot over another is obviously a more difficult task than defining hand preference. Clearly, preference in both unilateral and the bilateral contexts should be taken into consideration. The factors underlying foot preference may be related to the nature of the tasks (i.e., manipulative versus stabilizing) and their complexity, however, little research has focused on this aspect of footedness.

2. LATERAL PERFORMANCE: HANDS AND FEET

2.1 Measurement and expression of manual performance asymmetries

Performance measures are not susceptible to the inherent problems of hand preference inventories and thus present an objective alternative to such measures. Performance measures compare the relative performance of the two hands on a given task (Annett, 1985; Peters & Durding, 1978). Tasks that have been examined have compared the performance of the hands on their relative strength (e.g., Provins & Magliaro, 1993) and on their relative speed (e.g., Flowers, 1975). Because relative strength is dependent upon factors other than handedness (e.g., age, experience, and practice), comparing the hands on a test of relative speed is currently a more common practice. Researchers have compared the performance of the hands on peg moving (Annett, 1967), finger tapping (Peters, 1980), dot tapping (Tapley & M. P. Bryden, 1985), and manual aiming (Roy & Elliott, 1986). The performance of the preferred hand is generally found to be faster and more accurate than that of the non-preferred hand, especially for speeded tasks that require highly practiced elements (Peters, 1996). The reason for the performance superiority of the preferred hand, however, is a matter of debate. Three explanations have been considered: the visual-feedback hypothesis (Flowers, 1975), the motor-output hypothesis (J. Annett, M. Annett, Hudson, & Turner, 1979; Elliott & Chua, 1996), and the preferential-experience theory (Provins, 1997).

A number of studies have reported evidence supporting a visual-feedback explanation for manual asymmetries. In an indirect examination of the role of visual feedback, Roy (1983) found a greater preferred-hand advantage for manual aiming when speed rather than accuracy was stressed. He suggested that the speeded condition required the more efficient use of visual feedback, thus supporting the visual-feedback hypothesis. Stronger evidence for the visual-feedback hypothesis comes from work conducted by Todor and Cisneros (1985) who attempted to identify which phase of a movement is responsible for hand differences in an aiming task. Kinematic analysis revealed that the longer movement times of the non-preferred hand were mainly a result of the increased time spent after peak velocity, when visual feedback is thought to be most important. Yet, other research has provided evidence that the visual-feedback explanation cannot account entirely for the performance differences between the hands. For example, Todor and Doane (1978) found that the preferred hand did not perform significantly better in a condition requiring greater visual feedback. As well, Roy and Elliott (1986) found that the presence or absence of vision in a manual-aiming task did not affect the difference in performance between the two hands. Further work conducted by Carson, Chua, Elliott, and Goodman (1990) has replicated these findings.

Carson et al. (1990) proposed a motor-output explanation of manual asymmetries, suggesting that the right hand system is less variable in generating the forces necessary for a particular movement. Nonetheless, more recent work by these investigators (Carson, Elliott, Goodman, Thyer, Chua, & Roy, 1993) has failed to find any evidence supporting this theory. To test the force-variability hypothesis, Carson et al. (1993) added weights of different amounts to the limbs and compared the performance of the right and left hands. It was reasoned that if the left hand were more variable in specifying muscular forces, then the disadvantage of this system should increase under added mass. Note that moving more mass would require greater force, and thus greater force variability, resulting in greater endpoint variability (Schmidt, Zelaznik, Hawkins, Frank, & Quinn, 1979). However, the performance of the hands was not affected differentially by the added mass. Thus, the differences between the hands cannot be accounted for by a purely movement-output explanation (Carson et al., 1993).

A third hypothesis is that preferential experience or practice can account for the superiority of the preferred hand. Peters (1976) found that by the end of an extensive training period both hands were equivalent in tapping speed, indicating the contributing role of practice in the preferred-hand advantage. In contrast, Annett, Hudson, and Turner (1974) found improvement with both hands in a peg-moving task, though the difference between the hands did not disappear. Similarly, P. J. Bryden and Allard (1998) reported a significant difference in the performance of the two hands in peg moving at the end of the training period. Studies such as these suggest that practice can indeed influence the performance of the hands, but that the underlying asymmetry, in most cases, is still observable.

2.2 Measurement and expression of pedal performance asymmetries

As with the performance abilities of the two hands, the preferred foot is also the more proficient foot for many tasks. Perhaps the most studied foot performance task is foot tapping (see Peters, 1988, for a thorough review). Gardner (1941) not only found a small right-foot advantage for foot tapping, but also reported that the performance of the left foot was quite irregular. Stronger right-foot advantages have been reported (Malmo & Andrews, 1945; Peters & Durding, 1979a), and overall, it appears that males produce faster tapping rates than do females. Differences in the size of the right-foot advantage in foot tapping tend to be related to trial duration as well as the tapping apparatus (Peters, 1988). In addition, there is evidence that right handers show a clear, strong right-foot advantage, whereas left handers show a smaller, yet still significant right-foot advantage (Augustyn & Peters, 1986; Peters & Durding, 1979a, b).

Little work has been conducted on measures of foot performance other than foot tapping, except for the work conducted by Gardner (1941) more than fifty years ago. Gardner examined a number of tasks such as placing marbles in a container and moving a marble up an incline into a hole. Some of the tasks were also performed without the use of vision including placing pegs into a pegboard and grasping clothespins. Gardner showed a small right-foot advantage for all of the tasks investigated, with undoing a lock with the toes showing the largest difference between the feet. Overall, it appears that the preferred foot is also the most skilled foot in terms of performance tasks, nevertheless, it is clear that more work comparing the performance capabilities of the two feet needs to be done.

3. INTERRELATION BETWEEN PREFERENCE AND PERFORMANCE FOR THE HANDS AND FEET

3.1 The relation between preference and performance

Generally, there appears to be a strong, but by no means perfect, relation between self-reported hand preference and performance. Although single, global questions, such as "are you right or left handed?" show a poor concordance with behavioural measures (Coren, 1993b), larger inventories of hand preference tend to show a better relation with performance measures. For instance, Annett (1976), and Peters and Durding (1978) have shown a strong correlation between preference and performance on peg moving and finger tapping, respectively. Other researchers, however, have not found a strong relation between performance and preference. For instance, Rigal (1992) examined preference and performance in children (six to nine years) and found a correlation of only 0.54 between hand preference and overall performance on a series of unimanual tasks. However, performance on a writing test correlated highly with preference. Rigal also reports that the two performance tasks requiring the greatest amount of training (writing and aiming) significantly predict hand preference. More recently, Peters (1998) performed a thorough examination of the relation between hand performance and hand preference in adults. Participants completed several preference questionnaires and performed a number of unimanual tasks, including several dexterity tests such as peg moving, and finger tapping. Peters found the correlation coefficients to be significant between performance and preference, particularly when longer questionnaire versions were utilized, showing further evidence for the relation between preference and performance. Thus, individuals with stronger hand preferences display greater differences in performance between their two hands than those individuals with inconsistent hand preference.

3.2 The relation between hands and feet

Examinations of the relation between hand preference and foot preference seem to suggest a strong relation. Chapman et al. (1987) reported a correlation of 0.70 between a 13-item hand preference questionnaire and their behavioural measure of foot preference. Brown and Taylor (1988) reported a 90% agreement between the hand used to write and the foot used to kick, in a large sample of males. Dargent-Paré, De Agostini, Mesbah, and Dellatolas (1992) corroborated the strong relation between hand and foot preference in males and females across different ages. Clymer and Silva (1985) also showed strong agreement between measures of hand and foot preference. They found that 94.1% of right handers were right-footed, whereas only 5.9% were left-footed. In comparison, in the sample of left handers, 68.8% were left-footed and 31.1% were right-footed. It appears, then, that the relation between the hands and feet for measures of preference is strong.

In contrast, the relation between the hands and feet on performance measures is less clear. Peters and Durding (1979a) found only a very weak correlation between and hand- and foot-tapping speed. Augustyn and Peters (1986) compared foot- and finger-tapping speeds, and showed that, within handedness groups, the overall speeds of performance of the hands and feet were significantly correlated. However, when the difference in preferred versus non-preferred performance was compared for hands and feet, no significant correlation was observed. Thus, a large performance difference between the hands does not necessitate a large performance difference between the feet. Clearly, there is not as strong a relation between the hands and feet for performance as for preference measures. One possibility for the lack of a strong relation between hand and foot performance may be related to the highly skilled unimanual tasks examined in comparison to the relatively unskilled tasks examined in foot performance.

4. SKILLED BEHAVIOUR AND ITS CLASSIFICATION

Previous researchers have suggested that skill might be an underlying factor in both preference and performance (e.g., Steenhuis, 1996), based on evidence that the preferred hand is not only more often chosen to perform complex tasks, but also outperforms the non-preferred hand at such tasks (Flowers, 1975). Before reviewing the literature on the influence of skill on preference and performance behaviours, one should perhaps define "skill". At a simplistic level, skill can be used to denote either an act or a task, or as an indicator of the quality of performance (Magill, 1985). Skills that are acts or tasks are relevant to the present discussion, and can be defined as those acts or tasks that "require movement and must be learned in order to be properly performed" (Magill, 1985, p. 5). For the current review, Schmidt's (1991) definition will be used because it provides a much narrower definition of skill. He suggests that skills "generally involve achieving a well-defined environmental goal by (1) maximizing the achievement certainty; (2) minimizing the physical and mental energy costs of performance; and (3) minimizing the time used" (Schmidt, 1991; p. 5).

Having defined skill, one can then classify motor skills into general categories based on common components among different skills. First, motor skills have been classified based on the precision of the movement involved in the motor skill, resulting in a dichotomy of gross motor skills and fine motor skills. Gross motor skills are generally characterized as involving large musculature, wherein the precision of the movement is relatively unimportant (e.g., walking). Fine motor skills on the other hand require a high degree of precision, and thus involve the control of the small muscles of the body (Magill, 1985). Tasks involving fine motor skill are often examined in hand preference and performance tests, and include such tasks as writing and sewing. Another classification of motor skills is based on the distinctness of the beginning and ending of the movement. Discrete motor skills have a clearly defined beginning and end, whereas continuous motor skills are those that have an arbitrary beginning and endpoint. For example, hand- and foot-tapping tasks are considered discrete motor skills, whereas a tracking task would be considered a continuous motor skill. Motor skills have also been classified depending on the status of the environment. More specifically, closed skills are those that take place under fixed, unchanging environmental conditions. In other words, the stimulus waits to be acted upon. Lateral performance tasks, such as peg moving and finger tapping are predominantly closed skills. Open skills, conversely, are those that the "performer must act upon the stimulus according to the action of the stimulus" (Magill, 1985, p. 11). A good example of an open skill is playing soccer. Evidently, there are many ways to conceptualize skill. Handedness and footedness are typically measured using tasks that are considered fine, discrete motor skills, and which are often closed skills.

4.1 Skilled behaviour and lateral preference

Current evidence strongly suggests that hand preference is multifactorial (Healey, Liederman, & Geschwind, 1986; Steenhuis & M. P. Bryden, 1989). What the underlying dimension represents, however, is currently under debate. Healey et al. argue that the main dimension underlying hand preference is related to the musculature used to perform the task (i.e., proximal versus distal musculature). In contrast, Steenhuis and M. P. Bryden, based on a factor analysis of a comprehensive hand-preference inventory, reason that the level of skill required to perform a task distinguishes between tasks performed solely by the preferred hand and those performed by both hands. They found that individuals tended to report using their preferred hand more often for tasks requiring a high degree of manual skill. A fundamental characteristic underlying tasks requiring a high level of manual skill is, they suggest, that such tasks are composed of a relatively complex sequence of motor behaviours. Because the left hemisphere is thought to be the seat of the motor control system responsible for selecting and executing motor sequencing in speech and praxis, Steenhuis and M. P. Bryden argue that a strongly lateralized preference for such tasks makes inherent sense. For unskilled tasks, or those not requiring complex sequencing, the authors found a decrease in the number of participants choosing their preferred hand to perform the tasks. Further studies (Steenhuis, 1996) have clearly shown that the level of skill could be a dimension of hand preference, when hand preference is determined by responses on a questionnaire. Effects of skill on hand preference have also been noted when preference is measured directly (i.e., the participant is asked to perform a task, and hand use is observed). Steenhuis and M. P. Bryden (1999) made observations of hand use on a large number of tasks including writing, cutting, sewing, batting, and throwing. Implements were positioned to the left, midline, or right of the participant, and the hand used for picking up the implement was noted. They found that the preferred hand was used a greater proportion of the time for manipulating or using the implement as compared to simply picking it up. This effect was more noticeable for right-handed participants than for left-handed participants.

Using a similar paradigm, P. J. Bryden, Pryde, and Roy (1999a) recently examined the issue of task complexity and hand preference. The task involved reaching into different regions of hemispace in order to perform actions with objects located at each position. The actions performed with the object included: point, pick up, toss, sweep, and position. In accordance with the participants' hand preferences, the preferred hand was used more frequently on the various performance tasks. The distribution of hand use in hemispace indicated that preferred hand use was almost exclusive for actions carried out in right hemispace, whereas it was used only moderately for actions in left hemispace (less than 40% of the time). These trends were observed across all tasks, with no significant differences found between the tasks, suggesting that task complexity did not affect the frequency of preferred hand use. Performance on the preferential reaching task has also been examined in children (ages 3 –12) and adults using a simple tossing task and a very complex task requiring the object to be precisely oriented into a target goal (P. J. Bryden, Pryde, & Roy, 1999b; Pryde, P. J. Bryden, & Roy, 1999). Again, no differences in hand preference as a function of task were noted for any of the age groups. These studies are contrary to the empirical work of Steenhuis and M.P. Bryden (1989), and raise some questions concerning the role of skill in hand preference. It may well be that the lack of an effect of task complexity is related to the complexity of the task's goal. Consider the tasks examined by Steenhuis and Bryden: writing, cutting, sewing, batting, and throwing. Each of these tasks involves using an implement (pen or needle) that affords a particular, learned unimanual response in order to accomplish the goal (writing or sewing). For instance, one can pick up a pen with either hand, but if the goal is to write, it must be picked up and manipulated with the preferred hand. Note as well, that there are serious consequences of using the "wrong" hand in tasks such as writing and sewing. In contrast, the tasks examined by P.J. Bryden et al. (1999a) and Pryde et al. (1999) used a single implement (a dowel), which afforded no particular response to accomplish the goal (point or toss). In addition, there were few, if any, consequences of inaccurate performance.

Very little research has examined the issue of task complexity as a factor in foot preference. As described earlier, foot preference is generally described in terms of the mobilizing or stabilizing components (e.g., Peters, 1988). Recently, however, Hart and Gabbard (1996) argued that the preferred foot might be chosen to perform manipulative, mobilizing tasks primarily because such tasks are more difficult to perform than the task of stabilizing. In order to assess this prospect, Hart and Gabbard compared foot preference in simple and complex bilateral footedness tasks. The simple task required participants to tap with either foot while supporting themselves on one leg. In the complex task, visual cues were limited in order to make stabilizing more difficult. The results revealed that over 50% of individuals switched to the dominant leg to stabilize themselves in both simple and complex conditions. In other words, individuals who stated they were right-footed for kicking (i.e., mobilizing) instead balanced on their right foot to perform the tapping tasks. The same was true of left handers. There was a slight increase in the number of individuals who switched in the complex condition compared to the simple condition, but this was not a significant increase. The results suggest that foot preference may in part be dependent upon the context of the tasks. Therefore, it remains to be determined how task complexity influences the choice of the preferred foot. To some extent, the problem may in fact be related to the definition of foot dominance in bilateral and unilateral contexts.

In summary, it appears that skill and task complexity may underlie lateral preferences of both the hands and feet. Nonetheless, it is clear that the types of tasks that were evaluated influenced the findings. Arbitrarily defining tasks as either simple or complex is not enough. A method of assessing the difficulty of tasks is needed.

5. SKILLED BEHAVIOUR, TASK COMPLEXITY, AND LATERAL PERFORMANCE

Not only might skill be an important determinant of lateral preference, but it also appears to be an important factor in determining the degree of asymmetry in lateral performance. Instead of using the terms skilled and unskilled to refer to different classes of tasks, it may be more appropriate to discuss task complexity or task difficulty in relation to performance measures.

Very little research has been conducted concerning foot performance and task complexity. Kauranen and Vanharanta (1996) examined simple and choice reaction time for the upper and lower extremities. Reaction time for both feet was found to increase significantly as the number of response alternatives increased. The difference in reaction time between the feet was not significant for simple reaction time, though the difference reached significance for choice reaction time. The results suggest that the difficulty of the task (i.e., the number of response alternatives) influenced the performance difference between the feet. More research needs to be conducted on the performance capabilities of the two feet in a variety of tasks in order to reach any firm conclusions concerning the role of task complexity.

Task complexity, as a factor in manual asymmetries, has also not received much attention. In fact, little research on manual asymmetries and task complexity has varied the difficulty of the movement in quantifiable, measurable terms. Perhaps one of the first investigators to examine task complexity was Flowers (1975). He compared the performance of the two hands of both right and left handers on a simple and a complex task. The simple task was a rhythmical-tapping task, whereas the complex task was a manual-aiming task. Flowers argued that the simple task was essentially ballistic in nature because participants were unable to monitor visually, or to make visual corrections during the movements, whereas the manual-aiming task required participants to make visual corrections, and was closed-loop in nature. The results showed that there was a negligible difference in performance between the hands for the simple, ballistic task, whereas large hand differences were found in movement time and accuracy measurements for the complex task.

In another attempt to examine both simple and complex tasks, Provins and Magliaro (1993) compared performance of the two hands on a test of grip strength and on a handwriting task. As was expected, the performance difference between the hands was significantly larger for the handwriting task than for the grip-strength task. Borod, Caron, and Koff (1984) also examined performance across a broad range of tasks, which varied in complexity. The tasks considered included strength (e.g., grip strength), speed (e.g., dotting circles, marking targets), and accuracy (e.g., tracing spirals) measurements. Greater differences between the hands were found for more complex tasks (the accuracy in tracing spirals and signing one's name) than for simple tasks (grip strength). Likewise, Watson and Kimura (1989) showed a significant difference in performance between the hands for throwing, but not for intercepting (i.e., blocking).

Such studies (Borod et al., 1984; Flowers, 1975; Provins & Magliaro, 1993; Watson & Kimura, 1989), while purportedly examining task complexity and manual performance, have used arbitrary levels of task complexity. Evidently, a method of quantifying task complexity is required, if a systematic evaluation of task complexity as a variable in hand performance is to occur. Fitts (1954) has provided a relatively straightforward way of quantifying the difficulty of a task. Using a number of tasks including a peg-transfer task, a tapping task, and a disc-placing task, Fitts demonstrated that movement time increased linearly when the index of difficulty for a particular task also increased. He defined index of difficulty (ID) as log base two of two times the movement amplitude over the tolerance ($ID = \log_2 (2 \times \text{amplitude}/\text{tolerance})$). Tolerance, stated more simply, is the difference between the width of the target being captured and the width of the stylus. Since the formulation of this equation, Fitts' Law has been replicated for a large variety of tasks, including discrete aiming, moving objects to insert them into holes, moving a cursor on a screen and throwing darts (Shumway-Cook & Woollacott, 1995).

The use of Fitts' index of difficulty appears to be an effective way of quantifying the difficulty of a movement task, although only a few studies in the literature have explicitly applied Fitts' Law to their investigations of manual asymmetries and task complexity (Annett et al., 1979; Todor & Cisneros, 1985). Todor and Cisneros examined the performance of both hands across varying indices of difficulty (IDs were 5, 6 and 7 according to their calculations), using a Fitts' reciprocal-tapping task. As index of difficulty increased, the relative difference between the two hands increased. However, Todor and Cisneros calculated ID using only target width, and so did not take into account the size of the stylus. Calculating tolerance was important in this study because in one condition the size of the implement was actually larger than the size of the target, and so the results are slightly ambiguous.

A second study that investigated the performance of the two hands across task difficulty, and manipulated difficulty using Fitts' Law, was performed by Annett et al. (1979). The authors investigated a range of IDs from approximately 4.0 to 8.0 bits, using tolerance in their calculation of difficulty. They found that differences between the hands became greater as the hole-to-peg ratio decreased, that is, as the difficulty of the task increased. After filming participants performing the task Annett et al. found that the difference in movement time between the two hands was not attributable to insertion-time differences, nor were there differences in the speed of the two hands. Rather, they noted that the non-preferred hand simply made more errors than the preferred hand. They concluded that the essential difference between the hands was a result of the non-preferred hand being noisier or more variable in its output.

A more recent attempt to manipulate the difficulty of a task was that of van Horn and McManus (1994). The authors manipulated characteristics of the Annett pegboard (1967), the Bishop square-tracing task (Bishop, 1980), and the Tapley-Bryden dot-marking task (Tapley & M. P. Bryden, 1985), to alter the difficulty of each of these tasks. For the Annett pegboard task, van Horn and McManus manipulated the movement amplitude, the distance between the holes themselves, the diameter of the pegs, and the shape of the pegs, although Fitts' Law was not used to quantify task difficulty. Main effects were found for all manipulations of the Annett pegboard, such that increasing the task difficulty produced longer movement times. Yet none of these changes interacted significantly with the hand used to perform the task, indicating that the difference between the hands remained constant. The authors then proceeded to manipulate the difficulty of the Bishop square-tracing task and the Tapley-Bryden dot-marking task. In the Bishop square-tracing task, the size of square, and the distance between the two lines was manipulated, whereas the size and distance between the circles was manipulated in the Tapley-Bryden task. Again, none of the difficulty manipulations affected the magnitude of the between-hand difference. Thus, as the demands of the task increased, the movement time of both hands increased, but the difference between the hands remained constant.

The contradictory evidence found by van Horn and McManus (1994) indicates that the role of task complexity in manual asymmetries is far from understood. Currently, the author is executing a systematic evaluation of the role of task complexity or task difficulty in manual performance asymmetries (P. J. Bryden & Roy, 1999; P. J. Bryden, 1998, 2000). In an attempt to replicate the results of van Horn and McManus, P. J. Bryden and Roy (experiment 1) manipulated index of difficulty (Fitts, 1954) in the Annett pegboard and then observed the performance differences of the two hands across each level of task difficulty. Participants performed four different versions of the Annett pegboard in which the size of the pegs was manipulated. No interaction was found between hand performance and task difficulty, indicating a constant between-hand difference in performance times across increasing task difficulty and replicating the findings of van Horn and McManus. These findings have been replicated using both a unimanual tapping task (P. J. Bryden, 1998) and a manual-aiming paradigm (P. J. Bryden, 2000).

These studies provide evidence that manipulations of task difficulty within a single, unimanual task do not result in an increase in the preferred-hand advantage. Although manipulations of this kind can increase the difficulty of a task, as defined by Fitts' Law (1954), the degradation of task difficulty might not reflect differing degrees of skill, especially if one considers Schmidt's (1991) definition of skill. Recall that Schmidt (1991) defined skills as having well-defined environmental goals, in which achieving the goal is maximized, and both energy and time are minimized. By this definition, it could be argued that different indices of difficulty might encompass the same skills. In many ways, manipulations of index of difficulty within a single task produce conditions that have similar goals (e.g., place peg into hole, regardless of the size of the hole). As well, the performance at the different difficulty levels is also similar in the manner in which the goal is attained. By this argument, the different dimensions of skill may be related to the goal of the task, wherein the goal may ultimately drive the choice of the preferred hand.

Some evidence for the notion that manipulations of the goal of a task influence the magnitude of performance asymmetries comes from P. J. Bryden, Roy, and M. P. Bryden (1998). They compared the performance of the two hands on a number of unimanual performance tasks that differed in the complexity of the task goal, including two versions of the Annett pegboard (Annett, 1985), placing and removing pegs in the Grooved Pegboard (e.g., Thompson, Heaton, Matthews, & Grant, 1987), and a version of the Annett pegboard in which only the target hole's orientation was altered. Note that what differs between the different tasks is the complexity of the task goal. Analysis revealed that the performance difference between the hands varied with the type of task. Summarizing briefly, the smallest difference between the hands was noted for the task with the simplest goal (removing pegs from Grooved pegboard), and the largest difference between the hands was found for the task with the most complex goal (placing pegs in Grooved pegboard).

Once again, though, what constitutes a complex task or goal? Consider the "skills" in performing a "complex" task such as the place phase of the Grooved pegboard. In order for a goal to be achieved accurately, the appropriate musculature must be first selected and then coordinated into a complex sequence of movements. This complex sequence of movements also typically requires continuous monitoring using visual and proprioceptive feedback. Yet, which of these components is responsible for driving the choice of the preferred hand and its superiority in performance? In fact, previous research has shown little evidence to support the influence of any one of the above components of skill (i.e., visual feedback, motor output), in isolation, as underlying lateral preference or performance (Annett et al., 1979; Carson et al., 1993). Perhaps the most complex tasks are those that require multiple components of skill, whereas simple tasks are those that involve only one or two components. Clearly, before conclusions can be reached concerning the effects of task complexity and skill on hand performance, the different components of tasks used to evaluate hand performance need to be described.

6. CONCLUSIONS

The aim of this chapter was to review the influence of skill on the expression of lateral preference and performance. Evidence suggests that individuals have stronger preferences for more complex tasks. The dominant hand is chosen more frequently for tasks such as writing and throwing than for picking up objects. Likewise, the dominant foot is chosen more frequently when the focus of the task is manipulative. However, these findings could be dependent upon the tasks examined, or more specifically the goals of the tasks examined.

There is also some evidence that skill might play a role in lateral performance asymmetries. Manipulations of task complexity within a single task do not yield significant increases in the magnitude of the preferred-hand advantage. Yet, when comparisons are made across different tasks, with different goals, the preferred-hand advantage increases. The largest differences between the hands are found for highly learned tasks that require complex sequencing, visual feedback, and the precise control of motor output. Task complexity may also influence the performance of the dominant and non-dominant feet. Unfortunately, very little research has examined this issue.

Thus, rather than skill underlying preference and performance, one might argue that the preferred limb will be chosen and will out-perform the non-preferred limb on any highly learned task that requires one or all of the following: timing and coordination of musculature, complex movement sequencing, on-line visual control. By the same argument, the preferred limb would not be chosen as frequently, nor would it necessarily perform better at a task that does not encompass these requirements. Which components comprise skill, and drive performance differences between the hands (or feet) are yet to be determined. Future research should first define the important components comprising skill, and then determine how each of these components, in isolation and in interaction, affects the two limbs differentially.

Acknowledgements. The data presented here were collected at the University of Waterloo, and represent a portion of my doctoral dissertation in the Departments of Kinesiology and Psychology. I would like to thank Dr. Fran Allard, Dr. Jacqui Crebolder, Dr. Pat Rowe, and Linda Kalbfleisch, M.Sc., for their insightful comments on early drafts of this chapter.

7. REFERENCES

Annett, M. (1967). The binomial distribution of right, mixed, and left handedness. *Quarterly Journal of Experimental Psychology, 19,* 327-333.

Annett, M. (1972). The distribution of manual asymmetry. *British Journal of Psychology, 63,* 343-358.

Annett, M. (1976). A coordination of hand preference and skill replicated. *British Journal of Psychology, 67,* 587-592.

Annett, M. (1985). *Left, right, hand and brain: The right shift theory.* Hillsdale, NJ : Lawrence Erlbaum.

Annett, M., Hudson, P., & Turner, A. (1974). The reliability of differences between the hands in motor skill. *Neuropsychologia, 12,* 527-531.

Annett, J., Annett, M., Hudson, P. T. W., & Turner, A. (1979). The control of movement in the preferred and non-preferred hands. *Quarterly Journal of Experimental Psychology, 31,* 641-652.

Augustyn, C., & Peters, M. (1986). On the relation between footedness and handedness. *Perceptual and Motor Skills, 63,* 1115-1118.

Bishop, D. V. M. (1980). Handedness, clumsiness and cognitive ability. *Developmental Medicine and Child Neurology, 22,* 569-579.

Brown, E. R., & Taylor, P. (1988). Handedness, footedness, and eyedness. *Perceptual and Motor Skills, 66,* 183-186.

Bryden, M. P. (1977). Measuring handedness with questionnaires. *Neuropsychologia, 15,* 617-624.

Bryden, M. P. (1982). *Laterality: Functional asymmetry in the intact brain.* New York : Academic Press.

Bryden, M. P., Ardila, A., & Ardila, O. (1993). Handedness in native Amazonians. *Neuropsychologia, 31,* 301-308.

Bryden, P. J. (1998). *The origins of manual asymmetries: What is revealed by pushing task difficulty.* Unpublished doctoral dissertation, University of Waterloo, Waterloo, Ontario.

Bryden, P. J. (2000). Pushing task difficulty to the extreme. *Manuscript in preparation.*

Bryden, P. J., & Allard, F. (1998). Does extended practice affect the magnitude of manual asymmetries on a pegboard task? *Brain and Cognition, 37,* 44-46.

Bryden, P. J., Pryde, K. M., & Roy, E. A. (1999a). A behavioural measure of hand preference. *Manuscript submitted for publication.*

Bryden, P. J., Pryde, K. M., & Roy, E. A. (1999b). A developmental analysis of the relation between hand preference and performance: II. A performance-based method of measuring hand preference in children. *Paper submitted for the 1999 annual meeting of Theoretical and Experimental Neuroscience (TENNET X)* in Montreal, Canada.

Bryden, P. J., & Roy, E. A. (1999). Spatial task demands affect the extent of manual asymmetries. *Laterality, 4,* 27-37.

Bryden, P. J., Roy, E. A., & Bryden, M. P. (1998). Between-task comparisons: Movement complexity affects the magnitude of manual asymmetries. *Brain and Cognition, 37,* 47-50.

Borod, J., Caron, H. S., & Koff, E. (1984). Left handers and right handers compared on performance and preference measures of lateral dominance. *British Journal of Psychology, 75,* 177-185.

Carson, R. G., Chua, R., Elliott, D., & Goodman, D. (1990). The contribution of vision to asymmetries in manual aiming. *Neuropsychologia, 28,* 1215-1220.

Carson, R. G., Elliott, D., Goodman, D., Thyer, L., Chua, R., & Roy, E. A., (1993). The role of impulse variability in manual aiming asymmetries. *Psychological Research, 55,* 291-298.

Chapman, J. P., Chapman, L. J., & Allen, J. J. (1987). The measurement of foot preference. *Neuropsychologia, 25,* 579-584.

Clymer, P.E., & Silva, P.A. (1985). Laterality, cognitive ability and motor performance in a sample of seven year olds. *Journal of Human Movement Studies, 11,* 59-68.

Coren, S. (1993a). The lateral preference inventory for measurement of handedness, footedness, eyedness, and earedness: Norms for young adults. *Bulletin of the Psychonomic Society, 31,* 1-3.

Coren, S. (1993b). Measurement of handedness via self-report: The relation between brief and extended inventories. *Perceptual and Motor Skills, 76,* 1035-1042.

Dargent-Paré, C., De Agostini, M., Mesbah, M., & Dellatolas, G. (1992). Foot and eye preferences in adults: The relation with handedness, sex, and age. *Cortex, 28,* 343-351.

Dodrill, C. B., & Thoreson, N. S. (1993). Reliability of the lateral dominance examination. *Journal of Clinical and Experimental Neuropsychology, 15,* 183-190.

Elias, L. J., & Bryden, M. P. (1998). Footedness is a better predictor of language lateralisation than handedess. *Laterality, 3,* 41-51.

Elliott, D., & Chua, R. (1996). Manual asymmetries in goal-directed movement. In D. Elliott & E. Roy (Eds.), *Manual asymmetries in motor performance.* Boca Raton : CRC Press.

Fitts, P. M. (1954). The information capacity of the human motor system in controlling the amplitude of movement. *Journal of Experimental Psychology, 47,* 381-391.

Flowers, K. (1975). Handedness and controlled movement. *British Journal of Psychology, 66,* 39.

Gabbard, C., & Iteya, M. (1996). Foot laterality in children, adolescents and adults. *Laterality, 1,* 199-205.

Gardner, L. P. (1941). Experimental data on the problem of motor lateral dominance in feet and hands. *Psychological Records, 5,* 1-63.

Harris, A. J. (1958). *Harris tests of lateral dominance: Manual of directions.* New York : The Psychological Corporation.

Hart, S., & Gabbard, C.(1996). Bilateral footedness and task complexity. *International Journal of Neuroscience, 88,* 141-146.

Hart, S., & Gabbard, C. (1998). Examining the mobilizing feature of footedness. *Perceptual and Motor Skills, 86,* 1339-1342.

Healey, J. M., Liederman, J., & Geschwind, N. (1986). Handedness is not a unidimensional trait. *Cortex, 42,* 33-53.

Luria, A. R. (1965). *The higher cortical functions in man.* New York : Basic Books.

MacNeilage, P. (1993). Implications of primate functional asymmetries for the evolution of cerebral hemispheric specialization. In J.P. Ward and W.D. Hopkins (Eds.), *Primate laterality: Current evidence of primate behavioral asymmetries.* New York : Springer-Verlag.

Magill, R. A. (1985). *Motor learning: Concepts and applications* (3rd Ed.). Dubuque, Iowa: Wm. C. Brown Publishers.

Malmo, R. B., & Andrews, M. L. (1945). A recording device for foot-tapping, with results from polyneuropathic subjects. *American Journal of Psychology, 58,* 247-252.

McManus, I. C. (1997). Pulling together the laterality research of M. P. Bryden. Presentation at *Theoretical and Experimental Neuropsychology,* Montreal, Canada, June, 1997.

McManus, I. C., & Bryden, M. P. (1993). Handedness on Tristan da Cunha: The genetic consequences of social isolation. *International Journal of Psychology, 28,* 831-843.

Oldfield, R. C. (1971). The assessment and analysis of handedness: The Edinburgh Inventory. *Neuropsychologia, 9,* 97-113.

Peters, M. (1976). Prolonged practice of a simple motor task by preferred and non-preferred hands. *Perception and Motor Skills, 43,* 447-450.

Peters, M. (1980). Why the preferred hand taps more quickly than the non-preferred hand: Three experiments on handedness. *Canadian Journal of Psychology, 34,* 62-71.

Peters, M. (1988). Footedness: Asymmetries in foot preference and skill and neuropsychological assessment of foot movement. *Psychological Bulletin, 103,* 179-192.

Peters, M. (1996). Hand preference and performance in left handers. In D. Elliott & E. Roy (Eds.) *Manual asymmetries in motor performance* (pp. 99-120). Boca Raton : CRC Press.

Peters, M. (1998). Description and validation of a flexible and broadly usable handedness questionnaire. *Laterality, 3,* 77-96.

Peters, M., & Durding, B. M. (1978). Handedness measured by finger tapping: A continuous variable. *Canadian Journal of Psychology, 32,* 257-261.

Peters, M., & Durding, B. M. (1979a). Footedness of left- and right handers. *American Journal of Psychology, 92,* 133-142.

Peters, M., & Durding, B. M. (1979b). Left handers and right handers compared on a motor task. *Journal of Motor Behavior, 11,* 103-111.

Previc, F. H. (1991). A general theory concerning the prenatal origins of cerebral lateralization in humans. *Psychological Review, 98,* 299-334.

Provins, K. A. (1997). The specificity of motor skill and manual asymmetry: A review of the evidence and its implications. *Journal of Motor Behavior, 29,* 183-192.

Provins, K. A., & Magliaro, J. (1993). The measurement of handedness by preference and performance tests. *Brain and Cognition, 22,* 171-181.

Provins, K. A., Milner, A. D., & Kerr, P. (1982). Asymmetry of manual preference and performance. *Perceptual and Motor Skills, 54,* 179-194.

Pryde, K. M., Bryden, P. J., & Roy, E. A. (1999). A developmental analysis of the relation between hand preference and performance: I. Preferential reaching into hemispace. *Paper accepted for the 1999 annual meeting of Theoretical and Experimental Neuroscience (TENNET X)* in Montreal, Canada.

Raczkowski, D., Kalat, J. W., & Nebes, R. (1974). Reliability and validity of some handedness questionnaire items. *Neuropsychologia, 12,* 43-47.

Ransil, B. J., & Schachter, S. C. (1994). Test-retest reliability of the Edinburgh Handedness Inventory and global handedness preference measurements, and their correlation. *Perceptual and Motor Skills, 79,* 1355-1372.

Reib, M., Reib, G., & Freye, H. (1998). Some aspects of self-reported hand preference. *Perceptual and Motor Skills, 86,* 953-954.

Reitan, R. M., & Davidson, L. A. (1974). *Clinical neuropsychology: Current status and applications. The lateral dominance examination.* Washington, DC : Winston.

Rife, D. C. (1940). Handedness, with special reference to twins. *Genetics, 25,* 178-186.

Rigal, R. A. (1992). Which handedness: Preference or performance? *Perceptual and Motor Skills, 75,* 851-866.

Roy, E. A. (1983). Manual performance asymmetries and motor control processes: Subject-generated changes and response parameters. *Human Movement Science, 2,* 271-277.

Roy, E. A., & Elliott, D. (1986). Manual asymmetries in visually directed aiming. *Canadian Journal of Experimental Psychology, 40,* 109-121.

Schmidt, R. A. (1991). *Motor control and learning from principles to practice.* Champaign, IL : Human Kinetics Books.

Schmidt, R. A., Zelaznik, H. N., Hawkins, B., Frank, J. S., & Quinn, J. T. (1979). Motor output variability: A theory for the accuracy of rapid motor acts. *Psychological Review, 86,* 415-451.

Shumway-Cook, A., & Woollacott, M. (1995). *Motor control: Theory and practical implications.* Baltimore: Williams & Wilkins.

Steenhuis, R. E. (1996). Hand preference and performance in skilled and unskilled activities. In D. Elliott & E. Roy (Eds.), *Manual asymmetries in motor performance* (pp. 123-142). Boca Raton : CRC Press.

Steenhuis, R. E., & Bryden, M. P. (1989). Different dimensions of hand preference that relate to skilled and unskilled activities. *Cortex, 25,* 289-304.

Steenhuis, R. E., & Bryden, M. P. (1999). The relation between hand preference and hand performance: what you get depends on what you measure. *Laterality, 4,* 3-26.

Tapley, S. M., & Bryden, M. P. (1985). A group test for the assessment of performance between the hands. *Neuropsychologia, 23,* 215-222.

Thompson, L. L., Heaton, R. K., Matthews, C. G., & Grant, I. (1987). Comparison of the preferred and non-preferred hand on four neuropsychological motor tasks. *The Clinical Neuropsychologist, 1,* 324-334.

Todor, J. I., & Cisneros, J. (1985). Accommodation to increased accuracy demands by the right and left hands. *Journal of Motor Behavior, 17,* 355-372.

Todor, J. I., & Doane, T. (1978). Handedness and hemispheric asymmetry in the control of movements. *Journal of Motor Behavior, 15,* 539-546.

Van Horn, J., & McManus, I. C. (1994). Task difficulty and hand differences in peg moving, circle marking and square tracing. *Unpublished manuscript,* University College, London.

Watson, N. V., & Kimura, D. (1989). Right hand superiority for throwing but not for intercepting. *Neuropsychologia, 27,* 1399-1414.

Chapter 10

Examining the Notion of Foot Dominance

Carl Gabbard[1] and Susan Hart[2]
[1]*Texas A&M University, USA:* [2]*New Mexico State University, USA*

An increasing body of evidence suggests that the study of lower limb laterality has promise in broadening the window in which we view specific aspects of brain organization and function (see also Chapter 9, this volume). Recent testament of this is the hint that footedness is as good or better an indicator of language laterality as handedness (Day & MacNeilage, 1996; Elias & Bryden, 1998; Watson, Pusakulich, Hermann, Ward, & Wyler, 1993; Watson, Pusakulich, Ward, & Herman, 1998). Perhaps one of the more attractive characteristics of footedness that complements this line of reasoning is the notion that, compared to handedness, it is less likely to be influenced by dextral "social pressures" and "right-side-world" influences. For example, Peters (1990) states that activities in everyday life that require use of the feet (relative to use of the upper limbs) are fewer and usually less complex and practised. Therefore, it might be reasonable to suggest that the study of cerebral lateralization and footedness are complementary (e.g., Elias, Bryden, & Bulman-Fleming, 1998).

However, one aspect of footedness that warrants further consideration in research and clinical use is the fundamental question of *what is meant by a dominant foot.* For example, is the dominant limb the one used to manipulate an object, while the other foot lends postural control? Or, is it the limb providing the greatest degree of balance/postural support? As evidence for the dilemma, more recently Elias et al. (1998) concluded that they could

M.K. Mandal, M.B. Bulman-Fleming and G. Tiwari (eds.), Side Bias: A Neuropsychological Perspective, 249-265.

not be certain which component of footedness was predictive as a measure of cerebral lateralization. In this chapter we present information that addresses this issue. The chapter begins with a brief review of selected explanations of the phenomenon and comments relative to definition. This is followed by an argument based on empirical data suggesting that defining foot dominance requires a contextual perspective grounded in the understanding of its functional characteristics. The chapter ends with a brief discussion of the methodological implications for research and neuropsychological assessment.

1. PROMINENT THEORIES OF FOOTEDNESS

Theoretical explanations for describing the mystery of hand preference have been quite diverse and represent a respectable hodgepodge of biological and environmental perspectives (e.g., Annett, 1985; Corballis, 1997; Laland, Kumm, Van Horn, & Feldman, 1995; McManus & Bryden, 1992; Yeo & Gangestad, 1993). Although some cursory attempts have been made to use these models to explain footedness (e.g., Annett's right-shift theory), meaningful attention and debate on the issue have been sparse.

2. CEREBRAL LATERALIZATION AND MOTOR CONTROL

Although much is still a mystery about control of the upper and lower limbs, there is general agreement that the primary motor cortex of each hemisphere controls most aspects of *voluntary* movement primarily in the contralateral side of the body (e.g., Ganong, 1993; Hellige, 1993; Seeley, Stephens, & Tate, 1992). Opinions regarding the phenomenon associated with programming of limb preference however, remain divided. That is, although an individual can use both sides reasonably well, one hemisphere generally overrides the other with respect to preferential use and skill. And, for the vast majority of individuals, this is reflected by left-hemisphere control resulting in predominant use of the right limb for most tasks; an observation that is well established for the upper (e.g., Gilbert & Wysocki, 1992; Hugdahl, Satz, Mitrushina, & Miller, 1993) and lower (e.g., Porac, Coren, & Duncan, 1980; Gabbard & Iteya, 1996) limbs.

3. OPERATIONAL DEFINITION OF FOOTEDNESS

In perhaps the most comprehensive review of this manifestation of motor dominance, Peters (1988) offers an operational definition that describes footedness as used in several assessment inventories (e.g., Chapman, Chapman, & Allen, 1987; Coren, 1993; Reitan & Davison, 1974). Typically, foot preference for a particular task is characterized by its stabilizing and mobilizing (or manipulating) features. That is, one limb is used to manipulate an object or lead out, while the other foot has the role of lending postural (stabilizing) support (e.g., kicking a ball, stepping-up on a chair, letter tracing [with foot] while standing, and picking up a pebble). In this bilateral context, which provides a relatively clear division of functional limb action, consensus is that the mobilizing limb is the preferred (dominant) foot, whereas the foot that is used to support the actions of the preferred foot is defined as the nonpreferred limb. In this context, tasks that are more unilateral such as one-foot balance and hopping (on one limb) are questionable, because they do not provide clear bilateral role differentiation.

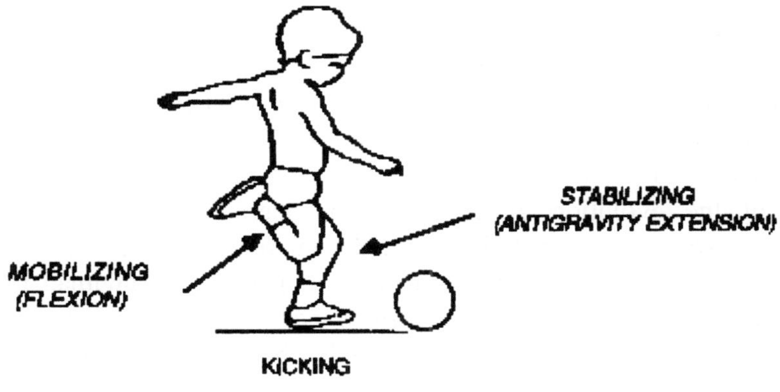

Figure 1. Kicking a ball illustrates the typical bilateral context with division of limb action

In a review of 10 inventories (Gabbard & Hart, 1996) noted in research and clinical work (e.g., Coren, 1993; Chapman et al., 1987; Dargent-Paré, De Agostini, Mesbah, & Dellatolas, 1992; Dodrill & Thoreson, 1993) two observations appeared most evident. First, virtually all of the items in each inventory were bilateral tasks, requiring actions of stabilization/mobilization.

The common exception (use of a predominately unilateral task) was "hop on one leg," noted in only two of the 10 inventories. And second, the task of "kicking a ball" appeared to be the predominant test of foot dominance as defined by the operational definition described and used in all inventories reviewed. Other common bilateral tasks, understood to be undertaken while standing on one foot were: pick up a pebble with toes, step on a bug, stamp one foot, step up on a chair, write name in the sand, arrange pebbles, and manipulate a golfball around circle.

The notion that kicking a ball is an ideal fit with the operational definition has been noted by Chapman et al. (1987), Peters (1988), and Porac and Coren (1981). Peters states that "the choice of foot for kicking is as compelling as the choice of hand for writing" (p. 183). The researcher also notes that "picking up a pebble" tends to be congruent with choice of kicking limb; a task used in three of the inventories reviewed.

Aside from the common operational definition described, the following selected explanations of footedness appear to have received the most attention in the literature.

4. THE 'POSTURAL ORIGINS' PERSPECTIVE

The notion of a complementary-role (bilateral) function of the feet was also given attention in the publication of MacNeilage's (1991) "Postural Origins" theory of primate neurobiological asymmetries. In the researcher's treatise of the evolution of cerebral hemispheric specialization of all primates, it was emphasized that an understanding of footedness and its relation to other functional asymmetries (in humans and nonhuman forms) can be gained by noting its postural significance. According to this theory, through evolution the right side of the body has become the "operative" side in higher primates (the dominant side?). And, "use of one leg for operations on the environment requires postural support of the body with the other leg" (p. 182). This is based in part on the premise that posture in early true primates was necessarily asymmetrical. Presumably, while clinging vertically in a tree habitat, the need developed to grasp (cling to) with one side (postural support) and either leap or feed with the other (contralateral) side; resulting in the evolution of manual and pedal asymmetry.

From MacNeilage's treatise, it is somewhat implied that, although complementary-role characteristics apply with most functions of the feet, the operative (right) side is dominant for most persons. This is supported further by the researcher's use of a single measurement (indicator) of footedness in the 1996 study (Day & MacNeilage); i.e., use of the foot selected for kicking. In reference to a point that will be commented on later, Maki (1990; a dissertation directed by MacNeilage) reported that foot preference for kicking (the operative limb) was not closely related to limb choice for tasks of unipedal stability (i.e., balancing on one foot and spinning around on one foot).

5. A NEURODEVELOPMENTAL EXPLANATION

From a neurodevelopmental perspective, Previc (1991) poses arguments that stimulate differing thoughts from the typical operational definition concerning foot dominance (see also Chapter 2, this volume). Tracing its origins to asymmetric prenatal development of the ear and labyrinth and position of the fetus (cephalic-leftward, right ear facing out) during the final trimester, Previc hypothesizes that in most humans, about two-thirds, there is a left-otolithic advantage that underlies a predisposition to use the left side of the body for postural control and, by elimination, the right side for voluntary (mobilizing) motor function.

Of pivotal importance to this theory is the ipsilateral relation that exists between the labyrinths and the antigravity extensor muscles (principally the gastrocnemius and soleus). According to Pompeiano (1985), although the labyrinths exert bilateral control over the antigravity reflexes, their excitatory influence is greatest for the ipsilateral muscle groups, such that stimulation to the left labyrinth produces greater extension of the left antigravity muscles and reduced extension on the right side. Therefore, according to Previc, there are lateral asymmetries in antigravity excitatory strength that arise from an imbalance in vestibular functioning. That is, there is a developmental advantage on the left side as a result of asymmetric prenatal development.

Complementing this is the notion that antigravity extension, including postural reflexes on the left side, emerges before voluntary motor control (flexion) on the right side of the body. The researcher illustrates this notion by describing the predominant sport-skill stances in a basketball lay-up shot and in a golf swing; i.e., for the right hander, the left leg's antigravity muscles (extension) are used for postural support, whereas the right leg is flexed. Greater extension (and presumably support) on the left side reflects greater strength of its vestibulospinal reflexes, a result of early fetal maturation. Kicking, the popular inventory item noted earlier, provides a particularly good example of the point.

Although Previc makes reference to the suggestion that in the common bilateral context there may not be a clearly dominant limb (one limb is used for postural support-extension, the other mobilizing-flexion), the theory predicts that most individuals, when placed in either a bilateral or unilateral (stabilizing only) context, would favour the left-side for postural control – a theory that we tested and discuss in a subsequent section of this chapter.

From the explanations provided, two reasonable arguments could be offered related to the notion of a dominant foot. The first is acceptance of the operational definition described by Peters (1988) as used in most inventories today. That is, in the bilateral context, the mobilizing or manipulating limb is the preferred (dominant) foot, whereas the foot that is used to support the actions of the preferred foot is the nonpreferred limb. It appears that several reports and experimental observations of bilateral-task behaviour (e.g., Gabbard & Iteya, 1996; Gentry & Gabbard, 1995; Spry, Zebas, & Visser, 1993; Whittington & Richards, 1987) support the general contention that humans are typically right-footed for actions of mobilization and left-sided for postural stabilization. Complementing this conclusion, in part, is the physiological observation of contralateral control and general left-hemispheric dominance for motor functions.

From another perspective, Previc's neurodevelopmental theory offers interesting explanations for foot dominance in the bilateral and unilateral context. The theorist's "initial" notion, that there is no clearly dominant limb in the bilateral context, seems reasonable. That is, one foot provides necessary postural support while the other executes voluntary (mobilizing) action; in essence, complementary role action. However, to agree with this proposition, one must assume that the neuropsychological demands placed on each action (mobilization and stabilization) are equivalent. With most bilateral tasks it seems that the greatest neurological demand is with mobilization (e.g., kicking a ball). Perhaps the most interesting tenet of Previc's theory is the notion that anti-gravity extension (postural support) on the left side of the body emerges before voluntary motor control (mobilization) on the contralateral (right) side. This suggests that for most individuals, the "dominant" foot for either bilateral or unilateral task behaviour is the left one. A point that also seems somewhat complementary to MacNeilage's assertion is that the preference for postural control evolved over time to the left side. This general claim, a dominant left-side preference for postural control in most persons, led us to examine further the functional characteristics of footedness.

6. EXAMINING THE NOTION OF FOOT DOMINANCE

As mentioned in the introduction, it appears that in order to grasp the notion of foot dominance, a different perspective is required: One that includes identifying and understanding the distinctive anatomical and physiological characteristics and behavioural features that are unique to this form of functional asymmetry. Actions of the lower limbs involve three alternative motor functions: stability, mobility, and the combination of stability/mobility. These functions are most typically described within the broad contexts of unilateral and bilateral behaviour. First, in the predominantly unilateral context are functions of *mobility* (motor action/flexion) and *stability* (postural control). Second, in the more commonly used bilateral context, are the complementary functions of *mobility/stability.* Another characteristic of note is what may be referred to as the "focal limb." That is, in virtually any context, the demands of the task require a greater focus of resources on a specific limb. For example, in the bilateral context of kicking, arguably the focal limb is the mobile foot. The act of postural support is relatively less demanding of the system. It is worthwhile to note that a similar argument holds for virtually all bilateral tasks found in current inventories used in research and clinical settings. Therefore, it is not surprising that the focal leg is typically on the same side as the mobilizing (manipulating) foot. Does this make it the dominant limb? Perhaps a reasonable answer is that it is dominant for mobility, but not stability. The question arises as to what the limb of choice would be if the demands of stability and mobility of a single task were equivalent or if the demands of stability were greater than those for mobility? In essence, these questions address the general notion of foot dominance in a particular context. Would research support the general claim as interpreted from the works of Previc and MacNeilage that the left limb is dominant for postural control? Or, can we find support for the traditional notion that foot dominance lies with the mobilizing limb, regardless of postural complexity?

With these questions in mind, we present research conducted in our lab and information reported by other investigators. One note of attention when interpreting some of the findings described; we acknowledge that when differentiating limb function, it is virtually impossible to attribute total postural control to one limb. That is, the other limb and perhaps parts of the upper body may very likely play some role in total body stabilization.

6.1 Mobility

Of the three behavioural features of footedness, comparatively less has been reported concerning the mobility-only unilateral condition. This would be a context in which resource demands for postural support are minimal, such as foot tapping or some other act of mobilization while one is seated or lying on one's back. Based on physiological data showing contralateral hemispheric control of voluntary lower limb actions (Ganong, 1993; Seeley et al., 1992), behavioural predictions for limb selection are rather intuitive. A recent study in our lab confirmed this prediction (Hart & Gabbard, 1998). That is, we found a strong relation between limb preference for mobilization used in a bilateral inventory (Coren, 1993) and for action in a unilateral-mobility context. In this experiment, participants were seated in an adjustable chair with material positioned at the midline and were asked to (1) draw their initials in a sandbox with one foot, and (2) use one foot to roll a golf ball around a circle as quickly and accurately as possible. Perhaps not too surprisingly, 98% of right-footers, who were right preferent for mobility with bilateral tasks, chose the right limb for both experimental tasks, whereas 2% displayed mixed responses. Left-footers, although not as lateralized as their right-footed counterparts, exhibited a similar profile: 84% chose the same preferred (left) limb for tasks, the remaining 13% were right preferent and 3% were mixed. Table 1 shows these results and those for the remaining contexts. As a point of reference, the * denotes concordance between the limb used in a bilateral assessment inventory and the foot selected for the experimental condition. For the remaining discussion, the comparison is for choice of stabilizing limb.

Table 1. Percentages of participants exhibiting foot preferences in selected contexts.

Footedness context			**Right footed**	**Left footed**
Unilateral-Mobility		R	98*	13
		L	0	84*
		M	2	3
Unilateral-Stability	Simple	R	62	31*
		L	15*	44
		M	23	25
	Complex	R	60	31*

		L	21*	41
		M	19	28
Unilateral-Stability/ Mobility	Simple (Stab.)	R	55	44*
		L	34*	40
		M	11	16
	Complex (Stab.)	R	58	50*
		L	36*	47
		M	6	3

* Concordance with limb selected in bilateral assessment inventory

6.2 Stability

In 1997 we published the findings of an experiment that tested Previc's notion that most individuals, when placed in either a bilateral or unilateral context of stability, would favour the left side for postural control (Hart & Gabbard, 1997), a theory that to some degree might be also interpreted from the work of MacNeilage. More specifically, this investigation compared foot preference for stabilizing while performing a one-leg static balance task with the stabilizing limb identified by bilateral inventory (Coren, 1993).

Briefly, groups of strong right- and left-footers were examined for preference and performance with a one-leg static balance task on a force plate in simple (lights on) and a complex condition (limited visual cues). Previous research verifies that postural stability is more difficult when visual cues are severely reduced. Participants were instructed to perform the first two trials in each condition with their preferred limb prior to which they were allowed a practice trial with each foot. Results indicated quite clearly that preference for stabilization in the bilateral context was independent of limb choice in the unilateral condition (Table 1). That is, the majority of individuals switched limbs for stabilization in the unilateral context. For example, in the simple condition, 62% of right-footers, i.e., left-preferent for stability in the bilateral context, "switched" to the right limb for stability in the unilateral context, with 23% and 15% having no preference or preferring the left limb, respectively. With the left-footed group, i.e., those who prefer the right foot to stabilize with bilateral tasks, 44% preferred the left foot, and 25% and 31% had no preference or preferred the right foot for the stability task, respectively. From another perspective, among right-footers there was only 15% agreement between contexts but there was a 31% concordance with left-footers. Preference data for the complex condition revealed similar results (Table 1). Although not critical to this discussion, in regard to performance on the forceplate (centre-of-pressure values), although there were no significant differences within the left-footed group, right-footers performed significantly better with their right limb.

These findings generated two general conclusions. Perhaps foremost was the observation that footedness (limb choice) for bilateral task behaviour was independent of behaviour in a unilateral context of stabilization. That is, one's preference for stabilization in a bilateral context may not be the same limb selected for postural control to perform a one-leg static balance. Complementing this notion is the suggestion that foot preference may be, in part, dependent on task complexity (see Chapter 9, this volume). And secondly, these results did not lend support to the general notion that, as a result of prenatal developmental processes (Previc) or evolution (MacNeilage), the majority of individuals are predisposed to postural control (stabilizing) on the left side. Most supportive of these conclusions was the relatively low percentage of left- and right-footed individuals who chose to stabilize on their left limb in the unilateral context. *In fact, the vast majority of right-footers and most of the left-footed group preferred the right leg, or had no preference (mixed) for performing the unilateral tasks.* It is interesting to note that the majority of both foot groups actually switched stabilizer preference. The implications of these findings are discussed in the last section of this chapter.

In regard to Previc's hypothesis, several possibilities could account for the discordance between the theory's predictions and the present findings. For example, although Previc's notion of prenatal asymmetrical stimulation of the labyrinth of the inner ear (favouring the left side) might be scientifically reasonable, at this point there is no evidence that this imbalance alone has an observable effect on behaviour. Further, there appears to be considerable discrepancy in the obstetric literature concerning the amount of time infants are positioned occipitally during the third trimester, with the proportion of time ranging from 20% to 60% (e.g., Cunningham, McDonald, & Gant, 1989; Taylor, 1976; Wren & Lobo, 1989). Solicited opinions of selected obstetric nurses and physicians suggest that, because of frequent movement of the head and neck (i.e., head-turning behaviour) during the last trimester, any quantifying of the degree of left/right otolithic stimulation would be difficult.

One could speculate further that perhaps a greater amount of prenatal stimulation to left-side sensory organs subserving excitatory influence to ipsilateral muscle groups does not simply produce a behavioural advantage for stabilizing preference on the left side. Unipedal stabilization involves a plethora of sensory mechanisms (labyrinth of the inner ear along with proprioceptors) exerting bilateral control and producing both excitatory and inhibitory influences to muscle groups involved in maintaining balance. The selection of a particular limb for stability is perhaps too complex to be predicted from a single physiological feature, considering the number and interactive combinations of sensory organs involved in maintaining upright equilibrium.

Other researchers have reported findings that lend support to those described here. In an unpublished doctoral thesis on the subject of footedness by Nonis (1996), the author found that in children 3 to 6 years of age, participants exhibited increasing "mixed" preference in performing the one-leg balance task (also known as hopping). As expected, participants were mainly right preferent for bilateral tasks such as kicking a stationary and a moving ball, and pick-up tasks. In another study, Katsarkas, Smith, and Galiana (1994) found no apparent biases for the right versus left foot in postural stability (sway) on one foot. Perhaps more interestingly, this was observed in normals and patients standing on the foot ipsilateral and contralateral to a brain lesion. Similar results with 5- to 9-year-olds were reported by Armitage and Larkin (1993) using unilateral hopping and one-leg balance. That is, differences between left- and right foot performance were not significant. Although these results do not complement the present findings, which note a right-side bias with increased complexity, the data are also less supportive of Previc's general theory.

6.3 Mobility / Stability

As noted earlier, in the bilateral context, consensus is that the mobilized foot is the dominant limb, with complementary evidence that humans are predominantly right-footed for actions of mobilization and left-sided for postural stabilization. However, with virtually all bilateral tasks used to assess foot preference (e.g., kicking a ball, tracing letters), actions of stabilization are arguably less demanding than those required for mobilizing, thus, posing the question of task complexity and limb selection. What would be the limb of choice if the demands of stability and mobility were equivalent? Or perhaps more practically, what if the stabilizing requirements in the bilateral context were greater than the need for the mobilizing effort? We addressed these questions by examining bilateral foot preference characteristics under conditions of varying relative complexity, such that in one condition the demands of mobility were greater than those for stability, and in the second condition the stabilizing demands for postural support were greater than those for mobilizing effort (Hart & Gabbard, 1996).

For this experiment, strong right- and left-footers (as determined by bilateral inventory) were required to stand on one foot on a nonactivated force platform and tap a telegraph key in two conditions of stabilizing complexity. In the first, participants were asked to tap the key five times with one foot while stabilizing with the other limb. In this simple condition, participants had optimal lighting to complete the task with no experimental perturbation of balance and were encouraged to tap as accurately as possible, such that the mobilizer served as the focal limb. As a deceptive technique used to enhance focus of attention on the mobilizing feature of the task, participants were told that tapping rhythmicity (consistency between taps) was being recorded.

The complex condition required the participant to perform the same task, but with limited visual cues, i.e., complete darkness with the exception of the glowing telegraph key, therefore, increasing the stabilizing demands of the task, resulting in a switch in the participant's focus of attention. To enhance this condition, participants were provided deception by being told that a performance measure for stability was recorded. Participants were allowed one practice trial with each limb prior to each condition; then, preference behaviour for the stabilizing limb was recorded.

Results of the first experimental condition, designed to increase mobility demands beyond those required for stabilizing effort, revealed a 34% concordance between bilateral inventory preference and the experimental task; that is, 55% of right-footers switched limbs for stability to perform the experimental condition. In other words, they favoured the right foot. Of the remaining sample, 11% had no preference. With left-footers, only 44% duplicated their stabilizing preference with the experimental task (right foot), whereas 16% exhibited no lateral bias and 40% favoured the left limb for stability, respectively (Table 1). A note: in this condition we predicted that behaviour (concordance) would be high given the similarity of focus of attention to the mobilizing limb. In retrospect, it appears that because participants were required to stabilize off the ground – on the forceplate – they gave more attention to stabilizing than we had anticipated.

In the second condition, designed to increase stabilizing demands beyond the effort required for the mobilizing response, both foot groups exhibited similar profiles. *That is, about half switched from stabilizing on their limb favoured with bilateral tasks, to preferring the opposite limb in the experimental condition.* Perhaps most evident from the data is the observation that in both simple and complex conditions, 50% or more of participants switched stabilizing limbs or showed no preference when comparing bilateral task behaviour to experimental conditions. Keep in mind that in virtually all bilateral task inventories, the act of stabilization is arguably less demanding than the mobilizing requirements. For example, in the simple and complex conditions, only 34% and 36% of right-footers choose to use the same limb for stabilization (the left). Left-footers exhibited a similar pattern in both conditions by showing concordance (same preference) values of 44% (simple) and 50% (complex) for stabilization.

7. CONCLUSIONS

So, what does this research tell us? From the findings discussed, including our studies and supportive investigations, we would argue the case that *foot preference may be in part dependent on the context of the task* and not tied predominantly to one general definition, such as the traditional notion that the mobilizing limb is that of the dominant foot. In addition, *it appears as if an individual's foot preference is dependent on the neurological demands of the task, such that the favoured limb is reserved for the arguably more difficult aspect of a behavioural action.* As noted earlier, in virtually all foot preference inventories the mobilizing action is arguably much more demanding than the more practised act of providing stabilizing support, such as the task of kicking a ball. In this paper we show rather convincing evidence that when the focus of attention is more on stabilization, most participants prefer to switch limbs. Fundamentally this observation appears to question the notion of a dominant limb.

In addition, the findings described appear to be rather nonsupportive of the theory of postural control suggesting that individuals are predisposed to favouring the left side for stability. Perhaps most evident of this is the finding that there was no overwhelming bias favouring the left side for stabilizing effort when relative complexity was a factor, that is, when the focus of attentional and physiological resources was placed on the stabilizing limb. With the research we described here, *a predominant portion of right- and left-footed participants chose the right limb or had no preference for stabilizing support.*

9.1 Implications

From these findings it seems that a global meaning for footedness as described by the common operational definition is perhaps too simplistic to truly assess individual footedness. For further consideration is the following. In their efforts to develop a foot-preference inventory, Chapman and colleagues examined 13 foot behaviours as the bases of overall footedness; all but three of the tasks fit the bilateral-context description. From the 13 tasks, an 11-item inventory was recommended; nine being bilateral activities. As one would expect, all of the items that correlated well with the overall footedness index were bilateral tasks; e.g., kicking a ball, writing name in sand with toes, and smoothing the sand with toes. Also not surprising was that the two lowest correlation values (0.48) were those of the two unilateral task items: hopping (stability only) and foot tapping while seated (manipulation only).

Similarly, as noted earlier in reference to the postural origins theory, Maki (1990) found a weak relation between bilateral kicking and the unilateral task of spinning on one foot. In addition, Armitage and Larkin (1993) found that when footedness was determined using three bilateral tasks, 65% of normal 8- and 9-year-olds were right-sided. However, when the footedness profile was recalculated based on an additional three tasks of unipedal action (hopping, balancing, and foot tapping) there was a significant decrease in right-sidedness (30% right-footed) and two-fold increase in mixed-footedness (70%).

Our point is that any definition of footedness should be considered in light of its specific contextual use, and certainly be given more definitive attention when used to examine any relation to brain function. It would seem that the statement by Elias et al. (1998), that they could not be certain which component of footedness was predictive as a measure of cerebral lateralization, underlines the issue.

10. REFERENCES

Annett, M. (1985). *Left, right, hand and brain: The right shift theory.* London: Erlbaum.

Armitage, M., & Larkin, D. (1993). Laterality, motor asymmetry and clumsiness in children. *Human Movement Science, 12,* 155-177.

Chapman, J.P., Chapman, L.J., & Allen, J.J. (1987). The measurement of foot preference. *Neuropsychologia, 25,* 579-584.

Coren, S. (1993). The lateral preference inventory for measurement of handedness, footedness, eyedness, and earedness: Norms for young adults. *Bulletin of the Psychonomic Society, 31,* 1-3.

Corballis, M C. (1997). The genetics and evolution of handedness. *Psychological Review,104,* 714-727.

Cunningham, G. F., MacDonald, P. C., & Gant, N. F. (Eds.). (1989). *Williams obstetrics* (18th ed.). Norwalk, CT: Appleton & Lange.

Dargent-Paré, C., De Agostini, M., Mesbah, M., & Dellatolas, G. (1992). Foot and eye preferences in adults: Relationship with handedness, sex, and age. *Cortex, 28,* 343-351.

Day, L., & MacNeilage, P. (1996). Postural asymmetries and language lateralization in humans (Homo Sapiens). *Journal of Comparative Psychology, 110,* 88-96.

Dodrill, C. B., & Thoreson, N. S. (1993). Reliability of the lateral dominance examination. *Journal of Clinical and Experimental Neuropsychology, 15,* 183-190.

Elias, L. J., & Bryden, M. P. (1998). Footedness is a better predictor of language lateralization than handedness. *Laterality, 3* (1), 41-51.

Elias, L. J., & Bryden, M. P., & Bulman-Fleming, M. B. (1998). Footedness is a better predictor than is handedness of emotional lateralization. *Neuropsychologia, 36,* 37-43.

Gabbard, C., & Hart, S. (1996). A question of foot dominance. *Journal of General Psychology, 123,* 289-296.

Gabbard, C., & Iteya, M. (1996). Foot laterality in children, adolescents and adults. *Laterality, 1,* 199-205.

Ganong, W. F. (1993). *Review of medical physiology* (16th ed.). Norwalk, CT: Appleton & Lange.

Gentry, V., & Gabbard, C. (1995). Foot Preference behaviour: A developmental perspective. *The Journal of General Psychology, 122,* 37-45.

Gilbert, A. N., & Wysocki, C. J. (1992). Hand preference and age in the United States. *Neuropsychologia, 30,* 601-608.

Hart, S., & Gabbard, C. (1997). Examining the stabilising characteristics of footedness. *Laterality, 2,* 17-26.

Hart, S., & Gabbard, C. (1998). Examining the mobilizing features of footedness. *Perceptual and Motor Skills, 86,* 1339-1342.

Hart, S., & Gabbard, C. (1996). Bilateral footedness and task complexity. *Internationa Journal of Neuroscience, 88,* 141-146.

Hellige, J. B. (1993). *Hemispheric asymmetry.* Cambridge, MA: Harvard University Press.

Hugdahl, K., Satz, P., Mitrushina, M., & Miller, E. N. (1993). Left handedness and old age: Do left handers die earlier? *Neuropsychologia, 31*,325-333.

Katsarkas, A., Smith, H., & Galiana, H. (1994). Postural instability on one foot in patients with loss of unilateral peripheral vestibular function. *Journal of Vestibular Research, 4,* 153-160.

Laland, K, N., Kumm, J., Van Horn, J. D., & Feldman, M. W. (1995). A gene-culture model of human handedness. *Behavior Genetics, 25,* 433-445.

McManus, I. C., & Bryden, M. P. (1992). The genetics of handedness, cerebral dominance and lateralization. In I. Rapin & S. J. Segalowitz (Eds.), *Handbook of neuropsychology, Vol. 6, Child neuropsychology* (pp. 115-144). Amsterdam : Elsevier.

MacNeilage, P. (1991). The "postural origins" theory of primate neurobiological asymmetries. In N. Krasnegor, D. Rumbaugh, M. Studdert-Kennedy, & R. Schiefelbusch (Eds.), *The biological foundations of language development* (pp.165-188). Hillsdale, NJ : Lawrence Erlbaum.

Maki, S.G. (1990). An experimental approach to the postural origins theory of neurobiological asymmetries in primates. *Unpublished doctoral dissertation*, University of Texas at Austin.

Nonis, K. P. (1996). A mixed longitudinal study of the development of lower limb preference and hopping performance in girls. *Unpublished doctoral dissertation*. The University of Western Australia.

Peters, M. (1988). Footedness: Asymmetries in foot preference and skill and neuropsychological assessment of foot movement. *Psychological Bulletin, 103*, 179-192.

Peters, M. (1990). Neuropsychological identification of motor problems: Can we learn something from the feet and legs that hands and arms will not tell us? *Neuropsychology Review, 1,* 165-183.

Pompeiano, O. (1985). Experimental central nervous system lesions and posture. In M. Igarashi & K. G. Nute (Eds.), *Proceedings of the symposium on vestibular organs and altered force environment* (pp. 1-23). Houston, TX: NASA Space Biomedial Research Institute.

Porac, C., & Coren, S. (1981). *Lateral preferences and human behavior.* New York: Springer-Verlag.

Porac, C., Coren, S., & Duncan, P. (1980). Lifespan age trends in laterality. *Journal of Gerontology, 35,* 715-721.

Previc, F. H. (1991). A general theory concerning the prenatal origins of cerebral lateralization in humans. *Psychological Review, 98,* 299-334.

Reitan, R.M., & Davison, L.A. (Eds.). (1974). Clinical neuropsychology: Current status and applications. *The lateral dominance examination.* Washington DC: Winston & Sons.

Seeley, R. R., Stephens, T. D., & Tate, P. (1992). *Anatomy and physiology.* St. Louis: Mosby.

Spry, S., Zebas, C., & Visser, M. (1993). What is leg dominance? In J. Hamill (Ed.), *Biomechanics in Sport XI: Proceedings of the XI Symposium of the International Society of Biomechanics in Sports.* Amherst, Mass.

Taylor, E. S. (1976). *Beck's Obstetrical Practice and Fetal Medicine* (10th ed.). Baltimore : Williams & Wilkins.

Whittington, J.E., & Richards, P.N. (1987). The stability of children's laterality preferences and their relationship to measures of performance. *British Journal of Educational Psychology, 57,* 45-55.

Wren, B. G., & Lobo, R. A. (Eds.). (1989). *Handbook of Obstetrics and Gynaecology.* St. Louis: Mosby.

Watson, G. S., Pusakulich, R. L., Hermann, B., Ward, J. P., & Wyler, A. (1993). Hand, foot, and language laterality: Evidence from Wada testing (Abstract). *Journal of Clinical and Experimental Neuropsychology, 15,* 35.

Watson, G. S., Pusakulich, R. L., Ward, J. P., & Hermann, B. (1998). Hand, foot, and language laterality: Evidence from Wada testing. *Laterality, 3*, 323-330.

Yeo, R. A., & Gangestad, S. W. (1993). Developmental origins of variation in human hand preference. *Genetica, 89,* 281-296.

Chapter 11

"Tell Me, Where is [this] Fancy Bred?": The Cardiac and Cerebral Accounts of the Lateral Cradling Bias

Oliver H. Turnbull[1] and Marilyn D. Lucas[2]
[1]University of Wales, UK: [2] University of the Witwatersrand, South Africa

"Tell me where is fancy bred.
Or in the heart or in the head?"
William Shakespeare, The Merchant of Venice, ii: 63

It has been some time since scientists seriously debated the relative merits of 'heart versus head' as the organ of mind. The relative importance of these organs was, of course, an important issue for a number of ancient civilisations (Finger, 1994, pp.14-15) - and produced a long-standing debate that featured figures as prominent as Democritus, Plato and Aristotle. The debate centred around the relative role of the heart (and to a lesser extent some other viscera) in our emotional lives. In contrast, the head was argued (at least by some) to factor more heavily in the intellectual domain. The issue remained important for several centuries, and its continued longevity is suggested by its (relatively) recent appearance in Shakespeare's quote, together with the existence of phrases such as having a 'broken heart', or hearing a 'heart-warming' story. However, the proposition that there is a cardiac basis of mind has not been seriously entertained for several hundred years. Nevertheless, there is one small corner of psychological science in which there has been a more recent debate on the relative merits of the

M.K. Mandal, M.B. Bulman-Fleming and G. Tiwari (eds.), Side Bias: A Neuropsychological Perspective, 267-287.

cardiac versus the cerebral theories. This is the investigation of the lateral cradling bias – a domain that has developed into a small but vigorous research area during the last few decades. The present chapter reviews this rather more narrow 'heart versus head' debate, presenting a series of arguments for and against each position. The result is, alas, less clear-cut than the outcome of the more general question of the heart as an organ of mind. Nevertheless, the cardiac versus cerebral debate has provided an interesting series of attempts to explain what, at first glance, appears to be a relatively simple phenomenon.

1. THE LATERAL CRADLING BIAS

There appear to be a wide range of situations in which humans show some form of lateral preference - in activities involving the hands, feet and eyes. However, these preferences almost always involve the right side of the body - most prominently demonstrated by the fact that the vast majority of humans are right handed. One clear exception is the finding that humans, especially female humans, prefer to cradle infants to the left side of the body midline (i.e., resting the infant's head on the left arm or shoulder). This preference has proved reliable in a wide range of contexts. It has been consistently reported regardless of age or parental status, and has been found in all human societies so far investigated (e.g., Bruser, 1981; Saling & Cooke, 1984). By investigating the artistic output of previous generations it has also been possible to show that the preference has existed in a wide range of historical contexts (e.g., Finger, 1975). The phenomenon is also found in a wide range of situations: leftward cradling has been observed in the real mother-infant interaction (e.g., Salk, 1960), when a doll is used to represent the baby (De Château & Andersson, 1976), and even when participants are asked to merely *imagine* cradling an infant (Nakamichi & Takeda, 1995; Harris, Almerigi, & Kirsch, 2000). Finally, a left-sided cradling preference has been reported even in some species of great ape (Salk, 1960; Manning & Chamberlain, 1990; Nishida, 1993).

In this context, a likely explanation for the leftward cradling bias might be that it is reducible to the well-established phenomenon of right handedness. Specifically, one might trust the explanation offered by most right-handed women when asked about their cradling preference, who usually suggest that they cradle the infant in their left arm in order to free their right hand for use in the many domestic tasks that have to be performed whilst holding an infant. However, many left-handed women are also left cradlers, and they tend to argue that they cradle the infant leftwards because this is their stronger, safer, arm. Thus, it seems that such arguments are after-the-fact accounts - designed to explain the origin of a bias that is beyond the conscious awareness of the people concerned. This assertion is backed by the findings of many studies of the cradling bias - which show no significant relation between handedness and cradling. It is true that the magnitude of the cradling bias is marginally affected by handedness, but left handers show almost as significant a leftward cradling preference as do right handers. For example, in the classic Salk (1960) study, leftward cradling was shown by 83% of right handers, and 79% of left handers.

2. SALK'S CARDIAC THEORY

The first closely argued explanation for the lateral cradling bias stressed the importance of the lateral position of the heart (Salk, 1960). Lee Salk appears to have first developed the theory when observing a mother rhesus monkey in the Central Park Zoo in New York (see Salk, 1973). Visiting the Zoo repeatedly, Salk observed the mother showing a left-sided cradling preference on 40 of 42 observed occasions. Salk reports wondering whether the phrase 'close to a mother's heart' was "more than just an expression" (Salk, 1973, p.24), and on this basis proposed a 'cardiac' theory - offering a mechanism that accounted for the role of the heart in the mother-infant relationship. Specifically, Salk suggested that the child developed an association, *in utero*, between the regular sounds of the mother's heartbeat and the experience of a secure and relatively stress-free environment. When, after birth, the child was distressed, it would be pacified by being re-exposed to these comforting heartbeat sounds. Salk also hinted at a more extreme version of this claim, by suggesting an emotional role for heartbeat sounds in humans of all ages - noting that "from the most primitive tribal drumbeats to the symphonies of Mozart and Beethoven there is a similarity to the rhythm of the human heart" (Salk, 1973, p.29).

Salk did, however, concede that the importance of heartbeat sound was greatest for infants. Indeed, Salk stressed the role of the immediate post-partum period heavily, and attempted to make an explicit link with the 'imprinting' theories then popular in explaining some animal behaviours (see especially Salk, 1961). We will focus on a number of features of the theory in this chapter - a theory which can be generally stated as: the heart is an organ positioned left of the body midline; mothers can best pacify their new-borns by placing them near the heart, especially in the immediate post-partum ('imprinting') period, and this would favour a left-sided cradling bias.

One of the empirical features of Salk's work was to quantify the magnitude of this lateral bias. The effect was certainly large, and (as noted above) almost entirely unaffected by handedness. Of 287 mothers, some 80% of participants cradled leftwards. We will not review the fate of this empirical aspect of Salk's theory in much detail, because a number of subsequent studies have confirmed the strength of the cradling bias as a phenomenon. However, we will briefly review the other aspects of Salk's empirical work, which was aimed at supporting the components of the theory that emphasised, firstly, the importance of the heartbeat sound and, secondly, the proposal that the immediate post-partum period was critical for 'imprinting'.

In this regard Salk presented a variety of sources of evidence, from the strictly empirical to the frankly anecdotal. The main sources of claimed evidence were (a) neonates exposed to heartbeat sounds gain more weight (and show other health benefits) and (b) older children fall asleep more easily when exposed to heartbeat sounds. As regards the 'imprinting' aspect of the theory, Salk repeatedly cited a range of arguments alluding to the importance of the heartbeat in our adult lives. Salk's most commonly cited arguments are the existence of phrases such as 'my heart longs for you' and being 'broken-hearted' (e.g., Salk, 1960, p.168). The other line of argument (cited above) is the importance of music (with its assumed heartbeat-like rhythm) in all cultures. These arguments cannot be taken as serious evidence for a specific imprinting claim - especially as they can be explained using a variety of other after-the-fact accounts [3]. However, the main empirical source of evidence for the 'imprinting' aspect of hypothesis appears to be that (c) mothers who are separated from their infants immediately post-partum do not show a leftward cradling bias. We will briefly review these three findings.

(a) *Weight gain, and other health benefits, in neonates*: The fact that heartbeat sounds are important for infants was bolstered by Salk's finding that infants exposed to heartbeat sounds after birth showed more weight gain

[3] For example, the heart is probably thought to be important in our emotional lives because the heart-beat is 'audible' when it increases its work-rate, such as at times when we are anxious. As for the importance of music, the bio-mechanics of operating percussion instruments (such as drums) promote the production of rhythmical, rather than non-rhythmical tunes. These simple arguments (and doubtless there are many others, all of limited scientific value) have no link with *in utero* exposure, or post-partum imprinting.

than those not exposed to such sounds. If one assumes that infants gain more weight when pacified (because they are not crying), then this clearly supports Salk's claim. In fact, the details of Salk's (1960) study are of particular interest in terms of experimental methodology. In our opinion (and also the opinion of several other investigators, see below) the studies exhibit a number of design features that might have hampered the gathering of well-controlled data - and perhaps introduced a degree of experimenter bias. For example, the original planned treatment condition was exposure to a 72 beats-per-minute heartbeat sound. Salk reports that an attempt to introduce a control condition of 128 beat-per-minute sound was terminated, almost as soon as it was introduced (Salk, 1960, pp.172-174), because there was "an immediate increase in the crying and restlessness of the infants" (Salk, 1973, p.29). Instead, a 'no-heartbeat-sound' control condition was used [4.]

Indeed, the results of heartbeat exposure were so dramatic that they appeared to require that we immediately institute a revolution in the nature of post-natal care (Salk, 1960). Infants showed dramatic increases (at greater than the 0.001 level) in weight-gain, despite no difference in food intake (babies were weighed each morning, apparently by the nursing staff, who seem to have been aware of the nature of the study). In contrast, the no-heartbeat babies were (surprisingly) reported as showing a median <u>loss</u> of 20 grams in weight in the course of the study (Salk, 1973, p.29).

In addition, the incidence of infant crying (measured as the percentage of time when any infant cried on the ward) was reduced from 60% to 38% in the heartbeat condition. Further support for the claim came from the fact that "clinical observation" revealed a "greater depth and regularity in breathing in heartbeat babies" (p.172), and that fewer babies were suspected of gastro-intestinal or respiratory upset. Numerical support for these claims was not offered.

Finally, there are anecdotal mentions of the benefits of heartbeat sounds. In the early stages of the experimental condition, the audio-tape developed a "slight hissing" sound, and it is reported that the nurses "on all shifts" found the babies more fussy and restless at these times (p.173). As for the reliability of the findings, Salk (1961 pp.742-743) claimed that the

[4] The fact that a control condition was so rapidly rejected might set a skeptic to worrying about whether the experimenter was 'overly committed' to the benefits of heart-beat sounds. The alternative is, of course, that Salk was entirely correct about the negative effects of non-heart-beat sounds - but this possibility should be considered in the light of later work, which failed to replicate this finding (see ii below).

results had been replicated by a Dr. R.C. Reed. Unfortunately, Salk's paper does not specify precisely which components of the original study were replicated, or offer any data to support the claim, except that a single infant's heart rate was apparently reduced from 146 to 114 beats per minute when he or she heard the heartbeat sound, and that the rate returned to 138 beats per minute when the sound was stopped.

(b) *Older children fall asleep more easily*: As regards older children, Salk (1960, p.174) reports that the use of heartbeat sounds was a highly successful treatment in one five-year-old with sleeping difficulties (reducing his time-to-sleep at nights from 2-3 hours to under 15 minutes). In a later study (Salk, 1961 pp.743-744) of children in a children's home (almost all aged 18-36 months) it was found that they fell asleep more quickly when exposed to heartbeat sounds. Indeed, they took less than half as long to fall asleep as when exposed to any other control condition (no sound, a metronome, or lullabies).

(c) *Post-partum separated mothers do not show a leftward cradling bias*: In Salk (1970, p.113) it was claimed that mothers who had experienced no post-partum separation from their infants (i.e., had held the child in the first 24 hours) showed a 77% leftward cradling bias, whereas mothers who had not held their children in the 24-hour post-partum period did not show a leftward cradling bias (the exact figure was 53% leftward). The figures were intermediate for a group who had been separated from their current child, but for whom this was not the first pregnancy. Salk argued that this initial 24-hour period was vital to the 'imprinting' process, and for the establishment of a "natural mother-infant interaction" (Salk, 1970, p.114).

3. THE FATE OF SALK'S THEORY

Salk's argument certainly represents an interesting proposal, with obvious implications for child care, especially in the immediate post-partum period. As a scientific theory it is also clearly open to empirical test in several areas, it has served to begin a series of cradling-bias investigations that has continued for several decades, and has been widely cited in the literature. However, the theory has not fared as well as might have been expected from the dramatic results of Salk's study. Firstly, the aspect of the theory that stresses the soothing effects of heartbeat sounds has not been as universally accepted as Salk's empirical evidence might have demanded. Certainly, hospitals (at least in our experience, and that of our colleagues internationally) do not routinely play heartbeat sounds to neonates, though the strength of the original findings might well have suggested that such methods would be warranted [5]. Because the 'medical uptake' aspect of the theory lies outside the domain of a chapter on lateral biases, and because we have no specialised knowledge of the fate of this aspect of the theory, we will not focus on this issue.

A second issue remains – that is, has Salk's explanation for the cause of the lateral cradling bias stood the test of time? It would appear that Salk's argument has rarely been examined systematically for its validity. Instead, the tendency has often been for authors to use the 'cardiac' findings as a platform for their own research. Nevertheless, some aspects of Salk's theory have been tested. However, such tests have generally not supported the original claims. There are also some other aspects of Salk's theory which are open to criticism, but appear not to have been mentioned in the literature. To the best of our knowledge, the full range of arguments against Salk's cardiac theory have yet to be systematically laid out. We will outline the basis of these several criticisms of the theory, before turning to the more recently popular, cerebral, theory of the lateral cradling bias.

There are, by our count, some nine lines of argument against Salk's theory – ranging from the strictly empirical to the entirely theoretical. It could be that none of these arguments is, individually, capable of discrediting the theory. However, as a group, they form a formidable argument against the cardiac theory of the cradling bias. We begin with the empirical data, and later pass on to the theoretical aspects of the case against the cardiac theory.

(i) Firstly, there is the question of the methodology of Salk's original 'weight-gain' study. Our comments above, relating to experimental design, were partly based on the assumption that variables such as weight are difficult to measure reliably in neonates. Related to this is the fact that the data appear not to have been collected 'blind'. Hence Salk's weight-gain findings may have been open to experimenter bias (albeit unconscious). This should also be interpreted in the face of a failure to replicate Salk's weight-

[5] However, heart-beat-sound devices can be found for private purchase. One of us (ML) actually bought such a device in the early 1980's (years before developing an interest in cradling). Several people have mentioned to us that specific heart-beat-sound devices (or conventional audio-tapes playing heart-beat sounds) can still be purchased from child-care specialists. Some such devices also produce a 'gurgling' sound, presumably on the argument that the infant would have also been exposed to these sounds *in utero*.

gain findings (Palmqvist, 1975), in which infants exposed to heartbeat sounds did not show significantly greater weight gain. Of course, this does not mean that they are not necessarily pacified by heartbeat sounds (though see ii below) – it may well merely be that pacification does not necessarily lead to weight gain. Nevertheless, these data suggest that one empirical link between cradling and the heart is open to question.

(ii) Secondly, Salk's theory rests of the specific importance of the maternal (72 beats per minute) heartbeat. Salk had used a regular rhythm heartbeat rate of 72 beats per minute to pacify infants, but rejected a rate of 128 beats per minute because it was almost immediately apparent that it was stressful for infants (Salk, 1960, 1973). Later investigations have attacked this finding from a number of perspectives. For example, it has been suggested that the infant may often be unable to hear the maternal heartbeat *in utero* (Querleu & Renard, 1981, cited in Stables & Hewitt, 1995, p.29). Weiland and Sperber (1970, p.157) claimed that the 72-beats-per-minute rhythm was not that of the average human heartbeat. On a more empirical front, Detterman (1978) was unable to replicate Salk's finding that heartbeat sound pacifies infants. In contrast, it has been shown that higher frequency tones, at both 150 and 500 cycles per second, *can* pacify infants (Birns, Blank, Bridger, & Escalona, 1965). At rates more comparable to that of the Salk study, Ockleford (1984) found that 144 beats per minute was the most soothing rate. It may be that *any* high intensity sound has a pacifying effect (Weiland & Sperber, 1970). On a more theoretical note, we should recall that the infant's *in utero* experience of a fast maternal heartbeat would have been linked with the experience of anxiety (as mediated by blood-borne increases in adrenalin). Thus, such higher-frequency rates should certainly not lead to the pacification of the infant. Finally, we note that the infant's heart rate is much faster than that of the mother, and that this heartbeat might also be heard by the infant *in utero*. Ockleford's (1984) report of a preferred 144 beats per minute is roughly that of an infant (rather than a normal-paced maternal) heartbeat. It is not entirely clear which rate of heartbeat the infant might link with the secure pre-partum environment. Thus, for a variety of reasons, the heartbeat aspect of Salk's theory seems questionable.

(iii) A further criticism relates to Salk's (1960) report of the lower incidence of infant crying when hearing the heartbeat sound - reduced from 60% to 38% in the heartbeat group. Crying was measured as the percentage of time when any infant cried on the ward – that is, only the presence of the crying sound was documented (based on a tape recorder, turned on automatically for 30 seconds every 7 minutes, with crying time was averaged from these 30-second intervals). With such a system, in which the actual number of crying babies is not measured, it is important that group size be matched. However, as Detterman (1978) pointed out, there were 112 infants in the control group, but only 102 infants in the heartbeat group. This discrepancy may well account for difference in amount of crying - because fewer babies on the ward means fewer babies crying at any given time. Indeed, because the number of babies was small ("an average of nine babies in the nursery at one time", Salk, 1960, p.173), and because one infant's crying can wake another, the effects of adding a single infant to the ward might well have been substantial.

(iv) A further line of argument against Salk is that non-rhythmical movements will pacify infants, as will non-body-contact vestibular-proprioceptive movement (Korner & Thoman, 1972). The former type of movement lacks the regularity that is so central to heartbeat, and the latter are movements in which the heart is 'out-of-earshot' of the infant. Thus, the potential role of hearing a regular and specific heart rate might be argued to be further undermined.

(v) Salk also attempted to marshall several lines of evidence to support the 'imprinting' aspect of the cardiac theory – that "the stimulus of holding an infant releases a certain maternal response" (Salk, 1973, p.27). We have mentioned above that we consider several of these proposals (such as feeling things 'from the bottom my heart') to be inappropriate forms of argument. The only empirical aspect to this claim relates to the ill-effects of post-partum separation for maternal 'bonding', when Salk reports that a strong left-sided preference was only true of experienced mothers. Again, there are several problems with this finding. Firstly, on a simple quantitative matter, De Château, Holmberg, & Winberg (1978) did find that the leftward preference was reduced in mothers who had been separated in the first 24 hours, but the rate was still some 60-70% leftward. That is, the leftward cradling bias was still present (Salk, 1970, found a rate of 53%). Secondly, as De Château et al. (1978) note, a strong leftward cradling bias is found even in women who have never had children. Indeed, we have found a strong effect in every study of nulliparous women that we have ever conducted, with the sample usually being female university undergraduates (Lucas, Turnbull, & Kaplan-Solms, 1993; Matheson & Turnbull, 1998; Turnbull et al., 1995; Turnbull & Lucas, 1996). In addition, a 79% leftward cradling bias has been reported in six-year-old girls (De Château & Andersson, 1976; see also Saling & Bonert, 1983). In summary, there seems no question but that a strong left-sided cradling bias exists in women who have never had children. Thus, Salk's findings imply that post-partum separation actually disrupts an existing left-sided cradling preference in women (cf. De Château et al., 1978). Finally, although there do appear to be advantages to additional mother-infant contact in the immediate post-partum period (e.g., De Château & Winberg, 1977), it is not clear that it is the sound of the maternal heartbeat that produces such advantages. A variety of explanations could be advanced to support the proposition that mothers should be allowed to cradle their babies soon after birth, not least because of the widespread belief in Western culture that immediate post-partum contact is beneficial. Such accounts certainly do not require that the infant hears the heartbeat of the mother.

(vi) A further point relates to the question of cardiac anatomy. Although the heart is widely thought of as lying to the left of centre, its medial aspect lies close to the midline. Thus, heartbeat sounds are detectable either side of the body midline (though it must be conceded that the left ventricular beat is the more prominent one). There is also the related matter of actually hearing the heartbeat sound. Bundy (1979) pointed out that unless the infant's ear was pressed against the mother's chest, the heartbeat sounds would be difficult to detect. This situation would apply especially when clothing is worn by the mother, which is the case for most cradling experiences (see iv, above). Of course, the infant does have skin-to-skin contact when breast-feeding, under which circumstances the heartbeat would be more audible. However, note that mothers must feed from both breasts (cf. Stables & Hewitt, 1995), so the circumstance under which the heartbeat is most likely to be heard is that under which women are least likely to be able to favour the left side.

(vii) The imprinting hypothesis has been explored by other researchers as an anxiety-reduction technique for mothers, rather than infants. Weiland and Sperber (1970; Sperber & Weiland, 1973) argued that in anxiety-provoking situations, individuals tended to hold objects leftwardly. Thus, they argued that the heartbeat was a distraction, rather than pacification, stimulus. This theory can be criticised for several reasons. Firstly, the cradling bias appears from six years of age (De Château & Andersson, 1976) when mothering anxieties seem unlikely. Secondly, it should follow from this assumption that mothers of sickly and premature infants would cradle more leftwardly than any other group. However, the reverse pattern occurs, for when mothers are separated from their newly born infants (through incubation or high care from others), the dominant cradling bias is disrupted (Salk, 1970).

(viii) It might also be argued that Salk presents an untestable hypothesis, because a control group of infants who have never been exposed to heartbeat sound prenatally cannot exist (Ginsburg, Fling, Hope, Musgrove, & Andrews, 1979). On this matter, it might perhaps be argued that congenitally deaf infants would not have heard the heartbeat sound - offering a potential control group. On this matter, we have collected data on cradling in the deaf, suggesting that they *do* show a leftward cradling bias (Turnbull & Matheson, 1996; Turnbull et al., under review). However, it might then perhaps be counter-argued that the acoustic stimulus could be heard by non-auditory means, simply as low frequency vibration. Thus, our findings cannot speak directly to the veracity of Salk's theory, because we cannot know the extent to which the deaf experience heartbeat 'sounds' *in utero*. However, it may well be that 'all mothers have a heart' aspect of the theory is not entirely untestable – if one is prepared to design an appropriate experiment, with suitable control groups.

(ix) Finally, the theory might be argued to be 'untestable' because the heart is on the left in all humans. There would be no control group of women with hearts on the opposite side, who would be argued to show a reversed cradling bias. Of course, it is not entirely true that such a control group do not exist. The medical phenomenon of reversal of the entire viscera (situs inversus) is extremely rare, and there are marginally more frequent cases of reversal of the heart and associated vasculature (dextrocardia). Furthermore, we note that the prevalence of rightward cradling is approximately 25% - so that such cardiac anomalies cannot be invoked to account for a phenomenon as common as rightward cradling. As for testing such a proposal, individuals with reversed heart positions are often difficult to trace, because they frequently have no medical disorders associated with their unusual cardiac anatomy. However, such an empirical investigation might be possible, and the attack of 'untestability' seems inappropriate on this count. Nevertheless, in the single case of dextrocardia in which the cradling bias has been tested, the mother cradled leftwards (Todd & Butterworth, 1998), i.e., counter to Salk's argument.

In summary, then, it appears that Salk's cardiac account of the basis of the lateral cradling bias can be attacked on a number of fronts. There are several ways in which the theory has been tested, and it has frequently been found wanting. Also, it is clear that the literature on cradling generated since Salk (1960) has been dedicated largely to testing theories other than Salk's. Instead, Salk's theory has been a catalyst, or springboard, for other lines of inquiry. It is to the most recent of these that this chapter now turns - an account of the cradling bias that is cerebral, rather than cardiac. The cerebral account of the cradling bias has dominated the theoretical agenda of work on the cradling bias in the 1990's. In some respects it seems a more scientifically 'sophisticated' account than Salk's cardiac theory - perhaps because it is not linked to ill-defined claims about feeling things 'from the bottom of your heart', and the role of music in adult life. Also, the literature on the cerebral theory is far less peppered with anecdotal evidence, which Salk often strayed towards. Finally, the cerebral theory's alignment with modern neuroscience has made it fit well with the general Zeitgeist. Nevertheless, as we shall see, the cerebral theory, like the cardiac theory before it, is open to a range of theoretical and empirical criticisms.

4. THE 'CEREBRAL' THEORY AND HEMISPHERIC ASYMMETRY

The cerebral, or more specifically the 'hemispheric asymmetry', account of the cradling bias has its origin in a body of evidence on the lateralization of emotion that became increasingly well-developed during the 1970's and 1980's. Specifically, the theory reminds us that the right hemisphere (in those with conventional cerebral dominance) is specialised for the perception and expression of emotion (see Campbell, 1982 for an early review). Based on this well-established finding, it appeared plausible that women might well experience a 'more optimal' emotional interaction with the infant if the infant were in the left hemi-space. This proposal was (to the best of our knowledge) first suggested by Lockard, Daley, and Gunderson (1979, p.236), but the argument also appears to have been independently suggested by Harris (1983), Kaplan (1985), and Manning and Chamberlain (1990; 1991). These accounts generally argue that cradling the infant leftwardly would allow the mother to interpret the infant's behaviour through her left visual and auditory field, as well as presenting the more expressive side of her face to the infant.

This account is a plausible approach to the problem of the origins of the cradling bias. It is based on established work in neuropsychology, it predicts a lateral bias in the correct (leftward) direction, and it does not predict (as the simple version of Salk's account does) that virtually all women should cradle leftwards (because they almost all have the heart on the left). Rather, the proportion of rightward cradlers should roughly mirror the percentage of individuals who show reversed or unusual hemispheric asymmetry of function - which is certainly greater than the percentage with situs inversus or dextrocardia (see vi, below). This account can even be credibly used to explain the nature of the lateral cradling bias in men. If men are less likely (as a population) to experience an emotional interaction when holding infants, this might explain why men are less likely to show the cradling bias, and also why men are more likely to show a leftward bias after becoming fathers (De Château, 1983).

One appealing aspect of the hemispheric-asymmetry hypothesis is that it is perhaps a more readily testable account than Salk's theory - because hearing a heartbeat sound *in utero*, and having the heart to the left of body midline, are almost universal in humans, and hence a control group would be difficult to find. However, as in the case of Salk's theory, empirical tests of the theory have not always proved supportive. Indeed, over the last few years we have been involved in a variety of attempts to test the hemispheric asymmetry account, all of which have not supported the theory. The various findings are briefly discussed below. As in the case of the criticisms of Salk's account, some are empirical and others are more theoretical.

(i) Firstly, we have made two clear attempts to test a possible link between the right-hemisphere advantages in visual abilities and the cradling bias (Lucas et al., 1993, Turnbull & Lucas, 1996). Essentially the same paradigm was used in both cases. In a group of healthy female undergraduates, we measured a visual function which was well-established as having a right-hemisphere (and hence left-visual-field) lateral bias. We also measured the direction of the participants' lateral cradling bias, and presumed that if the two were related then the variables would be correlated. In one case, the 'right-hemisphere' task was the perception of emotional expression on a chimeric faces task (Lucas et al., 1993) and in the other it was the lateral attentional bias (towards the left) on visual line bisection (Turnbull & Lucas, 1996). In both cases, we were able to replicate the findings of a left-visual-field (i.e., right-hemisphere) bias. We were also able to show the usual leftward cradling bias in both samples. However, in both studies, we did not find a significant correlation between the 'right-hemisphere' function and the lateral cradling bias (but see Harris et al., 2000).

(ii) We also attempted to test the hemispheric asymmetry theory by investigating other situations that might be tests of the lateral expression of emotion (Turnbull, Stein, & Lucas, 1995). We argued that perhaps embracing was another situation in which one had to make a choice of laterality, before having close contact with an emotionally important 'other'. One would presumably engage right-hemisphere emotional systems when embracing the other, meaning that one might better engage emotionally when the other individual was in the left hemispace [6]. We tested this by investigating embracing in a laboratory setting, and in a field-observation study (in the Arrivals lounge of an international airport). Both laboratory and field samples showed effectively the same results. There was no clear leftwards bias, in contrast to cradling. Instead, there was a weak rightwards bias. As in the case of cradling, the effect was sex-related, being strongest when women embraced women, intermediate in the mixed-sex case, and weakest when men embraced men. In effect, then, the evidence from embracing, a behaviour that might be argued to share the same basic hemispheric asymmetry features of cradling, fails to show the same direction of lateral bias.

(iii) Manning and Chamberlain (1991) also pursued the "right-hemisphere specialisation for emotional processing" hypothesis, and had shown that by covering the left eye of the cradler (and hence restricting visual access to the right hemisphere [7]) the frequency of leftward cradling was reduced. Although the cradling bias was reduced when the left eye was occluded, the group's dominant cradling bias continued to be significantly leftward. We investigated this phenomenon in both women and men (Matheson & Turnbull, 1998). In women, we found no significant effect of various blindfolding combinations on the cradling bias, though there was a small (but non-significant) trend in Manning and Chamberlain's reported direction - the leftward bias was 79% in the right eye-patch condition, and 67% with the left eye-patch. Importantly, however, the male sample was substantially more influenced by the experimental conditions of eye-occlusion than was the female group - the leftward bias was 71% in the right eye-patch condition, and a remarkable 25% with the left eye-patch. This suggests that visual information is far more important for those who do not display a clear lateral cradling bias (i.e., males). This would undermine the original Manning and Chamberlain (1991) claim.

(iv) In the middle 1990's, Sieratzki and Woll (1996) wrote a speculative article to The Lancet, suggesting that the emphasis should have been placed on the auditory, rather than the visual, modality. Their argument

[6] On a methodological point, one should note that embracing is a symmetrical activity - when I embrace you to my left, then I lie to your left.

[7] Of course, information from the left visual part of the visual world is, in principle, available through *both* eyes. However, the nose occludes the inferior portion of the contralateral visual field, so that patching the left eye occludes vision of the leftwardly cradled infant, as viewed with the right eye.

was that mothers wish to hear their child's vocal communication, as well as wishing to communicate verbally with their child. Importantly, this auditory interaction is of a prosodic nature, rather than formally linguistic (Fifer & Moon, 1994). The right hemisphere (in those with conventional cerebral dominance) is, of course, specialised for the reception and production of such prosodic information. Hence the child would be preferentially cradled in the left hemi-space. To test this hypothesis, we assessed hemispheric asymmetry for prosodic comprehension (using the dichotic-listening technique), to see whether this measure correlated with the cradling bias (Turnbull & Bryson, in press). By analogy with the studies cited under (i) above, we replicated both the left-ear advantage for most participants, and the leftward cradling bias. However, as with the previous studies, we found no significant relation between the two variables and hence could not support the claim that the perception of prosody is related to the direction of the cradling bias.

(v) The above criticisms by Sieratzki and Woll (1996) suggest that there is a particular role for specific senses in the development of the cradling bias. In this regard, it is notable that we have assessed the lateral cradling bias in both blind and deaf individuals (Matheson & Turnbull, 1998; Turnbull et al., under review; Turnbull & Matheson, 1996). The results showed a clear leftward preference amongst the blind, and an even more clear leftward preference amongst the deaf. Of course, these particular lines of evidence can be countered with the claim that the deaf participants might employ visual cues in developing their cradling bias, whereas the blind participants might employ auditory cues. Nevertheless, these data do suggest that attempts to tie the cradling bias to hemispheric asymmetries in particular sense modalities are likely to prove fruitless.

(vi) If the hemispheric asymmetry theory is correct, then the proportion of right cradlers should be roughly the same as the proportion of those with atypical cerebral dominance. The most reliable figures on hemispheric specialisation are those related to language - especially as measured by the Wada test. These data suggest that roughly 90-95% of humans have language represented in the left hemisphere, so that some 10% or less have 'anomalous' language dominance. This might be taken to imply that 10% or less have anomalous dominance for 'right hemisphere' functions such as the perception and expression of emotion. If this were true, the figures would not account for the roughly 25% of women who cradle rightwards (i.e., in the anomalous direction). However, it appears that the extent of hemispheric asymmetry for functions such as emotion may be much greater than the figure of 10% (or less) reported for language (see Gainotti, 1997, for review). This evidence is, at least, not directly contradictory of the hemispheric-asymmetry hypothesis.

In summary, both the 'cardiac' and the 'cerebral' accounts of the lateral cradling bias appear plausible, and each certainly accounts for a range of cradling-related phenomena. However, science requires that we do more than develop 'plausible' theories - it also requires that the theories be tested. Here is the opinion of Richard Feynman [8] on the matter:

"In general we look for a new law by the following process. First we guess it... then we compare the result with experiment or experience, compare it directly with observation, to see if it works. *If it disagrees with experiment it is wrong*. In that simple statement is the key to science. It does not make any difference how beautiful your guess is. It does not make any difference how smart you are, who made the guess, or what his name is - if it disagrees with experiment it is wrong" [emphasis added] (Messenger Lectures, 1964, from Gribbin & Gribbin, 1997, pp. 178-179).

So, although the cardiac and cerebral theories of the cradling bias appear to be plausible, perhaps even 'beautiful' theories, it seems that they have not fared well when directly tested. Of course, it may be that the theories could perhaps be reformulated in a better way, or perhaps the experimental tests of the theories have themselves been flawed. However, on the balance of evidence currently available, it appears that they are not adequate explanations of the cause of the lateral cradling bias. An alternative would be to come up with another theory - to make another, better, 'guess'.

In this regard, it might be useful to review precisely what an adequate theory of the lateral cradling bias would have to account for. A good theory would have to explain (a) why most women prefer to cradle leftwards, (b) why the figure is roughly 75%, (c) why the phenomenon is so stable across age, culture, and historical period, (d) why leftward cradling is also found in the great apes, and (e) why a leftward bias is not found in men (at least until they become fathers). One might also include a final feature of cradling, which we have not yet raised in our review - (f) that the direction of cradling

[8] Richard Feynman (1918-1988) made remarkable contributions to several areas of physics in an enormously long and productive scientific career. He would have been entitled to two (perhaps even three) Nobel prizes, and would be on any short-list for the 'physicist of the century'. In the end, he won a single Nobel, in 1965, for his work in quantum electrodynamics. He is also famous for his ability to communicate complex ideas clearly to both fellow physicists and the general public.

appears to be related to the quality of the mother-infant interaction. It seems worthwhile to briefly review this finding, and consider its implications for both the cardiac and cerebral theory, and for any future theory of the cradling bias.

During the late 1970's and early 1980's a series of papers were published that appeared to show that women who cradled infants right of the midline had a poorer quality of interaction with their children in the first weeks and months post-partum than their leftward-cradling peers (Bogren, 1984; De Château et al., 1978; 1982). For example, rightward cradlers are reported to show less body contact in cradling, take longer to relate to and accept the newborn, and have more contacts with child services in the first three years, etc. Rightward cradlers also scored more poorly on a number of other measures, arguably also related to the quality of the mother-infant interaction, such as how early in pregnancy they prepared for the birth, and how they felt about bodily changes in pregnancy. Notably, these effects appear to be independent of a range of socio-economic factors, and measures of pregnancy and birth complications.

It is difficult to see how these findings might be reconciled to either the cardiac, or the cerebral, theory of cradling. For example, the hemispheric asymmetry account argues that women cradle rightwards because they have reversed asymmetry of function for the perception and expression of emotion. Taken at face value, this argument cannot account for finding (f). It is difficult to see how a mere reversal of the 'normal' pattern of asymmetry would result in a poorer mother-infant interaction - unless we are to return to the medieval practice of labeling individuals with anomalous dominance (i.e., left handers) as devil-worshipers or witches.

On this point, the findings relating to the mother-infant interaction are surprising in the context of the many excellent mothers who are rightward cradlers. To bolster this point (and as Salk was so fond of the use of single-subject anecdotal evidence) one of us (OT) noticed some years ago that the late Princess Diana showed a strong rightward cradling bias (in some 75% of available press photographs and video clips, for example). The Princess was a women who had many faults, but one would be hard-pressed to claim that she was not a caring and affectionate mother. Indeed, we are finding it difficult to replicate the mother-infant interaction findings (which are also, in some respects, open to methodological criticism) in group studies in our laboratory (e.g., McKinnon, 1998). Indeed, it may be that the claims relating to rightward cradlers and the mother-infant interaction are unfounded. However, if they are genuine effects, then these findings represent yet another aspect of cradling bias research that must be accounted for by any appropriate theory.

In conclusion, it appears that the cause of the leftward cradling bias continues to elude us. It is certainly regrettable that such an apparently simple, and empirically robust, phenomenon does not have an acceptable explanation. Indeed, it is especially unfortunate in the context of our explanations for other examples of lateral asymmetry. If we have a good explanation for why most humans prefer to favour their right hand, foot and eye on complex tasks, then such a model should easily explain why the left side is favoured when women cradle infants. Conversely, understanding the cause of the lateral cradling bias might go some way towards helping us to understand the phenomenon of lateralization in general.

5. REFERENCES

Birns, B., Blank., Bridger, W., & Escalona, S. (1965). Behavioural inhibition in neonates produced by auditory stimuli. *Child Development, 36*, 639-645.

Bogren, L. Y. (1984). Side preference in women and men when holding their newborn child: Psychological background. *Acta Psychiatrica Scandanavica, 69*, 13-23.

Bruser, E. (1981). Child transport in Sri Lanka. *Current Anthropology, 22*, 288-290.

Bundy, R. S. (1979). Effects of head position on side preference in adult handling. *Infant Behavior and Development, 2*, 355-358.

Campbell, R. (1982). The lateralisation of emotion: A critical review. *International Journal of Psychology, 17*, 211-229.

De Château, P. (1983). Left-side preference for holding and carrying newborns: Parental holding and carrying in the first week of life. *Journal of Nervous and Mental Disease, 171*, 241-245.

De Château, P., & Andersson Y. (1976). Left-side preference for holding and carrying newborn infants. II: Doll-holding and carrying from 2 to 16 years. *Developmental Medicine & Child Neurology, 18*, 738-744.

De Château, P., Holmberg, H., & Winberg, J. (1978). Left side preference in holding and carrying newborn infants I: Mother's holding and carrying during the first week of life. *Acta Paediatrica Scandanavica, 67*, 169-175.

De Château, P., & Winberg, J. (1977). Long term effect on mother-infant behaviour of extra contact during the first hour post-partum. *Acta Paediatrica Scandanavica, 66*, 145-151.

De Château, P., Maki, M., & Nyberg, B. (1982). Left side preference in holding and carrying newborn infants III: Mother's perception of pregnancy one month prior to delivery and subsequent holding behaviour during the first post-natal week. *Journal of Psychosomatic Obstetrics and Gynaecology, 1*, 72-76.

Detterman, D.K. (1978) The effect of heart-beat sound on neonatal crying. *Infant Behavior and Development, 1*, 36-48.

Fifer, W.P., & Moon, C.M. (1994). The role of the mother's voice in the organisation of brain functions in the newborn. *Acta Paediatrica Supplement, 397*, 86-93.

Finger, S. (1985). Child holding preferences in Western art. *Child Development, 46*, 267-271.

Finger, S. (1994). *Origins of neuroscience: A history of explorations into brain function*. New York: Oxford University Press.

Gainotti, G. (1997). Emotion disorders in relation to unilateral brain damage. In T.E. Feinberg, & M.J. Farah (eds.), *Behavioral neurology and neuropsychology*. New York: McGraw Hill.

Ginsburg, H.J., Fling, S., Hope, M.L., Musgrove, D., & Andrews, C. (1979). Maternal holding preferences: A consequence of the newborn head-turning response. *Child Development, 50*, 280-281.

Gribbin, J., & Gribbin, M. (1997). *Richard Feynman: A life in science*. London: Viking/Penguin.

Harris, L.J. (1983). Laterality of function in the infant: Historical and contemporary trends in theory and research. In G. Young, S.J. Segalowitz, C.M. Corter, & S.E. Trehub (Eds.), *Manual specialization and the developing brain* (pp. 177-247). New York: Academic Press.

Harris, L.J., Almerigi, J.B., & Kirsch, E.A. (2000). Side preference in adults for holding infants: Contribitions of sex and handedness in a test of imagination. *Brain and Cognition, 43*, 246-252.

Harris, L.J., & Fitzgerald, H.E. (1985). Lateral cradling preferences in men and women: Results from a photographic study. *Journal of Genetic Psychology, 112*, 185-189.

Kaplan K.L. (1985). Lateral differences in breast sensitivity and cradling preferences in nulliparous females. *Unpublished masters dissertation*, University of the Witwatersrand, Johannesburg.

Korner, A.F., & Thoman, E.B. (1972). The relative efficacy of contact and vestibular-proprioceptive stimulation in soothing infants. *Child Development, 43*, 443-453.

Lockard, J.S., Daley, P.C., & Gunderson, V.M. (1979). Maternal and paternal differences in infant carry: U.S. and African data. *The American Naturalist*, 113, 235-246.

Lucas M.D., Turnbull O.H., & Kaplan-Solms K.L. (1993). Laterality of cradling in relation to perception and expression of facial affect. *Journal of Genetic Psychology, 154*, 347-352.

McKinnon, C. (1998). The lateral cradling bias and quality of interaction with infants. *Unpublished undergraduate dissertation,* University of Aberdeen.

Manning, J.T., & Chamberlain, A.T. (1990). The left-side cradling preference in great apes. *Animal Behaviour, 39*, 1224-1227.

Manning, J.T., & Chamberlain, A.T. (1991). Left-side cradling and brain lateralisation. *Ethology and Sociobiology, 12*, 237-244.

Matheson, E.A, & Turnbull, O.H. (1998). Visual determinants of the leftward cradling bias. *Laterality, 3,* 283-288.

Nakamichi, M., & Takeda, S. (1995). A child-holding thought experiment: Students prefer to imagine holding an infant on the left side of the body. *Perceptual and Motor Skills, 80*, 687-690.

Nishida, T. (1993). Left nipple suckling preference in wild chimpanzees. *Ethology and Sociobiology, 14*, 45-52.

Ockleford, E. (1984). Response to sound in pre- and full-term infants. *Journal of Infant and Reproductive Psychology, 2,* 92-96.

Palmqvist, H. (1975). The effect of heart-beat sound stimulation on the weight development of newborn children. *Child Development, 46*, 292-295.

Querleu, D., & Renard, X. (1981). Les perceptions auditives du foetus humain. *Médicine et Hygiène, 39*, 2102-2110.

Saling, M.M., & Bonert, R. (1983). Lateral cradling preferences in female preschoolers. *Journal of Genetic Psychology, 142*, 149-150.

Saling, M.M., & Cooke, W. (1984). Cradling and transport of infants by Southern African mothers: A cross-cultural study. *Current Anthropology, 25*, 333-335.

Salk, L. (1960). The effects of the normal heart-beat sound on the behaviour of the newborn infant: Implications for mental health. *World Mental Health, 12*, 168-175.

Salk, L. (1961). The importance of heart-beat rhythm to human nature: Theoretical, clinical and experimental observations. *Proceedings of the 3rd World Congress of Psychiatry*. Toronto: University of Toronto Press: 740-760.

Salk, L. (1970). The critical nature of the post-partum period in the human for the establishment of the mother-infant bond. *Nervous System, 31*, 110-115.

Salk, L. (1973). The role of heart-beat in the relations between mother and infant. *Scientific American, 228*, 24-29.

Sieratzki, J.S., & Woll, B. (1996). Why do mothers cradle babies to their left? *The Lancet, 347*, 1746-1748.

Sperber, Z., & Weiland, H. (1973). Anxiety as a determinant of parent-infant contact patterns. *Psychosomatic Medicine, 35*, 472-483.

Stables, D., & Hewitt, G. (1995). The effect of lateral asymmetries on breast-feeding skills: Can midwives' holding interventions overcome unilateral breast-feding problems? *Midwifery, 11*, 28-36.

Todd, B., & Butterworth, G. (1998). Her heart is in the right place: An investigation of the 'heartbeat hypothesis' as an explanation of the left side cradling preference in a mother with dextrocardia. *Early Development and Parenting, 7*, 229-233.

Turnbull, O.H., & Bryson, H.E. (in press). The leftward cradling bias and hemispheric asymmetry for speech prosody. *Laterality*.

Turnbull, O.H., & Lucas, M.L. (1996). Is the lateral cradling preference related to lateral asymmetries in attention? *Journal of Genetic Psychology, 157*, 161-167.

Turnbull, O.H., & Matheson, E.A. (1996). Left-sided cradling. *The Lancet, 384*, 691-692.

Turnbull, O.H., Rhys-Jones, S.L. & Jackson, A.L. (under review). The leftward cradling bias and prosody: An investigation of cradling preferences in the deaf community. *Journal of Genetic Psychology*.

Turnbull, O.H., Stein, L., & Lucas, M. L. (1995). Lateral preferences in adult embracing: A test of the 'hemispheric asymmetry' theory of infant cradling. *Journal of Genetic Psychology, 156*, 17-21.

Weiland, H., & Sperber, Z. (1970). Patterns of mother-infant contact: The significance of lateral preference. *Journal of Genetic Psychology, 117*, 157-165.

Chapter 12

Side Bias in Facial Expression

Hari S. Asthana[1], Braj Bhushan[2] and Manas K. Mandal[3]
[1]V.K.S. University, India: [2]Krishnamurti Foundation India, India: [3]Indian Institute of Technology, India

By side bias, we mean the asymmetry of functions of the paired or non-paired organs that are arranged symmetrically over the two sides (left, right) of the body. In other words, side bias refers to the phenomenon of differential involvement of one side of the body or of bodily structures (hand, foot, eye, ear and hemiface) in comparison to the other during behavioural functions.

Although the bias is more clearly evident in paired organs (such as hand, foot, eye, and ear), it is also distinguishable in similarly arranged areas of non-paired organs like the face. Ordinarily, people are less aware of the differential involvement of the two sides of the face. This behaviour (referred generally as facedness) is conceived in terms of the relative intensity of expression and the extent of movement on the left and right sides of the face (Borod & Koff, 1990). Functionally, facedness is different from other indices of side bias (handedness, footedness, eyedness and earedness). Whereas the latter indices provide important cues to understand subjective preference or proficiency in unimanual activities of sensory or motor origin, the former index provides interpersonal cues important to understand social interaction. By definition, facial asymmetry refers to the fact that the left and right sides of human face during rest or movement are not identical. The

M.K. Mandal, M.B. Bulman-Fleming and G. Tiwari (eds.), Side Bias: A Neuropsychological Perspective, 289-312.

asymmetry may be produced because of anatomical, physiological, neurological, psychological, pathological or socio-cultural factors (Gelder & Borod, 1990).

1. NEUROANATOMICAL SUBSTRATES OF FACIAL EXPRESSION

The muscles of the face are composed of two groups, the mastication muscles and the expressive or mimic muscles. On each side of the face, there are four muscles. These are the temprails, the masseter, and the internal and external pteryoid muscles (Rinn, 1984). Motor neurons of the brain that innervate facial muscles are of two types: the upper motor neurons that send impulses from the motor centre of the cortex to the brain stem and/or spinal cord and the lower motor neurons that send impulses from brain stem/spinal cord to the facial muscles (for details, see Rinn, 1984). The lower motor neuron tract that provides major motor innervation of the muscles of facial expression is known as the facial nerve or seventh cranial nerve (Brodal, 1957; Miehlke, 1973).

One of the most important aspects in the neuroanatomical basis of facial expression is the distinction between posed and spontaneous expressions of emotion. Posed expressions are those that are voluntarily produced by the individual himself or herself on request. Spontaneous expressions, on the other hand, are produced non-deliberately as a response to an environmental situation (Myers, 1976).

Neuroanatomical literature suggests relatively independent neuroanatomical pathways for posed and spontaneous facial expressions. For example, posed expressions are governed by the pyramidal tracts of the facial nerves that descend from the cortex. Spontaneous expressions, on the other hand, are governed by the extrapyramidal tracts of subcortical origin (Van Gelder, 1981, cited in Borod & Koff, 1991). Although there is some consensus on the fact that facial nerves of the pyramidal tracts are contralaterally distributed to the lower region of the face (Borod & Koff, 1984; Campbell, 1986), opinion differs on the precise nature of neuronal distribution in the upper face. With regard to spontaneous expression, there is little consensus about the neuronal pathways (crossed or uncrossed) that are distributed to the upper or lower regions of the face.

2. METHODOLOGICAL CONSIDERATIONS

Two common methods, judgmental and anatomical/electro-physiological, are available to measure facial behaviour. Most investigators have utilized the judgmental method to study facial asymmetry. In this method, observers' judgment is considered as the dependent and facial behaviour as the independent measure. Based on observers' judgments, facial behaviours are calibrated and inferences are drawn (Ekman, 1982). In the studies of facial asymmetry, the judgmental method is used with a variety of stimuli, for example, symmetrical composite faces, hemiregional composite faces, videotapes of whole faces and hemifaces.

2.1 Symmetrical composite faces from photographic stills

This method entails preparation of facial composites by cutting the original and mirror-reversed prints of each photograph along the vertical midline (e.g., Sackeim & Gur, 1978). The left-left (LL) composite is thus prepared by joining the left hemiface of normal orientation and its mirror image. Similarly, the right-right (RR) facial composite is prepared by assembling the right hemiface of normal orientation and its mirror image (see figure 1). Observers are asked to rate these composite photographs in terms of intensity of expression.

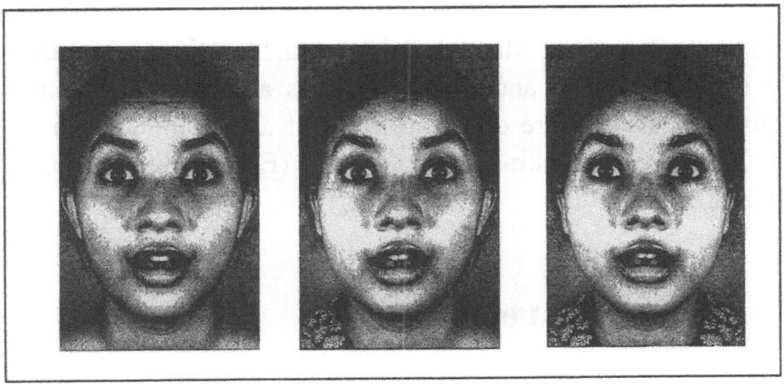

Figure 1. Examples of facial composites indicating asymmetries during expression of an emotion (Left to right: RR (right-side composite), LR (mirror reversal of normal orientation), LL (left-side composite). The mirror reversal of the original photograph was presented to avoid a possible left-to-right scanning bias.

2.2 Hemi-regional composite faces from photographic stills

To develop hemiregional composite photographs, first hemifacial composites are prepared. These composites are then cut along the horizontal mid-line bisecting the upper (forehead, brows, eyes and root of the nose) and lower (lower part of nose, cheeks, mouth and chin) regions of the face. The lower part of the left facial composite (LL) is then joined with the upper part of the right-right (RR) to produce the RR/LL hemiregional composite and vice versa to get the LL/RR composite. Observers are asked to rate/rank these hemiregional composite photographs in terms of expressed intensity.

2.3 Videotapes of whole faces and hemifaces

In this approach, participants' expressions are videotaped during different conditions of emotional expressions. These videotaped clips are then presented in either a dynamic or static mode for participants' rating in terms of intensity or muscular involvement.

The utilization of the electro-physiological/anatomical method in the study of facial asymmetry is reported rather less often. In the electro-physiological method, such as EMG recording (Girard, Tassinary, Kappas, & Bontempo, 1996; Girard, Tassinary, Kappas, Gosselin, & Bontempo, 1997; Schwartz, Fair, Salt, Mandel, & Klerman, 1976), muscle actions in the face are measured. The anatomical methods are based on distinct facial muscle movements that are coded in terms of action units, such as, FAST (Ekman, Friesen, & Tomkins, 1971), FACS (Ekman & Friesen, 1978) or MAX (Izard, 1983).

2.4 Measurement of muscle tone

This measurement technique requires the EMG leads to be placed over facial areas that are expected to be differentially active for the emotion studied. Surface electromyographic measurements are sensitive to differences among recalled emotions (Schwartz et al., 1976). Asymmetry in the EMG, of zygomatic muscles, is used as an indicator of facial asymmetry (Schwartz, Ahern, & Brown, 1979).

2.5 Facial Action Scoring Technique (FAST)

In this technique, experimenters observe separate areas of the hemiface for visible facial movements, which are then compared with standard photographic examples (Ekman, Friesen, & Tomkins, 1971). The difference in hemifacial movement with standard photographic examples substantiates the notion of facial asymmetry.

2.6 Facial Action Coding System (FACS)

The FACS (Ekman & Friesen, 1978) is designed to measure each visible facial behaviour in any context. This system isolates all anatomically based muscle measurements during emotional expression and action units are identified. Facial asymmetry is studied in terms of action units on each hemiface (Hager & Ekman, 1985).

2.7 Maximally Discriminative Facial Movement Coding System (MAX)

The MAX (Izard, 1983) was developed to objectively measure organized patterns of facial movements or appearance changes that signal human emotions. Unlike other anatomical coding systems (like FACS), MAX was designed to examine facial behaviour in young and preverbal children. Facial asymmetry is measured in terms of different codes representing appearance changes.

3. REVIEW OF LITERATURE

In 1872, Darwin pointed out that the two sides of human face are not equally expressive. However, the first systematic study of facial asymmetry was conducted by Lynn and Lynn (1938, cited in Borod, Haywood, & Koff, 1997). These authors used the term 'facedness' to characterize facial asymmetry during emotional or nonemotional expression. About the same time, Wolff (1943, cited in Sackeim, Gur, & Saucy, 1978) studied facial asymmetry from an emotional quality point of view. Wolff noted that the right side of the human face offers social or public expressions, whereas the left side of the face reveals hidden, and personalized feelings [For further readings on this subject matter, refer to Güntürkün, 1991].

3.1 Facial asymmetry during posed expressions of emotion

Asymmetry in facial expressions has been studied from still photographs and videotapes (see Table 1), and investigators have attempted to examine hemifacial asymmetry during posed facial emotion. Campbell (1978) used composite faces of nine different expressors depicting the happy (smile) emotion. He found that the left hemiface was judged to be expressing happiness more intensely than the right hemiface. Sackeim and Gur (1978) used composite faces posing six different emotions and a neutral expression. Participants rated each composite photograph on a 7-point scale of expressed intensity. The left-side composites were judged as expressing emotion more intensely than the right composites. In a somewhat different methodology, observers were asked to rate left-left and right-right facial composites with bipolar adjectives. Results revealed that left-left composites were rated as 'healthier, stronger, harder, more active, more excitable'. In contrast, right-right composites were rated as 'more sickly, weaker, more feminine, more passive and calmer' (Karch & Grant, 1978).

In the studies reported by Borod and her associates (Borod & Caron, 1980; Borod, Caron, & Koff, 1981), encoders were videotaped while posing expressions. Slow-motion replay was used to present each videotaped facial expression (whole face) and judges were asked to rate each expression for the degree of facial asymmetry. The left hemiface was judged as significantly more involved in facial expression than the right hemiface. Borod and her associates substantiated the notion on many occasions in later studies. These studies revealed that the left hemifacial composite was judged to have depicted greater affective tone than that of the right hemiface (Borod, Kent, Koff, Martin, & Alpert, 1988; Borod & Koff, 1983; Borod & Koff, 1990; Borod, Koff, & White, 1983; Borod, St. Clair, Koff, & Alpert, 1990). Support for the left hemifacial bias during expressions was obtained from other studies as well (for example, Asthana & Mandal, 1998; Baribue, Gurertt, & Braun, 1987; Mandal, Asthana, & Tandon, 1993; Moreno, Borod,

Table 1. Facial asymmetry during expression of emotion*#

Authors	Facial area	Mode of expression	Findings
Lynn & Lynn, 1938	Whole face	Spontaneous	No facial asymmetry for positive emotions[1]
Chaurasia & Goswami, 1975	Lower face	Spontaneous	Left-sided facial asymmetry for positive emotions[1]
Remillard et al., 1977	Lower face	Spontaneous	No facial asymmetry for positive emotions[1]
Campbell, 1978	Whole face	Posed	Left-sided facial asymmetry for positive emotions[1]
Sackeim & Gur, 1978	Whole face	Spontaneous	No facial asymmetry for positive emotions[1]
Campbell, 1979	Whole face	Posed	Left-sided facial asymmetry for positive emotions[1]
Schwartz et al., 1979	Lower face	Posed	No facial asymmetry for either positive or negative emotions
	Lower face	Spontaneous	Right sided for positive and left sided for negative emotions
Borod & Caron, 1980	Lower face	Posed	No facial asymmetry for positive and left-sided facial asymmetry for negative emotions
Strauss & Kaplan, 1980	Whole face	Posed	No facial asymmetry for either positive or negative emotions
Cacioppo & Petty, 1981	Whole face	Posed	No facial asymmetry for negative emotions[2]
	Whole face	Spontaneous	No facial asymmetry for negative emotions[2]
Ekman et al., 1981	Whole face	Spontaneous	No facial asymmetry for negative emotions[2]
Heller & Levy, 1981	Whole face	Posed	Left-sided facial asymmetry for positive emotions[1]
Rinn et al., 1982	Lower face	Spontaneous	Left-sided facial asymmetry for positive emotions[1]
Sirota & Schwartz, 1982	Lower face	Posed	No facial asymmetry for either positive or negative emotions
	Lower face	Spontaneous	No asymmetry for negative and right-sided asymmetry for positive emotions
Ladavas, 1982	Whole face	Posed	Left-sided facial asymmetry for adults but not for young participants

Moscovitch & Olds, 1982	Whole face	Spontaneous	Left-sided facial asymmetry for either positive or negative emotions
Borod, Koff, & White, 1983	Lower face	Posed	Left-sided facial asymmetry for either positive or negative emotions
	Lower face	Spontaneous	Left-sided facial asymmetry for either positive or negative emotions
Dopson et al., 1984	Whole face	Posed	Left-sided facial asymmetry for either positive or negative emotions
	Whole face	Spontaneous	Left-sided facial asymmetry for either positive or negative emotions
Sackeim et al. 1984	Facial regions	Posed	No consistent facial asymmetry for any region
Monserrat, 1984	Whole face	Spontaneous	Left-sided facial asymmetry for positive and no facial asymmetry for negative emotions
Hager & Ekman, 1985	Whole face	Posed	Left-sided facial asymmetry for positive and no asymmetry for negative emotions
Wemple et al., 1986	Whole face	Spontaneous	Left-sided facial asymmetry for negative emotions[2]
Baribeau et al., 1987	Whole face	Posed	Left-sided facial asymmetry for either positive or negative emotions
Sackeim & Grega, 1987	Whole face	Posed	Left-sided facial asymmetry for negative and no asymmetry for positive emotions
Borod et al., 1988	Whole face	Posed	Left-sided facial asymmetry for either positive or negative emotions
Schiff & Lamon, 1989	Lower face	Posed	Left-sided facial asymmetry for negative and no asymmetry for positive emotions
Borod et al., 1990	Whole face	Posed	Left-sided facial asymmetry for either positive or negative emotions
Schiff & MacDonald, 1990	Whole face	Spontaneous	Left-sided facial asymmetry for negative and right-sided facial asymmetry for positive emotions
Mandal & Singh, 1990	Whole face	Posed	Left-sided facial asymmetry for negative emotions[2]
Moreno et al., 1990	Lower face	Posed	Left-sided facial asymmetry for either positive or negative emotions

Kop et al., 1991	Lower face	Posed	No asymmetry for either positive or negative emotions
Brockmeier & Ulrich, 1993	Lower face	Spontaneous	Left-sided facial asymmetry for negative emotions and right-sided facial asymmetry for positive emotions
Mandal et al., 1993	Whole face	Posed	Left-sided facial asymmetry for either positive or negative emotions
Kowner, 1995	Whole face	Posed	Left-sided facial asymmetry for positive emotions[1]
Asthana & Mandal, 1997	Upper face	Posed	Right-sided facial asymmetry for either positive or negative emotions
	Lower face	Posed	Left-sided facial asymmetry for either positive or negative emotions
Asthana & Mandal, 1998	Whole face	Posed	Left-sided facial asymmetry for either positive or negative emotions
Yecker et al., 1999	Whole face	Posed	Greater right-sided facial asymmetry for approach and left-sided for withdrawal expressions

*The table is prepared based on Borod et al., 1997

@ Arranged in chronological order

1 Negative emotions were not examined

2 Positive emotions were not examined

Welkowitz & Alpert, 1990). The greater left hemifacial activity supports the proposition of greater right-hemispheric involvement in emotional expression. It has been documented that the fibre connections of each hemiface, especially the lower two-thirds part, are predominantly innervated by the contralateral hemisphere (Küypers, 1958).

Some studies, however, have failed to demonstrate systematic left-facedness during posed expression of emotion. Knox (1972) employed lateral half faces in a study in which observers were asked to judge which half face was expressing happiness more intensely. There was no significant difference in the judgment of the two sides of face. Others, however, have reported left facedness in the expression of happiness (Campbell, 1979; Hager & Ekman, 1985; Kowner, 1995). Cacioppo and Petty (1981) used composite faces posing a sad emotion. Participants reported no differences between the two sides of face. In one study, asymmetry was examined over 11 facial regions during emotional expressions. No consistent pattern of facial asymmetry was observed (Sackeim, Weiman, & Forman, 1984). Kappas and Michaud (1995) attributed facial asymmetry to factors that are independent of facial action, such as hair style, facial morphology, etc. They created composite photographs with the use of a digital camera and preserved the original orientation of facial outline and hair style. With such control, the asymmetry effect was found to be nonexistent. The absence of facial asymmetry is supported in other studies also (for example, Kop, Merckelbach, & Muris, 1991; Schwartz, Ahern, & Brown, 1979; Sirota & Schwartz, 1982; Strauss & Kaplan, 1980). Ladavas (1982) reported that left-sided asymmetries are found in older but not in younger groups (age 12-23 years) during posed expressions of emotions.

Conventionally, hemifacial asymmetry in expression of emotion has been examined either with symmetrical composite photographs or with chimeric faces. Although left-right as well as upper-lower facial differences in terms of behavioural pattern and neural innervation have been proposed by some investigators (for example, Ekman & Friesen, 1975; Rinn, 1984), no systematic attempt has been made to test the proposition empirically. We (Asthana & Mandal, 1997) conducted an experiment to examine the role of left-right as well as upper-lower regions of the face during posed emotional expressions with a modification in a composite-photograph methodology. For experimental purposes, two sets of composites, hemifacial and hemiregional, were prepared by bisecting the left-left and right-right composites into upper and lower regions to produce composites of upper right-right and lower left-left (RR/LL), and upper left-left and lower right-right (LL/RR) facial parts. Preparation of hemiregional composites thus provided an opportunity to examine the differential roles of upper and lower regions in addition to left-right hemifaces during an emotional expression. Altogether, six facial photographs (RL, LR, RR, LL, RR/LL, LL/RR) of an expression were presented to observers for rank-order judgment in terms of intensity of expressed emotion. Hemiregional composites of facial expression involving RR/LL were judged as most expressive followed by left facial composites. This finding indicated that the left hemifacial involvement was specific to the lower region and the right hemifacial involvement was specific to the upper regions during emotional expression. Given this observation, it is argued that the upper region of the face is probably controlled by ipsilateral rather than contralateral fibre connections.

3.1 Facial asymmetry during spontaneous expression

The phenomenon of facial asymmetry under spontaneous conditions (see Table 1) has not been studied as much as has posed expression. Borod and her associates (Borod & Koff, 1983, 1984; Borod et al., 1983; Borod et al., 1997) examined facial asymmetry in participants at the time they were watching affect-laden slides. The left hemiface was found to be significantly more involved than the right hemiface during expression of negative emotions (Brockmeier & Ulrich, 1993; Schiff & MacDonald, 1990; Wemple at al., 1986). Asymmetries were found equally often on both sides of the faces of viewers for positive emotions (Hager & Ekman, 1985; Lynn & Lynn, 1938; Sackeim & Gur, 1978). Dopson, Beckwith, Tucker, and Bullards-Bates (1984) examined facial asymmetry while participants were asked to remember a happy or sad experience of their lives. The left-sided asymmetry was found for both happy and sad emotions. Others documented similar asymmetry although the effect was stronger for negative emotions (Moscovitch & Olds, 1982). Studies on infants' facial expressions, however, documented a greater bias on the right side of the face during smiling and distress (Rothbart, Taylor, & Tucker, 1989).

Reviewing these studies, Borod and her associates (Borod et al., 1997) commented that spontaneous emotional expressions are elicited by a complex pattern of facial activity in which subcortical rather than cortical structures play a major role. Asymmetry of facial action is lost during spontaneous emotional expression because subcortical structures innervate the face with bilateral fibre projections (De Jong, 1979; Miehlke, 1973; see Borod et al., 1997).

3.2 Facial asymmetry and emotional valence

Although much of the current literature suggests left facedness, that is, superiority of the right hemisphere in the expression of facial emotion, some evidence (e.g., Ehrlichman, 1987; Sackeim, Greenberg, Weiman, Gur, Hungerbuhler, & Geschwind, 1982) suggested a differential facial involvement for emotion as a function of emotional valence. The valence hypothesis proposes that negative emotions are predominantly associated with the right hemisphere and positive emotions with the left hemisphere (see reviews by Leventhal & Tomarken, 1986; Tucker, 1981). Studies on the valence hypothesis carried out by Borod and associates (Borod & Caron, 1980; Borod & Koff, 1990; Borod_et al., 1983; Borod et al., 1988) indicated left facedness for negative emotions (a finding that was supported by Mandal

& Singh, 1990; Sackeim & Grega, 1987; Schiff & Lamon, 1989; Schiff & MacDonald, 1990; Wemple et al., 1986), and for positive emotions, both hemifaces were found equally expressive (a finding that was supported by Hager & Ekman, 1985; Remillard et al., 1977; Sackeim & Grega, 1987; Sackeim & Gur, 1978; Schiff & Lamon, 1989). On the other hand, some researchers have reported right facedness for positive emotions (Brockmeier & Ulrich, 1993; Schiff & MacDonald, 1990; Schwartz et al., 1979; Sirota & Schwartz, 1982). In a recent study on psychotic patients, similar observations were made. Approach expressions were more right-sided and withdrawal expressions were more left-sided (Yecker, Borod, Brozgold, Martin, Alpert, & Welkowitz, 1999).

Skinner and Mullen (1991) did a meta-analysis of 14 studies on facial asymmetry and concluded that the left hemiface expresses emotion more intensely than the right hemiface. This asymmetry was more pronounced for negative emotions than for positive emotions. Borod (1993) also reviewed studies on facial asymmetry and reported that "left-sided asymmetries were more frequent for negative (100%) than for positive (76%) emotional expression and that right-sided asymmetries were more frequent for positive than for negative emotional expression" (p.457). In a more recent review of 49 studies, Borod and her associates (Borod, Koff, Yecker, Santschi, & Schmidt, 1998) concluded that "when facial asymmetry was evaluated by trained judges and muscle quantification, facial expressions were left-sided, [and] the right cerebral hemisphere was implicated in emotional expression. However, when self-report experiential methods were utilized, the valence hypothesis received some support" (p. 1209).

3.3 Facedness and non-emotional factors

Certain non-emotional factors, such as hemifacial mobility, size, resting-state asymmetry, and handedness, have also been found to be operative in the elicitation of facial asymmetry (see Table 2).

3.3.1 Hemifacial mobility

If two hemifaces differ in degree of muscular activity, the hemiface with greater mobility might be perceived as more expressive (see Borod & Koff, 1984). Studies that examined asymmetries in hemiface mobility documented more mobility in the left than in the right hemiface (Campbell,

Table 2. Facial asymmetry during non-emotional expressions*

Authors	Facial activity	Findings
Chaurasia & Goswami, 1975	Unilateral facial movement	Greater mobility in left hemiface
Campbell, 1978	Neutral expressions	Left hemiface judged more miserable
Koff et al., 1981	Unilateral movement	Left facedness for unilateral facial actions
Ekman et al., 1981	Bilateral movement	Left hemiface judged more mobile
Alford & Alford, 1981	Unilateral movement	Greater mobility on the upper right hemiface
Stringer & May, 1981	Neutral expressions	Left hemiface judged more happy during neutral expression
Moscovitch & Olds, 1982	Unilateral movement	Greater mobility on the lower left hemiface
Campbell, 1982	Bilateral movement	Greater facial movement on left hemiface
Borod & Koff, 1983	Unilateral movement	Left hemiface rated as moving more extensively than right hemiface
Borod et al., 1988	Neutral expressions	Neutral expressions judged as more left-sided
Moreno et al., 1990	Neutral expressions	Neutral expressions judged as more left-sided
Graves & Landis, 1990	Speech production	Right-sided mouth asymmetry during verbal-list generation
Mandal et al., 1992	Neutral expressions	Left hemiface judged as more emotional than right hemiface
Hausmann et al., 1998	Speech production	Right-sided lip separation bias, especially for discrete words
Smith, 1998	Hemiface size	Facial size depends on cognitive specialization
Kowner, 1998	Neutral expression	Smaller relative to larger hemifaces were rated as exhibiting more intense expressions

* Arranged in chronological order

1982; Chaurasia & Goswami, 1975; Ekman et al., 1981; Koff, Borod, & White, 1981). In some studies, greater upper-left hemifacial mobility than upper-right mobility was found (Moscovitch & Olds, 1982) although the opposite trend has also been reported (Alford & Alford, 1981). The proposition that facial asymmetry is a function of left-hemispheric dominance for language processing was tested by Hausmann, Behrendt-Koerbitz, Kantz, Lamm, Radelt, and Güntürkün (1998). They found a right-sided lip-separation bias during verbal production, especially for discrete words. For continuous word production, sex specificity was observed with males showing the usual rightward bias and females showing no asymmetry. The effect of facial asymmetry on performance reading aloud was, however, found not to be significant (Gibson, Dancer, & Burl, 1996). Reviewing a number of studies, Graves and Landis (1990) commented that mouth asymmetry is clearly discernible during verbal word-list generation and verbal recall tasks and such asymmetry is less consistent with emotional expression.

3.3.2 Hemiface size

If the two hemifaces differ in size, the expression on the smaller hemiface might be perceived as being more intense than that on the larger hemiface (Koff et al., 1981). Literature on hemifacial size reported a larger right hemiface (Koff et al., 1981; Nelson & Horowitz, 1980). Some researchers, however, did not observe differences between the left and right hemifaces (for example, Mulick, 1965; Sackeim & Gur, 1980). Speculation that hemiface size might be related to hemifacial mobility was not supported (Koff et al., 1981). Recently, Keles, Diyarbakirli, Tan, and Tan (1997) reported that about 96% of right handers had larger left than right facial areas and about 68% of left handers had larger right than left facial areas. Kowner (1998) examined the effect of hemifacial size on attribution of personality and emotion. Hemifaces relatively smaller in size were perceived as more expressive and as having more positive than negative features than were larger hemifaces.

3.3.3 Resting state

The resting left hemiface is judged either more happy (Stringer & May, 1981) or miserable (Campbell, 1978) than the right hemiface. Lateral composites indicated asymmetry of the resting face (Bennett, Delmonico, & Bond, 1987), and these expressions were judged to be more left-sided (Borod & Koff, 1990; Borod et al., 1988; Moreno et al., 1990) and affect laden (McGee & Siknner, 1987). We (Mandal, Asthana, Madan, & Pandey, 1992) conducted a study to examine the asymmetrical nature of the resting facial state with the hypothesis that the left side of the face will be emotionally more involved than either the right side or the whole face. The left-side facial composites were judged by observers to be more emotional than the right-side composites or normal faces. Although Borod and Koff

(1990) attributed the left hemifacial involvement to non-emotional, peripheral factors (such as greater hemifacial mobility), we speculated that the bias is a revelation of the general affective state of the individual (facial leakage). Using an index of 'Fluctuating Asymmetry (deviation from bilateral symmetry in morphological traits with asymmetry values that are normally distributed with a mean of 0)', Shackelford and Larsen (1997, p. 456) suggested that facial 'Fluctuating Asymmetry' gives clues to understanding psychosocial and physiological distress. Recently, Smith (1998) conducted a study to examine hemiface size as a function of cognitive specialization in various university faculties belonging to different subject areas. The author found significant facial asymmetries as a function of academic faculty (humanities larger right hemiface, maths and physics larger left hemiface) and suggested differential muscular development, depending upon the nature of cognitive activity most often employed, as the mediating factor.

4. FACIAL ASYMMETRY AND HANDEDNESS

Considering facedness as an index of lateral dominance, Borod, Koff, and Caron (1983) hypothesised that there would be right facedness in right handers and left facedness in left handers during expression of emotion. They empirically tested this hypothesis and found an overall left facedness for both right handers and for left handers. The unrelatedness of facedness to handedness was supported by a later study (Borod et al., 1983). Sackeim et al. (1984) examined different facial regions during expressions of emotion and documented an effect of family history of sinistrality in facial asymmetry. The relation between unilateral facial movement and handedness has also been tested; it was observed that left handers "tended to move their right half face; right handers the left" (Borod et al., 1981). Chaurasia and Goswami (1975) found that handedness plays a significant role in facedness during unilateral movement of the body. About 59% right handers and only 29% left handers showed left facedness. Keles et al. (1997) also reported a relation between facial asymmetry and hemispheric dominance of speech.

5. GENERAL COMMENTS

We reviewed evidence to establish facial asymmetry as a form of side bias. Facial asymmetry (or facedness) is considered to be somewhat different from the traditionally accepted forms of side bias, such as handedness, footedness, eyedness and earedness. There are important distinctions between these forms of side bias and facedness. First, the traditional forms of side bias are measured on the basis of subjective preference to use one side of the body. Facedness, on the other hand, does not entail a component of conscious subjective preference. Any such preference for facial expression is purely sub-conscious. Second, manual proficiency or performance is considered a measure of lateral dominance for hand, foot, eye, and ear. Facial expressions, on the other hand, are rarely rated in terms of proficiency. Although expressiveness is considered as one component of facedness, the measure does not reflect the subjective ability to mobilize one side of the face. Third, motor asymmetry during the performance of a one-sided act (such as throwing a ball) is clearly observable for hand, foot, eye, and ear. Asymmetry of facial expression during social interaction is not clearly detectable. Bruyer and Craps (1985), however, have noted that "the human face is asymmetric with regard to facial expression .... and that the human perceiver takes such facial asymmetry into account when seeing a face" (p.55). According to these authors, we do not perceive asymmetry because the two sides of face are processed as one symmetrical structure. Others suggested that the face is perceived asymmetrical in expression because of left visual-field dominance (Asthana & Mandal, 1996; Borod & Koff, 1990). For example, the right hemiface is perceived more expressive in a face-to-face interaction due to leftward perceiver bias. Despite these theoretical speculations, facial asymmetries are not easily detectable unless manipulated experimentally. Kowner (1998) investigated resting asymmetrical faces versus their symmetrical hemifacial composites. He documented a null effect of a target person's facial asymmetry on observer's attributions. Finally, side bias in terms of hand, foot, eye, and ear is not evident while the individual is at rest. Experimental evidence has indicated that facial asymmetry is present even at the resting or neutral state of expression.

Although it is true that facedness differs from other forms of side bias in many respects, the bias has been found to be consistent in most individuals. The common assumption as to why a majority of individuals are left-faced has to do with the lateralized representation of affect in the cerebral hemispheres. It is believed that the right hemisphere is relatively superior to the left hemisphere in mediating the expression of emotion (Borod, 1992; Bryden, 1982; Mandal, Borod, Asthana, Mohanty, Mohanty, & Koff, 1999; Mandal, Mohanty, Pandey, & Mohanty, 1996; Mandal, Tandon, & Asthana, 1991). In essence, therefore, the left face is found more expressive as a function of contralateral motor control. For handedness, the left hemisphere, which usually represents speech, is considered to play an important role. Most individuals have left-hemispheric speech representation, and a vast majority of these are right handed (Bryden, 1982; Mandal, Asthana, Dwivedi, & Bryden, 1999). These two forms of side bias: facedness and handedness, are therefore not comparable as these subserve different kinds of psychological functions that are mediated by different hemispheres. Most studies have documented a weak association between handedness and facedness, especially for left handers. The inconsistency in facial asymmetry of the left handers can be attributed to their generally higher variability in functional lateralization, compared to right handers (Keles et al., 1997). The association between non-emotional facial movement and handedness was, nevertheless, found to be strong (Chaurasia & Goswami, 1975).

These findings indicate that facedness is an important area of study to understand behavioural asymmetry during the communication of emotion. Although the left hemiface was found to depict characteristic emotional signals, the role of the right hemisphere was also found to be important for displaying socially accepted emotional messages. We recently tested this proposition with the hypothesis that emotional display would be greatly modulated by the right side of the face due to greater voluntary control, which is contralaterally connected with the relatively less emotional side of the brain (left) for the facility of social interaction. The hypothesis was confirmed. It was found that right hemifacial expression was modulated as a function of culture. Conversely, the expressiveness in the left hemiface was found equally pronounced across cultures (Mandal, Harizuka, Bhushan, & Mishra, 2000). Thus, evidence suggests the differential involvement of the two hemifaces in social communication (see also, Gelder & Borod, 1990). Probably the right hemiface is dominant for expressing the non-emotional or social signals and the left hemiface is dominant for divulging the quality of emotional experiences. Such an assertion, originally proposed by Wolff (1943), needs further validation keeping constant such possible confounding effects as age and sex of the encoder, elicitation condition (posed, spontaneous), quality of experience, and the morphological/static characteristics of the face.

6. REFERENCES

Alford, R., & Alford, K. (1981). Sex differences in asymmetry in the facial expression of emotion. *Neuropsychologia, 19,* 605-608.

Asthana, H.S., & Mandal, M. K. (1996). Mirror-reversal of a face is perceived as expressing emotions more intensely. *Behavioural Neurology, 9,* 115-117.

Asthana, H.S., & Mandal, M.K. (1997). Hemiregional variations in facial expression of emotions. *British Journal of Psychology, 88,* 519-525.

Asthana, H.S., & Mandal, M.K. (1998). Hemifacial asymmetry in emotion expression. *Behaviour Modification (Special Issue: Facial expression of emotions), 22,* 177-183.

Baribeau, J., Guerette, R., & Braun, C. (1987). Hemifacial production of posed emotional expression in normal men and women. Paper presented at the meeting of *The International Neuropsychological Society,* Washington, D.C.

Bennett, H.L., Delmonico, R.L., & Bond, C. F. (1987). Expressive and perceptual asymmetries of the resting face. *Neuropsychologia, 25,* 681-684.

Borod, J.C. (1993). Cerebral mechanism underlying facial, prosodic, and lexical emotional expression: A review of neuropsychological studies and methodological issues. *Neuropsychology, 7,* 445-463.

Borod, J.C., & Caron, H. S. (1980). Facedness and emotion related to lateral dominance, sex, and expression type. *Neuropsychologia, 18,* 237-242.

Borod, J.C., Caron, H.S., & Koff, E. (1981). Asymmetry of facial expression related to handedness, footedness, and eyedness : A quantitative study. *Cortex, 17,* 381-390.

Borod, J.C., St. Clair, J., Koff, E., & Alpert, M. (1990). Perceiver and poser asymmetries in processing facial emotion. *Brain and Cognition, 13,* 167-177.

Borod, J.C., Haywood, C. S., & Koff, E . (1997). Neuropsychological aspects of facial asymmetry during emotional expression : A review of the normal adult literature. *Neuropsychology Review, 7,* 41-60.

Borod, J.C., Kent, J., Koff, E., Martin, C., & Alpert, M. (1988). Facial asymmetry while posing positive and negative emotions : Support for the right hemisphere hypothesis. *Neuropsychologia, 26,* 759-764.

Borod, J.C., & Koff, E. (1983). Hemiface mobility and facial expression asymmetry. *Cortex, 19,* 355-361.

Borod, J.C., & Koff, E. (1984). Asymmetries in affective expression : Behavior and anatomy. In N. Fox & R. Davidson (Eds.), *The psychobiology of affective development.* Hillsdale, NJ: Erlbaum.

Borod, J.C., & Koff, E. (1990a). Lateralization for facial emotional behaviour: A methodological perspective. *International Journal of Psychology, 25,* 157-177.

Borod, J.C., & Koff, E. (1990b). Facial asymmetry during posed and spontaneous expression in stroke patients with unilateral lesions. *Pharmacopsychoecologia, 4,* 15-21.

Borod, J.C., Koff, E., & Caron, H. S. (1983). Right hemispheric specialization for the expression and appreciation of emotion: A focus on the face. In E. Perceman, (Ed.), *Cognitive function in the right hemisphere.* New York : Academic Press.

Borod, J.C., Koff, E., & White, B. (1983). Facial asymmetry during posed and spontaneous expression of emotion. *Brain and Cognition, 2,* 165-175.

Borod, J.C., Koff, E., Yecker, S., Santschi, C., & Schmidt, J.M. (1998). Facial asymmetry during emotional expression: Gender, valence, and measurement technique. *Neuropsychologia, 36,* 1209-1215.

Brockmeier, B., & Ulrich, G. (1993). Asymmetries of expressive facial movements during experimentally induced positive versus negative mood states: A video- analytic study. *Cognition and Emotion, 7,* 393-405.

Brodal, A. (1965). *The cranial nerves.* Oxford : Blackwell.

Bruyer, R, & Craps, V. (1985). Facial asymmetry: Perceptual awareness and lateral differences . *Canadian Journal of Psychology, 39,* 54-69.

Bryden, M.P. (1982). *Laterality: Functional asymmetry in the intact brain.* New York: Academic Press.

Cacioppo, J.T., & Petty, R.E. (1981). Lateral asymmetry in the expression of cognition and emotion. *Journal of Experimental Psychology: Human Perception and Performance, 7,* 333-341.

Campbell, R. (1978). Asymmetries in interpreting and expressing a posed facial expression. *Cortex, 14,* 327-342.

Campbell, R. (1979). Left handers' smiles: Asymmetries in the projection of a posed expression. *Cortex, 14,* 571-579.

Campbell, R. (1982). Asymmetries in moving faces. *British Journal of Psychology, 73,* 95-103.

Campbell, R. (1986). Asymmetries of facial action: Some facts and fiction of normal movement. In R. Bruyer (Ed.), *The neuropsychology of facial perception and facial expression* (pp. 274-267). Hillsdale, NJ : Lawrence Erlbaum.

Chaurasia, B.D., & Goswami, H.K. (1975). Functional asymmetry in the face. *Acta Anatomica, 91,* 154-160.

Darwin, C. (1872). *The expression of the emotions in man and animals.* London: Murray.

De Jong, R.N. (1979). *The neurologic examination.* New York: Harper and Row.

Dopson, W.G., Beckwith, B.E., Tucker, D.M., & Bullards-Bates, P.C. (1984). Asymmetry of facial expression in spontaneous emotion. *Cortex, 20,* 243-251.

Ehrlichman, H. (1987). Hemispheric asymmetry and positive-negative affect. In D. Ottoson (Ed.), *Duality and unity of the brain. London:* MacMillan.

Ekman, P. (1982). *Emotion in the human face.* Cambridge, UK: University of Cambridge Press.

Ekman, P., & Friesen W.V. (1975). *Unmasking the face.* Englewood Cliffs, NJ: Prentice Hall.

Ekman, P., & Friesen, W.V. (1978). *The facial action coding system.* Palo Alto, California: Consulting Psychologists' Press.

Ekman, P., Friesen, W.V., & Tomkins, S.S. (1971). Facial affect scoring technique: A first validity study. *Semiotics, 3,* 37- 58.

Ekman, P., Hager, J.C., & Friesen, W.V. (1981). The symmetry of emotional and deliberate facial actions. *Psychophysiology, 18,* 101-105.

Gelder, R.S., & Borod, J.C.(1990). Neurobiological and cultural aspects of facial asymmetry. *Journal of Communication Disorders, 23,* 273-286.

Gibson, B., Dancer, J., & Burl, N. (1996). Null effects of facial asymmetry on intrafacial speech reading performance in men and women. *Perceptual and Motor Skills, 82,* 570.

Girard, É., Tassinary, L.G., Kappas, A., & Bontempo, D. (1996). A study of the ability to control precisely the contraction of muscles in different regions of the face. *Presented at the Thirty-Sixth Annual Meeting of the Society for Psychophysiological Research,* Vancouver, Canada.

Girard, É., Tassinary, L.G., Kappas, A., Gosselin, P., & Bontempo, D. (1997). The covert-to-overt threshold for facial actions: An EMG study. Presented at the *Thirty-Seventh Annual Meeting of the Society for Psychophysiological Research,* Cape Cod, Massachusetts.

Graves, R., & Landis, T. (1990). Asymmetry in mouth opening during different speech tasks. *International Journal of Psychology, 25,* 179-189.

Güntürkün, O. (1991). The Venus of Milo and the dawn of facial asymmetry research. *Brain and Cognition, 16,* 147-150.

Hager, J.C., & Ekman, P. (1985). The asymmetry of facial actions is inconsistent with models of hemispheric specialization. *Psychophysiology, 22,* 307-318.

Hausmann, M., Behrendt-Koerbitz, S., Kautz, H., Lamm, C., Radelt, F., & Güntürkün, O. (1998). Sex differences in oral asymmetries during word repetition. *Neuropsychologia, 36,* 1397-1402.

Heller, W., & Levy, J. (1981). Perception and expression of emotion in right handers and left handers. *Neuropsychologia, 19*, 263-272.

Izard, C.E. (1983). *The maximally discriminative facial movement coding system*. Instructional Resources Centre, University of Delaware, Delaware, USA.

Kappas, A., & Michaud, D.C. (1995). Are findings concerning the asymmetry of the perception of facial expressions due to methodological artifacts? *Presented at the 5th Annual Meeting of the Canadian Society for Brain, Behaviour and Cognitive Science*, Halifax, Nova Scotia.

Karch, G.R., & Grant, C.W. (1978). Asymmetries in perception of the sides of the human face. *Perceptual and Motor Skills, 47*, 727-734.

Keles, P., Diyarbakirli, S., Tan, M., & Tan, Ü. (1997). Facial asymmetry in right and left-handed men and women. *International Journal of Neuroscience, 91*, 147-160.

Knox, K. (1972). An investigation of nonverbal behavior in relation to hemispheric dominance. *Unpublished M.A. Thesis*, University of California, San Francisco.

Koff, E., Borod, J.C., & White, B. (1981). Asymmetries for hemiface size and mobility. *Neuropsychologia, 19*, 825-830.

Kop, W.J., Merckelbach, H., & Muris, P. (1991). Unilateral contraction of facial muscles and emotion: A failed replication. *Cortex, 27*, 101-104.

Kowner, R. (1995). Laterality in facial expression and its effect on attributions of emotion and personality: A reconsideration. *Neuropsychologia, 33*, 539-559.

Kowner, R. (1997). The perception and attribution of facial asymmetry in normal adults. *Psychological Record, 47*, 371-384.

Kowner, R. (1998). Perceiving and attributing hemispheric width in faces of normal adults. *Perceptual and Motor Skills, 87*, 567-572.

Küypers, H.G.J.M. (1958). Corticobulbar connections to the pons and lower brain stem in man. *Brain, 81*, 364-390.

Ladavas, E. (1982). The development of facedness. *Cortex, 18*, 535-545.

Leventhal, H. L., & Tomarken, A. J. (1986). Emotions: Today's problem. *Annual Review of Psychology, 37*, 565-610.

Lynn, J.G., & Lynn, D.R. (1938). Face-hand laterality in relation to personality . *Journal of Abnormal and Social Psychology, 38*, 250-276.

Mandal, M.K. (1987). Decoding of facial emotions, in terms of expressiveness, by schizophrenics and depressives . *Psychiatry, 50*, 371-376.

Mandal, M.K., Asthana, H.S., Dwivedi, C.B., & Bryden, M.P. (1999). Hand preference in the deaf. *Journal of Developmental and Physical Disabilities, 11*, 265-273.

Mandal, M.K., Asthana, H.S., Madan, S., & Pandey, R. (1992). Hemifacial display of emotion in the resting state. *Behavioural Neurology, 5,* 169-171

Mandal, M.K., Asthana, H.S., & Tandon, S.C. (1993). Judgment of facial expression of emotion in unilateral brain-damaged patients. *Archives of Clinical Neuropsychology, 8,* 171-183.

Mandal, M.K., Borod, J.C., Asthana, J.C., Mohanty, A., Mohanty, S., & Koff, E. (1999). Effects of lesion variables and emotion type on the perception of facial emotion. *Journal of Nervous and Mental Disease, 187,* 603-609.

Mandal, M.K., Harizuka, S., Bhushan, B., & Mishra, R.C. (2000). Cultural variations in hemifacial asymmetry of emotion expressions. *British Journal of Social Psychology,* in press.

Mandal, M.K., Mohanty, A., Pandey, R., & Mohanty, S. (1996). Emotion-specific processing deficit in focal brain-damaged patients. *International Journal of Neuroscience, 84,* 87-95.

Mandal, M.K., & Singh, S.K. (1990). Lateral asymmetry in identification and expression of facial emotion. *Cognition and Emotion, 4.* 61-70.

Mandal, M.K., Tandon, S.C., & Asthana, H.S. (1991). Right brain damage impairs recognition of negative emotions. *Cortex, 27,* 247-254.

McGee, A, & Skinner, M. (1987). Facial asymmetry and the attribution of personality traits. *British Journal of Social Psychology, 26,* 181-184.

Miehlke, A. (1973). *Surgery of the facial nerve.* Philadelphia: W.B. Saunders.

Monserrat, L. (1984). Facial asymmetry in the spontaneous expression of emotion: An index of functional brain asymmetry in the regulation of emotional expression . *Unpublished Doctoral Disseration,* New York University, New York.

Moreno, C.R., Borod, J.C., Welkowitz, J, & Alpert, M. (1990). Lateralization for the expression and perception of facial emotion as a function of age. *Neuropsychologia, 28,* 199-209.

Moscovitch, M., & Olds, J. (1982). Asymmetries in spontaneous facial expressions and their possible relation to hemispheric specialization. *Neuropsychologia, 20,* 71-81.

Mulick, J.F. (1965). An investigation of craniofacial asymmetry using the serial twin study method. *American Journal of Orthodontics, 51,* 112-129.

Myers, R.F. (1976). Comparative neurology of vocalization and speech: Proof of a dichotomy . *Annals of the New York Academy of Sciences, 280,* 745-757.

Nelson, C.A., & Horowitz, F.D. (1980). Asymmetry in facial expression. *Science, 209,* 834.

Remillard, G.M., Andremann, F., Rhi- Suci, A., & Robbins, N.M. (1977). Facial asymmetry in patients with temporal lobe epilepsy. *Neurology, 27,* 109-114.

Rinn, W.E. (1984). The neuropsychology of facial expression: A review of the neurological and psychological mechanisms for producing facial expression. *Psychological Bulletin, 95,* 52-77.

Rinn, W.E., Friedman, C., & Meller, P. (1982). An investigation of personality correlates of lateral gaze preference and facial asymmetry. A paper presented at the meeting of the *The International Neuropsychological Society,* Pittsburg, USA.

Rothbart, M.K., Taylor, S.B., & Tucker, D.M. (1989). Right-sided facial asymmetry in infant emotional expression. *Neuropsychologia, 27,* 675-687.

Sackeim, H.A., Greenberg, M., Weiman, A., Gur, R.C., Hungerbuhler, J., & Geschwind, N. (1982). Functional brain asymmetry in the expression of positive and negative emotions: Lateralization of insults in cases of uncontrollable emotional outbursts. *Archives of Neurology, 19,* 210-218.

Sackeim, H.A., & Grega, D.M. (1987). Perceiver bias in the processing of deliberately asymmetric emotional expressions. *Brain and Cognition, 6,* 464- 473.

Sackeim, H.A., & Gur, R.C. (1978). Lateral asymmetry in intensity of emotional expression. *Neuropsychologia, 16,* 473-481.

Sackeim, H.A., & Gur, R.C. (1980). Asymmetry in facial expression. *Science, 209,* 834-836.

Sackeim, H.A., Gur, R.C., & Saucy, M. (1978). Emotions are expressed more intensely on the left side of the face. *Science, 202,* 434-436.

Sackeim, H.A., Weiman, A.L., & Forman, B.D. (1984). Asymmetry of the face at rest: Size, area, and emotional expression. *Cortex, 20,* 165-178.

Schiff, B.B., & Lamon, M. (1989). Inducing emotion by unilateral contraction of facial muscles: A new look at hemispheric specialization and the experience of emotion. *Neuropsychologia, 27,* 923-935.

Schiff, B.B., & Macdonald, B. (1990). Facial asymmetries in the spontaneous response to positive and negative emotional arousal. *Neuropsychologia, 28,* 777-785.

Schwartz, G.E., Ahern, G.L., & Brown, S. (1979). Lateralized facial muscles response to positive and negative emotional stimuli. *Psychophysiology, 16,* 561-571.

Schwartz, G.E., Fair, P.L., Salt, P., Mandel, M.R., & Klerman, G.L. (1976). Facial muscle patterning to affective imagery in depressed and non-depressed subjects. *Science, 192,* 489-491.

Shackelford, T.K., & Larsen, R.J. (1997). Facial asymmetry as an indicator of psychological, emotional, and physiological distress. *Journal of Personality and Social Psychology, 72,* 456-466.

Sirota, A., & Schwartz, G.E. (1982). Facial muscle patterning and lateralization during elation and depression imagery. *Journal of Abnormal Psychology, 91*, 25-34.

Skinner, M., & Mullen, B. (1991). Facial asymmetry in emotional expression: A meta-analysis of research. *British Journal of Social Psychology, 30*, 113-124.

Smith, W.M. (1998). Hemispheric and facial asymmetry: Faces of academe. *Journal of Cognitive Neuroscience, 10*, 663-667.

Strauss, E., & Kaplan, E. (1980). Lateralized asymmetries in self-perception. *Cortex, 16,* 289-293.

Stringer, P., & May, P. (1980). Attributional asymmetries in the perception of moving, static, chimeric, and hemisected faces. *Journal of Nonverbal Behavior, 5,* 238-252.

Tucker, D.M. (1981). Lateral brain function, emotion, and conceptualization. *Psychological Bulletin, 89,* 19-46.

Van Gelder, R. (1981). *Neuroanatomy and facial asymmetry.* Internal publication of the ILO, Department of Psychology, Free University of Amsterdam.

Van Gelder, R., & Borod, J.C. (1990). Neurobiological and cultural aspects of facial asymmetry. *Journal of Communication Disorders, 23,* 273-286.

Wemple, C.Y., Safer, M.A., & Notarious, C.T. (1986). Display rules and facial expression : intensity, asymmetry and type. Paper presented at the *Convention of the American Psychological Association,* Washington, D.C.

Wolff, W. (1943). *The expression of personality.* New York: Harper.

Yecker, S., Borod, J.C., Brozgold, A., Martin, C., Alpert, M., Welkowitz, J. (1999). Lateralization of facial emotional expression in schizophrenic and depressed patients. *Journal of Neuropsychiatry & Clinical Neuroscience, 11,* 370-379.

Chapter 13

Asymmetries in Portraits: Insight from Neuropsychology

Michael E.R. Nicholls
University of Melbourne, Australia

On casual inspection, the human face appears to be very nearly symmetrical. Although individual features, such as a part in the hair, or a mole on the skin may vary within an individual, these individual differences cancel one another when they are considered as part of a larger population. Despite this apparent symmetry, it appears that we choose to portray ourselves asymmetrically. Giovanni Bellini's portrait of Leonardo Loredan, Doge of Venice (c. 1501) provides a good example of four asymmetries that typically appear in portraiture. These four examples include asymmetries in: (a) the expression of the face, (b) the way the head is turned, (c) the direction of lighting and (d) the horizontal position of the eyes.

As is frequently the case with portraits of this era, the expression of the Doge is mobile and difficult to define. Differences in expression between the left and right sides of the face can be seen by covering one or other side of the face. If one concentrates on the right side of the face (the Doge's right), the portrait appears to depict a stern and authoritarian character. The right side of the face is lit more severely than the left, accentuating a frowning aspect of the right eye and eliminating any hint of a smile from the right side of the mouth (Campbell, 1990). The left side of the face, which is more softly lit, depicts a gentler character. Whereas the eyelid was raised clear of the pupil in the right eye, giving it a glaring expression, the left eyelid lies in

M.K. Mandal, M.B. Bulman-Fleming and G. Tiwari (eds.), Side Bias: A Neuropsychological Perspective, 313-329.

a more natural and relaxed position. The left side of the mouth is also slightly upturned giving the Doge a faint hint of a smile. The attempt to combine benevolence with authoritarianism within the one portrait may reflect Leonardo Loredan's position as Chief Magistrate of Venice. What is particularly interesting in the present context, is that the left was portrayed as the benevolent side whereas the right was portrayed as authoritarian. Other portraits of this era, such as the Mona Lisa, show a similar asymmetry in the expression of emotion. The Mona Lisa's renowned smile is stronger on the left side of her face than on the right (McMullen, 1977).

Figure 1. Giovanni Bellini: Leonardo Loredan, Doge of Venice (c.1501). Reprinted with permission from the National Gallery, London.

Bellini's portrait is arranged so that the Doge looks towards the left hand side of the painting. As a result, the left side of the Doge's face is turned toward the viewer and features more prominently than the right. A preference for featuring the left side of the face appears to be typical of portraiture. McManus and Humphrey (1973) reviewed 1,474 painted single-subject portraits from Western Europe dating from the 16th to the 20th Centuries. They found that 68% of female and 56% of male portraits were painted so that the left side of the face was more visible than the right. Other researchers have confirmed this leftward bias. Conesa, Conesa, and Miron

(1995) sampled 4,180 single-subject paintings, photographs, etchings and drawings from the 14th to the 20th Centuries. Of the portraits that were not symmetrical, 54% featured the left side of the face more prominently than the right. Similar results have been reported by Nicholls, Clode, Wood, and Wood (1999) for a catalogue of portraits belonging to the National Portrait Gallery (Yung, 1981). Portraits were sampled sequentially from the catalogue and were classified according to whether either the left or right side of the nose was visible. Of the 361 portraits sampled, 57% of the male and 78% of the female portraits were arranged so the left side of the face was more prominent than the right. These results not only confirm the leftward bias, but also demonstrate that the bias is stronger for portraits of females than for males. Gordon (1974) has reported a similar gender bias in portraits by Goya, with female sitters being more likely to show their left face than male sitters.

The lighting for the Doge's portrait also comes from the left hand side of the painting (the Doge's right). A preference for light falling from the left side of the painting is common in portraiture. Coles (1974) analysed the direction of illumination in portraits painted by Rembrandt and Reynolds. Of the 315 portraits sampled from Rembrandt, 94% were illuminated from the left side of the painting. Reynolds' portraits showed a similar bias, with 121 portraits out of a total of 188 showing a leftward bias. Grüsser, Selke, and Zynda (1988) reported that the leftward bias for the illumination of portraits was strongest for portraits painted between the 15th and 19th Centuries and declined for 20th Century portraits. Interestingly, they also reported a preference for leftward illumination in paintings other than portraits. Grüsser et al. (1988) discussed the possibility that asymmetries in lighting are related to the hand used by the artist to paint the portrait. They noted that right handed artists, such as Lucas Cranach (1472-1553) preferred to illuminate their paintings and drawings from the left hand side of the painting. In contrast, sinistrals such as Leonardo da Vinci and Hans Holbein the Younger (1497-1543) show a preference to illuminate their work from the right hand side of the painting. Differences in lighting preference between sinistral and dextral artists could be related to the organisation of the artist's studio. A right-handed artist would most probably place the subject to his or her left. This placement would allow the artist to observe the model over the left hand, which holds the palette rather than having to look over the right arm, which is painting. In order to illuminate both the subject and the painting, the light source would need to be placed behind the artist and to the right of the subject. This arrangement would produce a painting with a light source coming from the left-hand side. For sinistral artists, the reverse would be true: they would place the subject to their right and the light source would fall on the left-hand side of the subject. This may explain why the preference for lighting from the left declined during the 20th Century. Artificial lighting sources might have eliminated the need for a single light source, such as a window, to illuminate both the subject and the painting.

Grüsser et al. (1988) noted that the leftward bias of illumination was stronger for portraits than for other paintings. This suggests illumination asymmetries have a special role in portraits. From the evidence reviewed above, it appears that most portraits feature the left side of the face. To achieve this pose, the model must turn his or her head slightly to the right. To illuminate the front of the face, the light source would need to come from the left-hand side of the painting (i.e. from the model's right). Leftward illumination might also serve to place the left side of the face in relief and thus accentuate the features located on that side of the face (see Fig. 1).

The Doge's portrait is also arranged so that one of the eyes is placed along the central vertical axis of the painting. Tyler (1998) has noted a tendency for one eye to be centred in portraiture. He examined the horizontal position of the eyes, nose and mouth in the portraits produced by 625 painters over the past 600 years. Although the nose and mouth showed a weak tendency to be placed in the centre of the painting, one of the eyes was nearly always placed very close to the horizontal centre of the canvas. Tyler (1998) did not report which eye was placed in the centre of the painting. Inspection of Bellini's portrait reveals that the left eye is placed closer to the centre of the painting than is the right. It is likely that centreing the left eye would achieve a balance in the arrangement of the painting, with the mass of the side of the face in the left half of the painting being balanced by the mass of the front of the face in the right half of the painting. Given that most portraits feature the left side of the face, it seems reasonable to assume that most portraits will also align the left eye with horizontal midline of the painting.

To test whether the left eye is centred more than the right, Nicholls et al. (1999) measured the relative horizontal placement of the left and right eyes in 137 portraits. Single-subject portraits of adults that included only the upper torso or shoulders and the head were sampled from a catalogue of Renaissance portraits (Campbell, 1990). Of the portraits that were sampled, 57% showed more of the left side of the face, 36% showed more of the right side and 7% showed no bias. For the measurement of eye position, nine portraits were excluded because they were profile portraits and the position of one eye could not be measured. The data collected from the remaining 129 portraits are shown in Figure 2. The data are sharply unimodal with half of all portraits featuring the left or right eye in the middle (± 2.5%) of the painting. This central tendency confirms the results reported by Tyler (1998). Of the portraits for which one eye fell in the middle of the painting, 63% of these featured the left eye. A measure of the deviation of the eyes from the centre of the painting was calculated by dividing the vertical displacement of the left and right eyes by the total width of the painting. This measure revealed that the average horizontal deviation of the left eye from the centre (7.03%) was less than that of the right eye (9.05%). A matched-pairs t-test revealed that this difference was statistically significant [$\underline{t}$ (128) = 2.26, $p < 0.05$]. Thus, it would appear that, not only is one eye centred in portraits, but that this is usually the left eye.

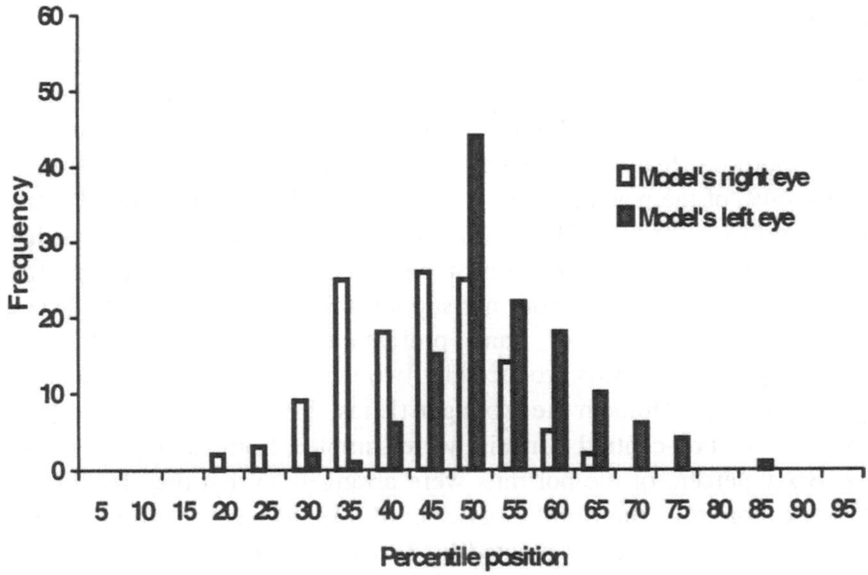

Figure 2. Relative horizontal position of model's left and right eyes within 129 Renaissance portraits

Asymmetries in head turning, direction of illumination and eye position probably stem from a desire to construct portraits that feature the left side of the face more prominently than the right. The existence of such an asymmetry raises the question of the causal mechanisms that might underlie the bias. One obvious explanation might be that the bias is related to the hand preference of the artist. The large majority of the population uses their right hand for writing, drawing or painting. Although left handers are thought to be over-represented amongst artists (Mebert & Michel, 1980), the difference between artistic and non-artistic populations is slight (Lanthony, 1995). Dextrality on behalf of the artist could facilitate the production of portraits that feature the left side of the face. These portraits are arranged on the canvas so that the outline of the nose, mouth and chin fall in the left half of the painting. These features, which fall in a convex arc relative to the right arm, can be drawn using smooth abductive movements. Abductive movements are both faster and more accurate than adductive movements, which work against the natural swing of the arm (Bradshaw, Bradshaw, & Nettleton, 1990). For sinistral artists, abductive movements of the left hand would facilitate the production of portraits with the outline of the face falling in the right hand side of the canvas. As a result, sinistral artists should prefer to produce portraits that feature the right side of the face.

Shannon (1979) investigated asymmetries for profile drawing in groups of left- and right-handed children. Children who were right handed preferred to draw portraits featuring the left side of the face 81% of the time. In contrast, 44% of the portraits drawn by left-handed children featured the left side of the face. Although Shannon's data support the proposition that hand preference affects the construction of portraits, an analysis of the portraits made by professional artists does not support this proposition. Nicholls et al. (1999) tested the effect of hand preference in established artists by examining painted portraits produced by two well-known left-handed artists; Raphael and Hans Holbein the Younger (Elias, 1998; Grüsser et al., 1988). A total of 111 non-centred portraits were sampled from catalogues of their work. Sixty percent of the portraits were arranged so that they featured the left cheek more than the right. Grüsser et al. (1988) have reported a similar leftward bias in 87 portraits painted by Holbein. Thus it would appear that sinistral artists show the same bias as dextrals; both preferring to produce portraits featuring the left side of the face rather than the right

Another reason for suspecting that dextrality is not the central cause of the left bias comes from a study of photographic portraits. Labar (1973) examined photographs from two school yearbooks and categorised them according to whether the left or right side of the face was directed toward the camera. Photographs featuring the left side of the face accounted for 63% of all photographs. It is interesting to note that the participants selected the photo that was to be included in the yearbook from a number of poses taken by the photographer. This self-selection suggests that the leftward bias is the result of sitter's preference rather than the photographer's direction. The fact that the leftward bias exists in photography, in which there is no mechanical constraint on rendering the image, suggests that dextrality is not the central cause of the leftward bias.

An interesting alternative explanation for the leftward bias has been put forward by Corballis and Beale (1976). They noted that portraits, particularly those painted during the Renaissance, were frequently painted in pairs. A good example of such a pair of paintings is a set painted by Lucas Cranach the Elder in 1502 that depict Johannes Cuspian and his first wife, Anna Putsch. These paintings were designed to be hung as a pair, with the female being placed on the left and the male on the right. Placing females to the left of males is entrenched in many cultures and can still be seen in formal ceremonies such as Christian weddings. To achieve a balanced composition between the pair of paintings, the portraits were drawn so that they faced one another. As a result, the female sitter would turn to her right and face her husband, thus revealing her left cheek. The male, on the other hand, would turn to his left, thus exposing the right side of his face. An inspection of a catalogue of Renaissance portraits (Campbell, 1990) supports this claim. Of the five husband and wife portrait pairs included in the book, four out of five females show the left cheek and all of the males show the right.

Although portraits designed to be hung in a pair would only account for a fraction of all painted portraits, it could be argued that these portraits influenced the arrangement of other portraits. However the data do not fit with this proposition. If portraits of males are generally hung to the right, and therefore show the right cheek, why do the majority of stand-alone portraits of males show the left cheek (e.g., Nicholls et al., 1999)? The fact that a leftward bias exists for both genders suggests that some other factor is at work. It is possible, however, that the inclusion of male paired-portraits could account for the reduced leftward bias that has been observed for portraits of males in general (e.g., McManus & Humphries, 1973).

Asymmetries in the aesthetic/perceptual appeal of paintings could also account for the leftward bias in portraiture. Arnheim (1956) and Gaffron (1950) have noted that aesthetic judgements are affected by left/right asymmetries in the arrangement of a painting [see Gordon (1981) for a review]. Mead and McLaughlin (1992) tested the role of stimulus asymmetry in aesthetic preference by asking participants to evaluate original and left-right mirror-reversed versions of 61 paintings. They found that participants' aesthetic evaluations of paintings changed dramatically in response to mirror-reversals. Paintings that were perceived to have a high aesthetic quality were those that: (a) contained more weight in the viewer's left side and (b) suggested a motion from the viewer's left to their right (see also, McLaughlin and Kermisch, 1997).

Asymmetries have also been reported for the perception of faces (see also Chapter 12, this volume). Mattingley, Bradshaw, Nettleton, and Bradshaw (1994) have investigated asymmetries in face perception using chimeric facial stimuli that consist of a left-right mirror-reversed pair of photographs. Each face comprised a 'happy' and a 'neutral' half with the happy expression being placed on the left side in one face and on the right in the other. When asked to select the face that appeared to be happier, participants tended to select the face with the happy expression on the left-hand side. Hoptman and Levy (1988) have reported a similar leftward attentional bias in the perception of photographic and cartoon faces. Asymmetries in the perception of faces could play an important role in the perception of portraits. Portraits that feature the left side of the face are usually arranged so that the eyes, mouth and nose fall in the middle, or to the left, of the painting (see Fig. 1). The leftward placement of these features could facilitate the process of face recognition more so than if the features were placed in the right half of the painting.

A number of theories have been put forward to account for asymmetries in the aesthetic/perceptual appeal of paintings in general and of portraits specifically. Gaffron's (1950) 'glance curve' theory suggests that the typical Western observer begins evaluating a painting by attending to the features contained in the bottom left-hand corner. As the evaluation continues, the glance progresses towards the top right-hand corner of the painting. The curve is three-dimensional, and as the glance moves from left to right, it also recedes from the front to the back of the painting. Paintings that contain more mass in their left side, and which suggest a left-to-right movement, have an arrangement that allows spontaneous recognition as the glance moves across the painting. As a result, these paintings are rated as more aesthetically pleasing than those that require deviations from the normal path of the glance curve. Gaffron (1962) suggested that the glance-curve could also account for the leftward bias in portraiture. For portraits that feature the right side of the face, the path of the glance curve would pass over the right cheek, ear and the forehead. Inspecting information-laden features such as the mouth, eyes and nose (Bruce & Young, 1998) would require deviations from the glance-curve path. In contrast, portraits that feature the left side of the face place the eyes, nose and mouth within the path of the glance curve. As a result, portraits that feature the left side of the face should be judged as more aesthetically pleasing than portraits that feature the right cheek.

A more recent version of the glance theory suggests that the leftward bias in face recognition is generated by the left-to-right reading habits that predominate amongst readers of English (Chokron, Bernard, & Imbert, 1997). Manning, Halligan and Marshall (1990) suggested that these scanning habits lead to an overrepresentation of the leftward extent of a stimulus compared to the right because the scan always starts on the left, but can be terminated before the rightward end is reached (cf. Kim, Anderson, & Heilman, 1997). The effect of scanning habit has been investigated by comparing readers of languages with different scanning directions. Sakhuja, Gupta, Singh and Vaid (1996) compared readers of Hindi (left-to-right) and Urdu (right-to-left) on a chimeric-face recognition task. When asked to select the face that appeared to be happier, readers of Hindi typically selected the face with the happier expression on the left side. Readers of Urdu, in contrast, selected the face with the positive expression on the right. Eviatar (1997) has reported a similar reduction in the leftward bias for face recognition for readers of Hebrew.

Left-to-right scanning biases reflect a bias in information processing that is not specific to face processing. That is, left-to-right scanning causes an overestimation of the leftward extent of a stimulus irrespective of what comprises the stimulus. It is possible, however, that processes specific to face recognition cause the leftward bias in face perception. It is well known that lesions to the posterior regions of the right hemisphere are likely to cause symptoms of prosopagnosia, an inability to recognise familiar faces (Benton, 1990). Faces are also recognised more readily when they are presented to the left visual field rather than the right visual field (Levine & Koch-Weser, 1982). This left-visual-field advantage presumably reflects the fact that information presented in the left visual field has direct access to the face-processing mechanisms located in the right hemisphere. Material presented in the right visual field lacks this direct access, and therefore must cross from the left to the right hemisphere via the corpus callosum. Paintings that feature the left cheek are arranged so that more facial features fall in the left half of the painting. Assuming that viewers tend to fixate in the middle of the painting, the left half of the painting would be projected via the left visual field directly to the face-recognition mechanisms located in the right hemisphere. It is possible that this arrangement would facilitate the perception of the portrait more so than if the portrait were arranged so the majority of the face fell in the right half of the painting.

Although plausible mechanisms exist that can account for a leftward aesthetic/perceptual bias, a number of problems exist for this theory. One such problem stems from the data collected from self-portraits. It seems reasonable to assume that an artist would want a portrait of himself or herself to be as aesthetically pleasing and perceptually accommodating as a portrait of anyone else. Therefore, self-portraits should be arranged so that they feature the left cheek more prominently. Contrary to this prediction, self-portraits are commonly arranged so that the right cheek is turned towards the viewer (Latto, 1996). Humphrey and McManus (1973) first noted a rightward bias for self-portraits. Of 57 self-portraits painted by Rembrandt, only 9 featured the left cheek more prominently than the right. It is possible that this rightward bias could be confounded by the fact that the model was always male; the gender that is associated with a reduced leftward bias (McManus & Humphrey, 1973). Nicholls et al. (1999) controlled for the effect of gender in self-portraits by testing male and female portraits separately. Self-portraits were obtained from a number of sources (Buscombe, 1978; Goldscheider, 1937; Meskimmon, 1996; Rubenstein, 1982; Yung, 1981) and were classified according to whether the left or right side of the nose was more visible. Of the 219 self-portraits that were sampled, 61% of the male and 67% of the female portraits turned the right cheek. Thus, it would appear that self-portraits, irrespective of whether they are male or female, feature the right side of the face more prominently than the left.

Why do self-portraits show the right cheek whereas portraits of others show the left cheek? Humphrey and McManus (1973) and McManus (1979) suggested that the rightward bias for self-portraits, at least in Rembrandt's case, was related to the perceived kinship of the sitter. Close kin, such as himself, male relations and males in general, were painted so that the right side of the face was prominent. Females, who were less socially like Rembrandt, were portrayed with the left side of the face turned toward the viewer. Gordon (1974) has applied Humphrey and McManus (1973) kinship theory to 295 portraits painted by Goya. Although Gordon (1974) found that females were more likely to show the left cheek than males, there was no difference in the profile chosen by the kin and non-kin of Goya. These results suggest that the kinship theory might apply to the work of some artists, but not others.

An alternative explanation for the rightward bias in self-portraits is related to the techniques used to produce them (Latto, 1996). Self-portraits are traditionally painted using a mirror. The use of mirrors in self-portraits is clearly demonstrated in a portrait by Johannes Gumpff (Goldscheider, 1937).

The painting depicts three views of Gumpff painting a self-portrait: his back, his reflection in the mirror, and his image on a canvas. In order to achieve a reflection that appears to feature the right cheek, the artist must turn his or her left cheek to the mirror. Thus, the artist poses as the majority of other models do; with the left cheek facing towards the artist. Because left and right are reversed in the mirror, however, the image in the mirror that is transcribed to the canvas appears to show the right cheek. In actual fact, it is the left cheek that is turned toward the viewer.

This then raises the possibility that the leftward bias in portraiture is generated by a desire to portray features contained on the left side of the face. Is there something special about the left side of a human face? Research demonstrates that when people express an emotion, the muscles on the left side produce a more intense expression than those on the right side (Borod, Haywood, & Koff, 1997; see also Chapter 12, this volume). This asymmetry has been clearly demonstrated by Sackheim, Gur and Sancy (1978). They divided photographs of faces into left and right halves, mirror-reversed them, and then rejoined them to form left-left and right-right composites. When participants were asked to select the image that appeared 'happiest' or 'saddest', they tended to select the image that was a left-left composite. This asymmetry in expression presumably reflects the fact that the left side of the face is controlled by the right cerebral hemisphere which is dominant for the expression of emotion (Benton, 1990).

Benjafield and Segalowitz (1993) investigated differences in expression between the left and right sides of face by measuring viewer's impressions of four left- and four right-sided portraits drawn by Leonardo da Vinci. To control for the effect of orientation, the portraits were shown to participants in original and mirror-reversed versions. The portraits were rated along ten semantic differential scales that loaded on three scales: evaluation, potency and activity. Portraits of females were rated as being less potent than portraits of males. Portraits featuring the right side of the face were judged as being more potent and active than left-facing portraits, irrespective of whether the image was in its original form, or was mirror-reversed. This suggests that the viewer's impression of a portrait is determined by which side of the face the model shows and not by the orientation of the portrait as it appears on the canvas (cf. McLaughlin & Murphy, 1994). These results led Benjafield and Segalowitz to conclude: "Astute observers such as Leonardo, Rembrandt and Goya may have observed that the left side of the face is generally more emotionally expressive, and thus focussed on the right, more reserved side to connote power and self-control" (p. 30).

A desire to portray the emotive qualities of the left side of the face might explain why the majority of portraits feature that side of the face. A preference to feature the left side of the face could also account for the rightward bias observed for self-portraits. Artists, like the majority of other models, might wish to have the emotional, left side of their face portrayed. By turning their left cheek to the mirror, the artist produces a right-facing image that is transcribed to the canvas. Laterality of emotional expression might also explain the gender difference in paintings wherein females are more likely than males to show a left bias. Research indicates that males are less inclined to portray emotion than women (Wagner, Buck, & Winterbotham, 1993). Thus, women might be more likely to present their left (emotive) cheek when sitting for their portrait. Males, on the other hand, might be more inclined to turn their right (impassive) cheek. This desire to portray, or conceal, emotion may explain why Nicholls et al. (1999) found no leftward bias for portraits of scientists. A total of 127 portraits of scientists belonging to the Royal Society were sampled from a catalogue produced by Robinson (1980). A measure of turning bias revealed that an equal proportion of scientists turned their head to the left or right. The popular conception of scientists as logical rationalists would suggest that they might prefer to hide their emotive side by turning their right cheek.

The model or the artist could make the decision as to which side of the face is portrayed. Models might intuitively turn their left cheek when posing for an emotive portrait and turn their right cheek when trying to appear impassive or powerful. Alternatively, artists, when trying to portray emotion, might draw upon their experience and direct the model to turn one or other cheek. Nicholls et al. (1999) tested the proposition that models have an intuitive knowledge of which cheek best reflects or conceals emotion. Participants were randomly allocated to two conditions. In the 'emotional' condition, participants were given a script that asked them to imagine that they were posing for a portrait for their family. The script encouraged them to express their warmth and love for their family. In the 'impassive' condition, participants were asked to imagine that they were successful scientists who were having their portraits made for the Royal Society. The script informed them that the portrait was a great honour, but that they should avoid looking smug or proud. After thinking about the role, participants were asked to pose in front of a camera. They were asked to pose without directly facing the camera. Analysis of turning bias revealed no difference in the frequency of left and right poses and no difference in posing behaviour between the genders. Given the stronger leftward bias reported for portraits of females (Gordon, 1974; McManus & Humphrey,

1973), one might have expected a gender difference to arise in posing behaviour. The fact that no effect was found can most probably be attributed to the roles adopted during the study that over-rode any pre-existing gender differences in the desire to portray emotion.

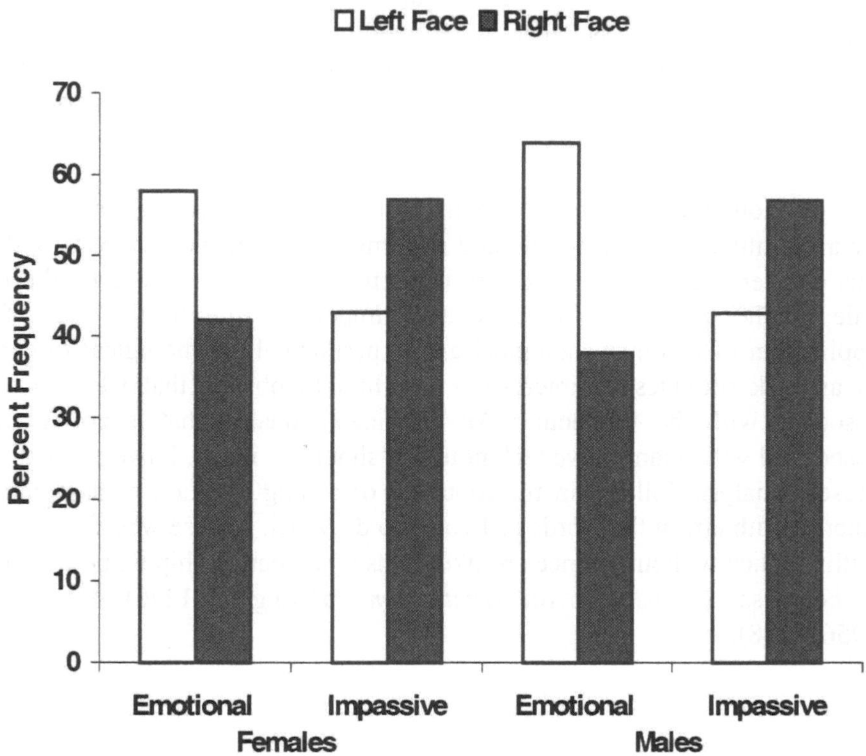

Figure 3. Frequency of left and right head turns as a function of emotional condition and gender

There was a significant interaction between emotional condition and turning direction. In the emotional condition, participants tended to turn their left cheek whereas in the impassive condition, participants turned their right cheek (see Fig 3). These results suggest that models have an intuitive knowledge of which side of their face can be used most effectively to express emotion. This knowledge could be ontogenetic or phylogenic in origin. Throughout a person's life, someone might learn the most effective way of expressing an emotion based on the reaction that the expression has evoked in the past. Alternatively, emotional expressions might be controlled by innate brain mechanisms. Darwin (1872/1965) first proposed the idea that emotions in humans and other animals have a common origin. Ekman (1980) has supported this proposition by demonstrating that members of an isolated tribe in New Guinea produced facial expressions of emotion, such as happiness and sadness, that were readily recognised by Westerners. Knowing to turn the left cheek when expressing an emotion, and to turn the right cheek when hiding emotion, may be part of the innate repertoire of facial expression discussed by Darwin.

Although it is likely that lateral asymmetries in portraits are determined by a multitude of idiosyncratic and systematic factors, the data reviewed in this chapter suggest that the desire to portray features contained on the left side of the face plays a particularly important role in the bias. The application of neuropsychological and experimental psychological research to artwork provides an interesting insight into objects that we normally associate with the 'aesthetic': An indefinable quality that is not usually associated with quantitative techniques. It should be noted, however, that the present analysis follows in the footsteps of a long tradition of associating science with art. In the words of Leonardo da Vinci: "Those who fall in love with practice without science are like pilots who board a ship without rudder or compass" (Translation of: *Treatise on Painting* (c. 1270). McMahon, 1956, p. 48).

REFERENCES

Arnheim, R. (1956). *Art and visual perception.* London: Faber & Faber.

Benjafield, J., & Segalowitz, S.J. (1993). Left and right in Leonardo's drawings of faces. *Empirical Studies of the Arts, 11*, 25-32.

Benton, A. (1990). Facial Recognition. *Cortex, 26*, 491-499.

Borod, J. C. Haywood, C. S., & Koff, E. (1997). Neuropsychological aspects of facial asymmetry during emotional expression: A review of the normal adult literature. *Neuropsychology Review, 7*, 41-60.

Bradshaw, J.L., Bradshaw, J.A., & Nettleton, N.C. (1990). Abduction, adduction and hand differences in simple and serial movements. *Neuropsychologia 28*, 917-931.

Bruce, V., & Young, A. (1998). *In the eye of the beholder: The science of face perception.* Oxford: Oxford University Press.

Bruce, V., & Young, A. (1998). *In the eye of the beholder: The science of face perception.* Oxford: Oxford University Press.

Buscombe, E. (1978). *Artists in early Australia and their portraits.* Eureka Research: Sydney.

Campbell, L. (1990). *Renaissance portraits.* Yale University Press: New Haven.

Chokron, S., Bernard, J.M., & Imbert, M. (1997). Length representation in normal and neglect subjects with opposite reading habits studied through a line extension task. *Cortex, 33,* 47-64.

Coles, P.R. (1974). Profile orientation and social distance in portrait painting. *Perception, 3,* 303-308.

Conesa, J., Conesa, C., & Miron, M. (1995). Incidence of the half left profile pose in single subject portraits. *Perceptual and Motor Skills, 81,* 920-922.

Corballis, M.C., & Beale, I. L. (1976). *The psychology of left and right.* Lawrence Erlbaum: Hillsdale, New Jersey.

Darwin, C. (1872/1965). *The expression of emotions in man and animals.* Chicago: University of Chicago Press.

Ekman, P. (1980). *The face of man: Expressions of universal emotion in a New Guinea village.* New York: Garland STPM Press.

Elias, L.J. (1998). Secular sinistrality: A review of popular handedness books and world wide web sites. *Laterality, 3,* 193-208.

Eviatar, Z. (1997). Language experience and right hemisphere tasks: The effects of scanning habits and multilingualism. *Brain and Language 58,* 157-173.

Gaffron, M. (1950). Right and left in pictures. *Art Quarterly,13,* 312-331.

Gaffron, M. (1962). Perceptual experience: An analysis of its relation to the external world through internal proceedings. In S. Koch (Ed.), *Psychology: A study of science.* Vol 4. McGraw Hill: London.

Goldscheider, L. (1937). *Five hundred self-portraits from antique times to the present day.* Phaidon Press: Vienna.

Gordon, I.E. (1974). Left and right in Goya's portraits. *Nature, 249,* 197-198.

Grodon, I.E. (1981). Left and right in art. In D. O'Hare (Ed.), *Psychology and the arts.* (pp 211-241). Harvester Press: Sussex.

Grüsser, O., Selke, T., & Zynda, B. (1988). Cerebral lateralisation and some implication for art, aesthetic perception and artistic creativity. In I. Rentschler, B. Herzberger, & D. Epstein (Eds.), *Beauty and the brain: Biological aspects of aesthetics* (pp 257-293). Birkhäuser Verlag: Berlin.

Humphrey, N., & McManus, I.C. (1973). Status and the left cheek. *New Scientist, 59,* 437-439.

Hoptman, M.J., & Levy, J. (1988) Perceptual asymmetries in left- and right handers for cartoons and real faces. *Brain and Cognition, 8,* 178-188.

Kim, M., Anderson, J.M., & Heilman, K.M. (1997). Search patterns using the line bisection test for neglect. *Neurology, 49,* 936-940.

LaBar, M. (1973). Turning the left cheek examined using modern photography. *Nature, 245,* 338.

Latto, R. (1996). Turning the other cheek: Profile direction in self-portraiture. *Empirical Studies of the Arts, 14,* 89-98.

Lanthony, P. (1995). Left handed painters. *Revue-Neurologique, 5,* 165-170.

Levine, S.C., & Koch-Weser, M.P. (1982). Right hemisphere superiority in the recognition of famous faces. *Brain and Cognition, 1*, 10-22.

Manning, L., Halligan, P.W., & Marshall, J.C. (1990). Individual variation in line bisection: A study of normal subjects with application to the interpretation of visual neglect. *Neuropsychologia, 28,* 647-655.

Mattingley, J.B. Bradshaw, J.L. Nettleton, N.C., & Bradshaw, J.A. (1994). Can task specific perceptual bias be distinguished from unilateral neglect? *Neuropsychologia, 32,* 805-817.

McLaughlin, J.P., & Kermisch, J. (1997). Salience of compositional cues and the order of presentation in the picture reversal effect. *Empirical Studies of the Arts, 15,* 21-27.

McLaughlin, J.P., & Murphy, K.E. (1994). Preference for profile orientation in portraits. *Empirical Studies of the Arts, 12,* 1-7.

McMahon, A.P. (1956). *Treatise on painting: (Codex urbina's latinus 1270) / by Leonardo Da Vinci.* Princeton, N.J.: Princeton University Press.

McManus, I. C. (1979). Determinants of laterality in man.: *Unpublished PhD thesis,* University of Cambridge.

McManus, I. C., & Humphrey, N. K. (1973). Turning the left cheek, *Nature, 243,* 271-272.

McMullen, R. (1977). *Mona Lisa: The picture and the myth.* New York: Da Caapo Press.

Mead, A. M., & McLaughlin, J. P. (1992). The roles of handedness and stimulus asymmetry in aesthetic preference. *Brain and Cognition, 20,* 300-307.

Mebert, C.J., & Michel, G.F. (1980). Handedness in artists. In J. Herron (Ed.), *Neuropsychology of left handedness* (pp.273-279). New York: Academic Press.

Meskimmon, M. (1996). *The art of reflection: Women artists' self-portraiture in the twentieth century.* Scarlet Press: London.

Nicholls, M.E.R., Clode, D., Wood, S.J., & Wood, A, G. (1999). Laterality of expression in portraiture: Putting your best cheek forward. *Proceedings of the Royal Society (section B), 266,* 1517-1522.

Robinson, N. H. (1980). *The Royal Society catalogue of portraits.* Royal Society: London.

Rubenstein, C. S. (1982). *American women artists from early Indian times to the present.* G. K. Hall: Boston.

Sackheim, H.A., Gur, R.C., & Saucy, M.C. (1978). Emotions are expressed more intensely on the left side of the face. *Science, 202,* 434-436.

Sakhuja, T., Gupta, G.C., Singh, M., & Vaid, J. (1996). Reading habits affect asymmetries in facial affect judgements: A replication. *Brain and Cognition, 32,* 162-165.

Shannon, B. (1979). Graphological patterns as a function of handedness and culture. *Neuropsychologia, 17,* 457-465.

Tyler, C.W. (1998). Painters centre one eye in portraits. *Nature, 392,* 877-878.

Wagner, H. L. Buck, R., & Winterbotham, M. (1993). Communication of specific emotions: Gender differences in sending accuracy and communication measures. *Journal of Nonverbal Behavior, 17,* 29-53.

Yung, K. K. (1981). *National Portrait Gallery: Complete Illustrated Catalogue 1856-1979.* National Portrait Gallery: London.

Chapter 14

Attentional and Intentional Factors in Pseudoneglect

Gina M. Grimshaw[1] and Jocelyn M. Keillor[2]
[1]California State University, San Marcos & University of California San Diego, USA : [2]Defence and Civil Institute of Environmental Medicine, Canada

Perhaps one of the most extreme examples of sidedness occurs when neurological patients fail to report, respond, or orient to stimuli presented to the side of the brain opposite their lesion, a deficit that is most frequent and severe following right-hemisphere lesions (Brain, 1941; Gainotti, Messerli, & Tissot, 1972). This hemispatial neglect is most commonly examined at the bedside by having the patient bisect a line printed on a sheet of paper. Patients with right-hemisphere lesions typically bisect the line substantially to the right of true centre (Heilman, Watson, & Valenstein, 1993). Interestingly, when neurologically intact participants bisect horizontal lines, they place the bisection point slightly to the left of true centre, a phenomenon known as pseudoneglect. This leftward bias can also influence behaviour outside the lab, as participants who display pseudoneglect can be more likely to bump into objects on their right sides (Turnbull & McGeorge, 1998). It has been argued that such a leftward bias does not exist at the population level and is instead the result of sampling error in a number of small-N studies (Mozer, Halligan, & Marshall, 1997). However, a recent meta-analysis of 73 studies and over 2,000 participants reports that pseudoneglect is a robust phenomenon, and although the magnitude and direction of deviations are subject to both individual differences and many experimental parameters, it is clear that, overall, people tend to bisect lines to the left of their true midpoint (Jewell & McCourt, 2000).

M.K. Mandal, M.B. Bulman-Fleming and G. Tiwari (eds.), Side Bias: A Neuropsychological Perspective, 331-346.

A number of mechanisms have been proposed to account for pseudoneglect, including both perceptual/attentional and premotor/ intentional processes.[3] For example, leftward deviations might occur because an attentional bias toward the left end of the line produces a magnification of its leftward extent (Milner, Brechman, & Pagliarini, 1992) or because an intentional bias exaggerates movement into left hemispace (Luh, 1995). Attentional and intentional factors have similarly been implicated in hemispatial neglect, and a considerable body of research has examined the relative roles of each. In this chapter we review the methods that have been used to examine the roles of attention and intention in hemispatial neglect, and evaluate the evidence for attentional and intentional influences in pseudoneglect.

1. DISSOCIATION OF ATTENTIONAL AND INTENTIONAL FACTORS IN HEMISPATIAL NEGLECT

Although hemispatial neglect is conceptualized primarily as an attentional disorder, both attentional and intentional systems can be affected. These factors are often confounded in studies of line-bisection performance because both lead to rightward deviations. Thus a number of techniques have been developed that allow one to dissociate atttentional and intentional processes. One way to determine if hemispatial neglect is attentional or intentional in origin is to eliminate either the perceptual or motor component of the task and see if the neglect remains. Using this approach, Heilman, Bowers, and Watson (1983) blindfolded five patients with hemispatial neglect (to eliminate visual perceptual biases) and asked them to point to a position in space along their bodies' midline with their right hands. They produced mean displacements of over 8 cm to the right of centre, demonstrating *directional hypokinesia*, a difficulty in moving into or toward contralesional hemispace. This finding illustrates that neglect can be intentional in origin.

In another study that used the elimination method, Liu, Bolton, Price, and Weintraub (1992) compared two patients, one with a right peri-Sylvian

[3] Although we use the terms attentional and intentional throughout this chapter, we do so to refer to broad classes of processes that operate primarily at input (visual/perceptual/attentional/representational) or output (intentional/premotor/motor).

infarction and the other with a right frontal hemorrhage. Both demonstrated left-sided neglect on standard bedside tasks (letter cancellation and clock drawing). However, only the patient with frontal damage was able to name objects on the left side of an array (a task with no motor component), but only the patient with peri-Sylvian damage was symmetrical in performing a tactile exploration task (with no visual perceptual component).

Although tasks that eliminate perceptual or motor components demonstrate that neglect *can* be either attentional or intentional, it is possible that many cases of neglect involve both factors. In order to assess the relative contributions of attentional and intentional influences, it is necessary to examine both factors directly in the same experiment. Butter, Rapcsak, Watson, and Heilman (1988) seem to be the first to have pitted attentional and intentional factors in direct opposition. They used an eye-movement paradigm in a patient with lesions of right prefrontal and premotor cortex. In the uncrossed condition, the patient was to fixate centrally, and then move his eyes to the examiner's finger (30 degrees left or right). As expected, the patient neglected the target when it was on the left. However, in the crossed condition, the patient was to move his eyes in the direction opposite to the examiner's finger. In this condition the patient failed to respond correctly to targets in either visual field, indicating both attentional and intentional neglect. However, after six months he still had difficulty moving leftward in response to right-sided targets, but could accurately move rightward in response to left-sided targets, indicating that the attentional neglect had resolved, and the intentional neglect remained.

In a multiple-case study, Coslett, Bowers, Fitzpatrick, Haws, and Heilman (1990) videotaped patients while they performed a line-bisection task. Patients were not able to watch their own hands, but received feedback through a videomonitor. Thus, attentional and intentional factors were decoupled. The line was placed either in left or right hemispace, to measure intentional factors, and the monitor was placed in left or right hemispace, to assess attentional factors. In this situation, a main effect of line position (performance worse on the left, regardless of monitor placement) would reflect intentional neglect, and a main effect of monitor position (performance worse with feedback on the left, regardless of line placement) would reflect attentional neglect. One of their four patients produced data that could not be interpreted in this way, as he had great difficulty with the task, and bisected the line to its left when feedback was from the left, and to its right when feedback was from the right, and seemed to be experiencing an attentional grasp of the monitor rather than responding to the line itself. However, the other three patients revealed a nice dissociation between attentional and intentional neglect. Two patients with right anterior lesions were affected by line position but not by monitor placement. These patients therefore had an intentional neglect. A patient with a posterior lesion demonstrated a significant effect of monitor location but no effect of line placement, reflecting an attentional neglect. These results concur with the anterior/posterior intentional/attentional dissociation found by Liu et al. (1992).

Another approach uses the landmark task developed by Milner and colleagues (Harvey, Milner, & Roberts, 1995; Milner et al., 1992; Milner, Harvey, Roberts, & Forster, 1993;). In order to remove the motor component involved in bisecting the line, participants examine pre-transected lines and indicate whether the landmark is closer to the left or right end of the line. Critical trials are those in which the line is centrally bisected. An intentional component is reintroduced into the task by having patients respond by pointing to the end of the line nearest the landmark. Note that attentional neglect will bias patients to point to the left, but intentional neglect will bias patients to point to the right. Harvey et al. (1995) used this task with eight neglect patients, seven of whom had a strong bias to point left, and one of whom had a strong bias to point right.[4] Lesion analysis partially supported the anterior/posterior distinction, with five of six patients with attentional neglect having parietal lesions. However, one patient with attentional neglect had a subcortical lesion, and the patient with intentional neglect had a large lesion that included both frontal and parietal areas.

Although all these tasks allow one to diagnose neglect as primarily attentional or intentional, the motor response is so different from that involved in normal line bisection that it does not permit a quantitative analysis of the relative contributions of attentional and intentional influences to line-bisection performance itself. Halligan and Marshall (1989) used a computerized line-bisection task that put attentional and intentional factors in direct opposition in a patient with a profound left-sided neglect subsequent to a right temporoparietal lesion. The patient used the computer's mouse to drag a cursor along the line to the bisection point. The mouse could start at either the left or right end of the line. In the congruent

[4] Harvey et al. (1995) also included a cueing manipulation in their study. In control subjects, left-end cues increased rightward responses, indicating that the cue acted attentionally to magnify the left end of the line. However, in patients, left-end cues increased leftward responses, indicating that the cues acted intentionally to bias motor responses in their direction.

condition, the cursor and the mouse were directly coupled so that leftward movement of the mouse produced leftward movement of the cursor. The patient bisected lines substantially to the right of true midpoint, an effect that was magnified when the cursor started at the right end of the line. In the incongruent condition, the mouse and the cursor were decoupled so that rightward movement of the mouse produced leftward movement of the cursor. If the neglect was intentional, the bisection point should now fall to the left of midpoint. Performance was unaffected by this manipulation, suggesting that the neglect was entirely attentional in nature.

Bisiach, Geminiani, Berti, and Rusconi (1990) used a similar procedure with a physical apparatus. They used a pulley sytem in which the patient responded by moving a pointer mounted on the top pulley string to bisect the line. When the patients grasped the top string to make the bisection (the congruent condition), rightward movements of the string produced rightward movements of the pointer. However, when they grasped the bottom string (the incongruent condition), rightward movements of the string produced leftward movements of the pointer. If hemispatial neglect were entirely attentional, patients should produce identical rightward deviations in both cases. If neglect were entirely intentional, they should produce rightward deviations with the top string, and leftward deviations of the same magnitude with the bottom string. To the extent that neglect is a combination of both factors, the bisection point with the bottom string should fall between these two extremes. For example, if attentional and intentional influences are equivalent, bisections with the bottom string should fall on the true midpoint. If the primary bias is attentional, the bisection point should fall on the same side of midpoint in both conditions, but if the primary bias is intentional, the bisection point should fall on the opposite side of midpoint in the incongruent condition.

Bisiach et al. (1990) reported the results of 15 consecutive cases of right-hemisphere damage with clinical symptoms of neglect. Thirteen of the 15 had some evidence of intentional influence on line-bisection performance, although only two of these actually reversed the direction of their deviations, and none produced evidence of purely intentional neglect. A lesion analysis found that patients with frontal lesions had greater intentional involvement than those with purely post-Rolandic lesions.

Most recently, Adair, Na, Schwartz, and Heilman (1998; see also Na, Adair, Williamson, Schwartz, Haws, & Heilman, 1998) have used a video-flipping technique to decouple attentional and intentional effects. Unlike the video apparatus used by Coslett et al. (1990), this procedure places the monitor at midline so that participants do not have to look in the direction opposite their hands for visual feedback. In the congruent condition, the camera is mounted under the translucent table, so that the left side of the line appears on the left side of the video display. In the incongruent condition, the camera is mounted above the table, so that the image is reversed on the monitor. If neglect were attentional, participants should reverse the direction of their deviations when the image is flipped, however if neglect is intentional, no differences should be observed between video conditions. Adair et al. used a quantitative approach in which patients were classified as having a primary attentional neglect if the bisection errors reversed direction in the incongruent condition, and an intentional neglect if they did not. The authors also identified a secondary bias. For example, if the magnitude of the errors decreased, but did not reverse direction, the patient would have a primary intentional neglect with a secondary attentional neglect. The authors found that 14 of 26 participants had primary attentional neglect, and 12 had primary intentional neglect. Those with anterior lesions were more likely to demonstrate intentional neglect, and those with posterior lesions were more likely to demonstrate attentional neglect. Of the 23 patients who demonstrated a secondary bias, 11 had biases that were concordant (i.e., both attentional and intentional biases in a rightward direction), but 12 had biases that were discordant. Although this finding might be expected by chance (measurement error will cause deviations between the congruent and incongruent conditions) the authors argue that this is not the case, because the direction of the secondary bias was systematic, with attentional patients displaying concordant secondary biases, but those with intentional neglect displaying discordant secondary biases.

In summary, both attentional and intentional biases contribute to neglect, and these effects are separable. There is some evidence for an anatomical dissociation between the two, with attentional neglect associated with posterior lesions and intentional neglect associated with frontal lesions, although this relation is not perfect (see Bisiach, Ricci, Lualdi, & Colombo, 1998, for a large-scale lesion study).

2. ATTENTIONAL AND INTENTIONAL FACTORS IN PSEUDONEGLECT

Although most of the clinical studies reviewed above included control participants, pseudoneglect is obviously not as profound as hemispatial neglect, and studies designed for patients have neither enough trials nor enough participants to reliably detect any lateral biases. Control participants are also matched to patients for age and therefore do not reflect the primarily young population of undergraduates that are employed in studies of pseudoneglect. Age differences have been reported in line-bisection performance, with older adults exhibiting a significant rightward bias on line bisection (Jewell & McCourt, 2000).

Of the studies that have specifically examined pseudoneglect in a young population, several have examined attentional factors by eliminating the motor component of the response through the use of the landmark task, either in free vision or with tachistoscopic presentation (to eliminate motor factors associated with eye movements). Pseudoneglect is observed with landmark tasks (McCourt & Jewell, 1999; McCourt & Olafson, 1997; Milner et al., 1992) indicating that attentional factors do contribute to pseudoneglect. However, as with hemispatial neglect, the fact that pseudoneglect remains in the absence of a motor response does not rule out a role for intentional factors in normal bisection performance. Interestingly, the meta-analysis of Jewell & McCourt (2000) finds that pseudoneglect is actually greater with landmark tasks than with manual bisection tasks. This effect may arise because several of the landmark studies used tachistocopic presentation, which eliminates ocular scanning of the line that may be used to correct for purely perceptual biases.

Dellatolas and colleagues have compared performance on manual bisection tasks with that on the landmark task in the same participants in an effort to provide a more direct comparison. Dellatolas, Vanluchene, and Coutin (1996) found pseudoneglect on a manual paper-and-pencil task that was not related to deviations on a paper version of the landmark task. This suggests a motor component in line-bisection performance in standard paper-and-pencil tasks that is not present on the landmark task.

Interestingly, pseudoneglect was not observed on a computer version of the task, although in this case deviations in manual bisection and landmark tasks were correlated. In a similar study of performance in children, Dellatolas, Coutin, and De Agostini (1996) observed that children demonstrated pseudoneglect on the landmark task at all ages (on paper). However, on the manual bisection task young children bisected to the left of midpoint with the left hand, and to the right of midpoint with the right hand. This effect was different in older children, who demonstrated pseudoneglect with both hands. The authors again support the role of motor or intentional factors in pseudoneglect, particularly in young children.

In a novel approach, Luh (1995) evaluated the perceptual/attentional component of pseudoneglect by comparing it with other perceptual/attentional asymmetries. No correlations were observed between deviations on a line-bisection task and asymmetries on a chimeric faces or a dot-enumeration task, leading to the conclusion that pseudoneglect is not attentional in origin. However, it should be noted that significant correlations between perceptual asymmetries are rare in any case (Boles, 1998). In a second study, pseudoneglect was not observed on a manual computerized bisection task (see Dellatolas, Vanluchene & Coutin, 1996, who also failed to find pseudoneglect on a computerized manual bisection task). The author argues that computerized bisection eliminates the intentional component because the mouse movements required to perform the task are very small. Although this may minimize biases in motor output, an intentional bias might be expected to affect the *planned* movement, which would be the movement of the cursor, and not the movement of the mouse.

The experiment of Schwartz, Adair, Na, Williamson, and Heilman (1997) is the only reported study to directly assess attentional and intentional components of pseudoneglect in the same task. They used the video-flipping method used by Adair et al. (1998) to examine line-bisection performance in normal participants . Lines were 230 mm in length. As with the Adair et al. study above, they classified participants as having attentional or intentional primary bias. Population-level pseudoneglect was not observed on the task, with 11 of 24 participants producing rightward deviations and 13 producing leftward deviations in the congruent condition. However, 18 participants reversed the direction of their bias in the incongruent condition, reflecting an attentional influence on bisection performance. Of the 13 participants displaying pseudoneglect, 4 had a primarily intentional bias.

In our laboratory, we have examined the relative contributions of attentional and intentional factors to pseudoneglect using the pulley procedure (Keillor, Grimshaw, Bryden, & Cocivera, 1995). The pulley task (Bisiach et al., 1990) permits such an examination in a somewhat more natural task than video procedures afford, in that participants do not view reversed visual feedback of their moving limb, and all feedback comes directly from the line itself and not via a video monitor. In the pulley procedure, the subject's hand is obscured by a screen, but the bisecting pointer and the line are clearly visible and at no time does the hand obscure the view of the line. Recall that in the congruent condition (moving the pointer with the top string of the pulley) both attentional and intentional factors produce leftward errors. However, in the incongruent condition, attentional factors produce leftward errors, and intentional factors produce rightward errors. An intentional effect is therefore revealed as a difference in the magnitude or even direction of errors between the congruent and incongruent conditions.

This study also considered how attentional and intentional factors interact with other factors that have been demonstrated to affect bisection performance, including line length, hemispace, and response hand. Both leftward deviations in normals and rightward deviations in patients are greater with longer lines (Bisiach, Bulgarelli, Sterzi, & Vallar, 1983; Halligan, Manning, & Marshall, 1990; Manning, Halligan, & Marshall, 1990). In patients, neglect can even reverse itself with very short lines, a phenomenon known as the crossover effect (Halligan & Marshall, 1988; Tegner & Levander, 1991). Response hand and hemispace also influence pseudoneglect, although the literature is not consistent on the magnitude or even direction of these effects. Most studies find that pseudoneglect is greatest in left hemispace (Luh, 1995; Milner et al., 1992) and when responding with the left hand (Dellatolas et al., 1996; Scarisbrick, Tweedy, & Kuslansky, 1987). Both of these effects are supported by the meta-analysis of Jewell and McCourt (2000), and could reflect either a general right-hemisphere activation that occurs when working on the left side of space (Kinsbourne, 1977), or a right-hemisphere dominance for attention (Heilman et al., 1993).

Participants were 30 right-handed undergraduate students who were tested using a modified version of the pulley apparatus used by Bisiach et al. (1990; see Figure 1).

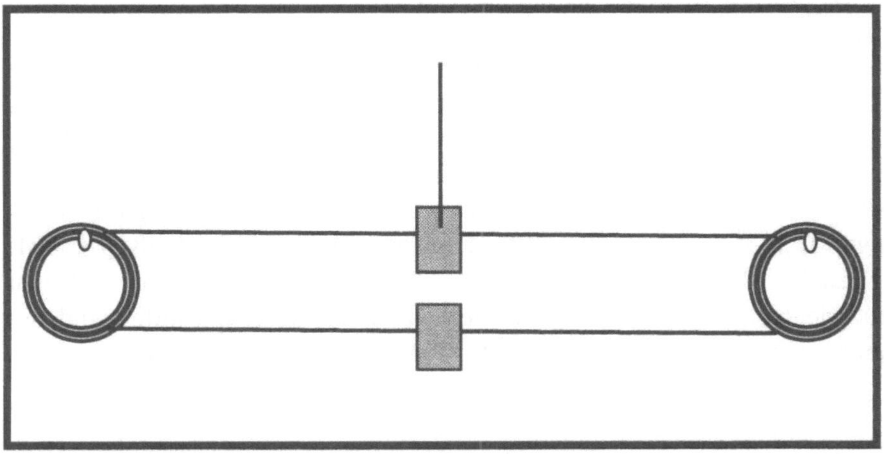

Figure 1. Schematic diagram of the pulley apparatus used to dissociate attentional and intentional influences on line-bisection.

The device consisted of a board, 1 meter in length, upon which a loop of fishing line was stretched between two pulleys that were mounted 80 cm apart. Wooden finger grips were attached to both the top and bottom portions of this loop, and a pointer protruded from the grip that was on top. The board was mounted on a 52° angle, and a shelf extended over the strings of the pulley so that participants could see the pointer but not their hands. Four pegs were positioned across the top of the board so that the printed bisection lines could be individually affixed in left hemispace, at midline, and in right hemispace. For each trial, the experimenter mounted the corners of a piece of paper from a pair of pegs so that the stimulus line was centred beneath them. The starting point of the pointer alternated between the left and right ends of the pulley. Three line lengths were used (50, 125, and 200 mm). Participants were instructed to bisect the line by moving the pointer to its centre using either the grip on the top string (congruent condition) or the grip on the bottom string (incongruent condition). There were thus 36 conditions (3 lengths x 3 positions x 2 congruencies x 2 hands). Three trials of each condition were presented in a pseudorandom order for a total of 108 trials.

Displacements from the objective midpoint were measured in millimeters, with leftward displacements receiving negative values. Mean deviations were expressed both in millimeters and as a percent of the total line length. An overall pseudoneglect was observed, with a mean leftward displacement of 0.62 mm, $t(29) = -3.29$, $p = .003$, or 0.45%, $t(29) = -3.02$, $p = .005$.

Deviations (in mm) were analyzed in a $2 \times 2 \times 3 \times 3$ (Hand × Congruency × Length × Hemispace) within-participants analysis of variance (ANOVA). For mm deviations, an interaction was observed between Length and Hemispace, $F(4, 116) = 40.56$, $p < .001$. For the shortest lines, pseudoneglect was largest in left hemispace, but for the longest lines, pseudoneglect was largest in right hemispace. No effect of Hemispace was observed for the medium line length. There were also main effects of Length, $F(2, 58) = 12.46$, $p < .001$, Hemispace, $F(2, 58) = 10.48$, $p < .001$, and Hand, $F(1, 29) = 4.43$, $p = .044$. The effect of Hand was such that pseudoneglect was larger with the left hand, although this effect did not interact with hemispace or line length. There were no effects of congruency, suggesting that errors in line bisection were entirely attentional in nature.

Given the large effect of Length, and the fact that many researchers have found a linear relation between line length and magnitude of pseudoneglect, deviations were expressed as a proportion of line length. This removed a great deal of the variance from the analysis, and allowed the examination of more subtle effects. The results are depicted in Figure 2. The effect of Hand disappeared, suggesting that it was mainly mediated by the longer lines (although the Hand x Length interaction failed to reach significance), as did the main effects of Length and Hemispace. However, the interaction between Length and Hemispace remained, $\underline{F}(4, 116) = 46.05$, $\underline{p} < .001$. Most importantly, an interaction was now observed between Congruency and Length, $\underline{F}(2, 58) = 5.45$, $\underline{p} < .007$. A congruency effect was observed for the shortest lines, $\underline{F}(1, 29) = 10.44$, $\underline{p}$ .003, but not for the long or medium line lengths, $\underline{F}(1, 29) = .47$, ns.

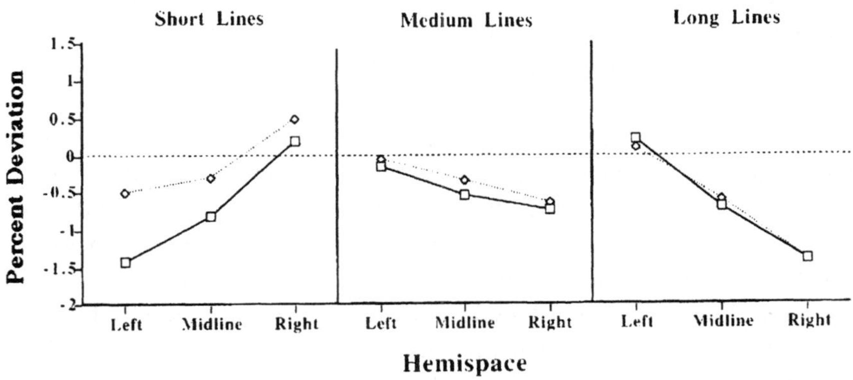

Figure 2. Percent deviations for 5 cm, 12.5 cm, and and 20 cm lines as a function of congruency and hemispace. Congruent trials are open squares and incongruent trials are open diamonds. There was an effect of congruency (reflecting an intentional component) for short lines only. There was also an interaction of line length and hemispace. For short lines, pseudoneglect was greatest in left hemispace, but for longer lines, pseudoneglect was greatest in right hemispace. Hemispace was not related to the magnitude of pseudoneglect on the medium-length lines.

Individual subject analyses revealed that 19 of 30 participants had pseudoneglect in the congruent condition (collapsed across all other factors). Of those 19 participants, only 3 reversed the direction of their error in the incongruent condition, reflecting a primary intentional pseudoneglect. Of the 11 participants with reversed pseudoneglect (rightward biases), 2 reversed direction in the incongruent condition. Participants with attentional biases were considered to have a secondary intentional bias if their deviations were smaller in the incongruent condition. Thirteen of the 16 participants with attentional neglect demonstrated a secondary intentional bias, but only 5 of the 9 participants with reversed attentional pseudoneglect demonstrated a secondary bias, as would be expected from chance deviations between congruent and incongruent conditions.

Because a congruency effect was observed for short lines only, individual subject performance was analyzed for these lines. Twenty of 30 participants demonstrated pseudoneglect, half of whom exhibited reversed pseudoneglect. Of those, 7 demonstrated primarily intentional neglect, whereas none of the participants with reversed pseudoneglect demonstrated a primary intentional bias. Of the 13 participants with attentional pseudoneglect, 11 demonstrated some evidence of a secondary intentional bias. Of the 10 participants with attentional reversed pseudoneglect, only 4 demonstrated a secondary bias, as would again be expected by chance.

Overall, these findings suggest that pseudoneglect is primarily attentional in nature, unless lines are very short. For line lengths of 125 and 200 cm, there is no difference in the deviations produced with either string of the pulley. The congruency effect that was observed for 5 cm lines indicates that intentional factors play a role in the bisection errors on these very short lines. This congruency effect was additive across hemispace, suggesting that the effect of hemispace on bisection errors arises relatively early in processing (attentionally) and not intentionally.

The interaction of line length and hemispace also supports the hypothesis that bisections on very short lines are qualitatively different than those on longer lines, in that deviations on longer lines were greater in right hemispace, but deviations on short lines were greater in left hemispace. This finding is consistent with the crossover effect observed in patients, which suggests that they too process short lines in a qualitatively different way. The most likely explanation is that short lines can be viewed in a single fixation, and thus their representation does not have to be formed over multiple saccades.

3. CONCLUSIONS

Findings from a number of different methodologies converge to suggest that attentional factors contribute to pseudoneglect, the strongest evidence coming from studies of the landmark task, in which pseudoneglect is observed in the absence of a motor response. Intentional influences on performance are more difficult to observe, and even more difficult to quantify. Schwartz et al. (1997) find that intentional factors might dominate for about 1/3 of participants who demonstrate pseudoneglect, although the estimate from our study is lower. This difference may arise because of a difference in criterion – in the Schwartz et al. study a change in direction of errors between conditions reflected attentional neglect, but in our study a change in direction reflected intentional neglect. Statistically, intentional influences were harder to detect in our study. For the shortest lines, we did find that intentional factors dominated for about 1/3 of participants, although almost all participants demonstrated some attentional effect with the short lines. The qualitative differences that have been observed between short and longer lines in a number of domains (intention, hemispace, and crossover effects) suggest that scanning patterns that are necessary for longer lines may be very important in the production of attentional biases.

The line-bisection task provides a model for the examination of perceptual/attentional and premotor/intentional effects on motor behaviour. A leftward bias can be demonstrated in both attentional and intentional systems, a bias that may have very real implications for our perceptions and actions in space.

Acknowledgements. The studies reported in this chapter were carried out at the University of Waterloo under the supervision of Phil Bryden, and were funded by the Natural Sciences and Engineering Research Council of Canada. The authors wish to thank Tracy Cocivera for assistance in data collection and analysis, Helena Phylactou and Kristin Kwasny for assistance in document preparation, Eric Roy for interesting discussion of these issues and Phil Bryden for financial, intellectual, and personal support.

4. REFERENCES

Adair, J. C., Na, D. L., Schwartz, R. L., & Heilman, K. M. (1998). Analysis of primary and secondary influences on spatial neglect. *Brain and Cognition, 37,* 351-367.

Bisiach, E., Bulgarelli, C., Sterzi, R., & Vallar, G. (1983). Line-bisection and cognitive plasticity of unilateral neglect of space. *Brain and Cognition, 2,* 32-38.

Bisiach, E., Geminiani, G., Berti, A., & Rusconi, M. (1990). Perceptual and premotor factors of unilateral neglect. *Neurology, 40,* 1278-1281.

Bisiach, E., Ricci, R., Lualdi, M., & Colombo, M. (1998). Perceptual and response bias in unilateral neglect: Two modified versions of the Milner landmark task. *Brain and Cognition, 37,* 369-386.

Boles, D. (1998). Relationships among multiple task asymmetries: I. A critical review. *Brain and Cognition, 36,* 253-267.

Brain, R. (1941). Visual disorientation with special reference to the lesions of the right cerebral hemisphere. *Brain, 64,* 244-272.

Butter, C. M., Rapcsak, S. Z., Watson, R. T., & Heilman, K. M. (1988). Changes in sensory inattention, directional motor neglect and "release" of the fixation reflex following a unilateral frontal lesion: A case report. *Neuropsychologia, 26,* 533-545.

Coslett, H. B., Bowers, D., Fitzpatrick, E., Haws, B., & Heilman, K. M. (1990). Directional hypokinesia and hemispatial inattention in neglect. *Brain, 113,* 475-486.

Dellatolas, G., Coutin, T., & Agostini, M. (1996). Bisection and perception of horizontal lines in normal children. *Cortex, 32,* 705-715.

Dellatolas, G., Vanluchene, J., & Coutin, T. (1996). Visual and motor components in simple line-bisection: An investigation in normal adults. *Cognitive Brain Research, 4,* 49-56.

Gainotti, G., Messerli, P., & Tissot, R. (1972). Qualitative analysis of unilateral spatial neglect in relation to laterality of cerebral lesion. *Journal of Neurology, Neurosurgery and Psychiatry, 35,* 545-550.

Halligan, P. W., Manning, L., & Marshall, J. C. (1990). Individual variation in line-bisection: A study of four patients with right hemisphere damage and normal controls. *Neuropsychologia, 28,* 1043-1051.

Halligan, P. W., & Marshall, J. C. (1988). How long is a piece of string? A study of line-bisection in a case of visual neglect. *Cortex, 24,* 321-328.

Halligan, P. W., & Marshall, J. C. (1989). Line-bisection in visual-spatial neglect: A case study. *Neuropsychologia, 29,* 517-522.

Harvey, M., Milner, A. D., & Roberts, R.C. (1995). An investigation of hemispatial neglect using the Landmark task. *Brain and Cognition, 27,* 59-78.

Heilman, K. M., Bowers, D., & Watson, R. T. (1983). Performance on hemispatial pointing task by patients with neglect syndrome. *Neurology, 33,* 661-664.

Heilman, K. M., Watson, R. T., & Valenstein, E. (1993). Neglect and related disorders. In K. M. Heilman, & E. Valenstein (Eds.), *Clinical neuropsychology* (3rd ed., pp. 279-336). New York : Oxford University Press.

Jewell, G., & McCourt, M. E. (2000). Pseudoneglect: A review and meta-analysis of performance factors in line-bisection tasks. *Neuropsychologia, 38,* 93-110.

Keillor, J., Grimshaw, G. M., Bryden, M. P., & Cocivera, T. (1995). Lateral biases in line-bisection in normals: A perceptual or a pre-motor phenomenon. *Poster presented to the International Neuropsychological Society,* Seattle, WA, February 8, 1995.

Kinsbourne, M. (1977). Hemi-neglect and hemisphere rivalry. *Advances in Neurology, 18,* 41-49.

Liu, G. T., Bolton, A. K., Price, B. H., & Weintraub, S. (1992). Dissociated perceptual-sensory and exploratory-motor neglect. *Journal of Neurology, Neurosurgery, and Psychiatry, 55,* 701-706.

Luh, K. E. (1995). Line-bisection and perceptual asymmetries in normal individuals: What you see is not what you get. *Neuropsychology, 9,* 435-448.

Manning, L., Halligan, P. W., & Marshall, J. C. (1990). Individual variation in line-bisection: A study of normal subjects with application to the interpretation of visual neglect. *Neuropsychologia, 28,* 647-655.

McCourt, M.E., & Jewel, G. (1999). Visuospatial attention in line bisection: Stimulus modulation of pseudoneglect. Neuropsychologia, 37, 843-857.

McCourt, M. E., & Olafson, C. (1997). Cognitive and perceptual influences on visual line-bisection: Psychophysical and chronometric analyses of pseudoneglect. *Neuropsychologia, 35,* 369-380.

Milner, A. D., Brechmann, M., & Pagliarni, L. (1992). To halve and halve not: An analysis of line-bisection judgements in normal subjects. *Neuropsychologia, 30,* 515-526.

Milner, A. D., Harvey, M., Roberts, R. C., & Forster, S. V. (1993). Line-bisection errors in visual neglect: Misguided action or size distortion? *Neuropsychologia, 31,* 39-49.

Mozer, M., Halligan, P., & Marshall, J. (1997). The end of the line for a brain-damaged model of unilateral neglect. *Journal of Cognitive Neuroscience, 9,* 171-190.

Na, D. L., Adair, J. C., Williamson, D. J. G., Schwartz, R. L., Haws, B., & Heilman, K. M. (1998). Dissociation of sensory-attentional from motor-intentional neglect. *Journal of Neurology, Neurosurgery, and Psychiatry, 64,* 331-338.

Scarisbrick, D. J., Tweedy, J. R., & Kuslansky, G. (1987). Hand preference and performance effects in line-bisection. *Neuropsychologia, 25,* 695-699.

Schwartz, R. L., Adair, J. C., Na, D., & Williamson, D. J. G. (1997). Spatial bias: Attentional and intentional influence in normal subjects. *Neurology, 48,* 234-242.

Tegner, R., & Levander, M. (1991). The influence of stimulus properties on visual neglect. *Journal of Neurology, Neurosurgery, and Psychiatry, 54,* 882-887.

Turnbull, O. H., & McGeorge, P. (1998). Lateral bumping: A normal subject analog to the behaviour of patients with hemispatial neglect? *Brain and Cognition, 37,* 31-33.

Subject Index